软件开发源码 精讲系列

Spring IoC
源码精讲

王 涛 ◎著

清华大学出版社
北京

内 容 简 介

本书从 Spring IoC 核心技术的使用开始讲解，具备充分的 Spring IoC 使用说明，并且对 Spring IoC 核心源码进行分析。

本书分为 22 章，主要内容包括 SpringXML 模式中 XML 文档的验证、XML 资源读取、Spring 默认标签的解析、自定义标签解析、各类资源对象注册和生命周期，以及 Spring 注解模式中的注解元数据、注解模式的启动和注解模式下 Bean 的注册等。本书可以帮助读者快速掌握 Spring IoC 容器的基本使用及 Spring IoC 相关的源码逻辑。

本书的源码分析在大部分情况下遵循测试用例优先，尽可能保证源码现象可复现。

本书适合具有一定 Java 编程基础和对 Spring 框架有基本开发能力的读者。

本书封面贴有清华大学出版社防伪标签，无标签者不得销售。

版权所有，侵权必究。举报：010-62782989，beiqinquan@tup.tsinghua.edu.cn。

图书在版编目(CIP)数据

Spring IoC 源码精讲/王涛著. —北京：清华大学出版社，2022.9
（软件开发源码精讲系列）
ISBN 978-7-302-60131-9

Ⅰ.①S… Ⅱ.①王… Ⅲ.①JAVA 语言－程序设计 Ⅳ.①TP312.8

中国版本图书馆 CIP 数据核字(2022)第 025864 号

责任编辑：	安　妮　薛　阳
封面设计：	刘　键
责任校对：	郝美丽
责任印制：	丛怀宇

出版发行：清华大学出版社
网　　址：http://www.tup.com.cn，http://www.wqbook.com
地　　址：北京清华大学学研大厦 A 座　　邮　编：100084
社 总 机：010-83470000　　邮　购：010-62786544
投稿与读者服务：010-62776969，c-service@tup.tsinghua.edu.cn
质量反馈：010-62772015，zhiliang@tup.tsinghua.edu.cn
课件下载：http://www.tup.com.cn，010-83470236

印 装 者：三河市铭诚印务有限公司
经　　销：全国新华书店
开　　本：185mm×260mm　　印　张：25.5　　字　数：624 千字
版　　次：2022 年 9 月第 1 版　　印　次：2022 年 9 月第 1 次印刷
印　　数：1～1500
定　　价：99.80 元

产品编号：093121-01

前言

Spring 框架是目前全球 Java 开发领域中最受欢迎的开发框架之一。现在主流的 Java 开发领域中有大量的项目基于 Spring 或者对接 Spring 进行适配。

笔者初识 Spring 是在 2015 年的一个项目中,当时项目使用的是 Spring 4.1 版本,该版本的功能虽然已经比较强大,但是各类配置文件的处理会比较烦琐。随着 Spring 版本的升级迭代,基于 SpringXML 的开发方式逐渐减少,Spring 注解模式开发逐渐增多,目前 Spring 注解模式开发已经成为主流技术。笔者作为 Spring 的使用人员,对于 Spring 中的一些实现细节十分感兴趣,并付诸实践记录了一些源码的流程,同时想把这些经验分享给更多的人,便有了本书。

Spring 框架升级迭代的速度相对其他一些框架而言更快,本书中所采用的 Spring 框架版本是 5.2.3.release,从 Spring IoC 层面来看,这些升级(新增 spring.factories 文件支持和新增 spring.components 文件支持)有些会影响 Spring IoC 相关内容,但是涉及 Spring IoC 相关的处理流程内容变化较少。如果有大版本的升级并改动了 Spring IoC 相关内容也属于正常情况,请读者耐心查看更新记录找到变化的内容再对其进行分析,从而完善对 Spring IoC 的相关认识。

本书的组织结构和主要内容

本书共分为 22 章。

第 1~8 章主要围绕 SpringXML 相关技术进行分析,将会对传统的 SpringXML 开发进行说明,并对 SpringXML 相关的各类配置解析进行分析,内容如下。

第 1 章 对 Spring 框架的使用进行说明,并讲述 Spring IoC 中的核心类。

第 2 章 对 Spring IoC 资源读取及注册相关内容进行分析,主要包含 SpringXML 资源文件的读取、注册和解析。

第 3 章 对自定义标签相关内容进行分析,包含如何编写一个自定义标签和自定义标签的处理过程分析。

第 4 章 对别名注册和 BeanDefinition 注册进行分析,包含别名注册环境搭建、别名注册流程和 BeanDefinition 注册流程的分析。

第 5 章 对 bean 标签解析进行详细分析,包含 bean 标签解析环境搭建和解析流程分析。

第 6 章 对 Bean 生命周期相关内容进行分析,包含 Bean 的初始化、Bean 的属性设置和 Bean 的摧毁相关分析。

第 7 章 对 Bean 的获取进行分析,包含获取 Bean 的环境搭建和获取 Bean 流程分析。

第 8 章 对 SpringXML 模式下容器的生命周期进行分析,包含容器的启动和容器的关闭(暂停)分析。

第9～14章主要围绕Spring注解相关技术进行分析,将会对Spring注解相关内容进行分析,包含且不限于注解Bean加载、配置类解析和常用注解分析,内容如下。

第9章 对Spring注解模式进行分析,包含Spring注解模式的环境搭建和启动流程分析。

第10章 对Spring配置类解析进行分析,包含常见注解的解析。

第11章 对Spring中负责配置解析的类进行分析,主要分析对象是ConfigurationClassPostProcessor。

第12章 对Spring注解模式下的Import注解相关源代码进行分析,主要分析对象是DeferredImportSelectorHandler。

第13章 对Spring注解模式下配置类中的Bean读取进行分析,包含测试环境搭建和读取流程分析。

第14章 对Spring中常见的元数据进行说明和演示。

第15～22章主要围绕Spring IoC辅助工具进行分析,将会对Spring中常见的一些辅助工具进行分析,包含且不限于占位符解析和Spring事件模式,内容如下。

第15章 对Spring中的事件处理机制进行演示和源码分析。

第16章 对Spring中的占位符解析进行源码分析。

第17章 对Spring中关于对象转换相关源码进行分析并简单实现了一个转换服务。

第18章 对Spring中的消息进行源码分析,包含Spring消息环境搭建和处理流程分析。

第19章 对Spring中资源解析器进行分析,包含资源解析器环境搭建和资源解析过程分析。

第20章 对Spring中BeanName的生成策略进行分析。

第21章 对Spring中的条件注解进行分析,包含条件注解的环境搭建和注解解析流程分析。

第22章 对Spring中的排序注解进行分析。

本书配套源代码可以扫描右侧二维码获取。

源代码

本书面向对象

本书面向具备Java编程能力的读者和对Spring IoC具有使用经验和有兴趣的读者,通过学习本书将学到Spring的基础使用以及一些高级功能,如Spring事件和转换服务等,以及Spring IoC的基础实现逻辑。

致谢

在此非常诚挚地感谢所有SpringFramework项目的创建者和开发者,感谢他们所做的基础性工作和对开源项目的热情,没有他们就没有本书的诞生。

由于编者水平有限,书中不当之处在所难免,欢迎广大同行和读者批评指正。

王 涛

2022年6月

目 录

第 1 章 Spring 容器环境搭建及基本使用 ················· 1

1.1 Spring 容器环境搭建 ················· 1
1.2 Spring 基本代码编辑 ················· 2
1.3 Spring IoC 核心类 ················· 5
小结 ················· 6

第 2 章 IoC 资源读取及注册 ················· 7

2.1 XML 文档验证 ················· 7
2.1.1 认识 XML 验证模式 ················· 7
2.1.2 Spring 中 XML 的验证 ················· 8
2.2 Document 对象获取 ················· 9
2.3 BeanDefinition 注册 ················· 10
2.3.1 doRegisterBeanDefinitions 流程 ················· 12
2.3.2 parseBeanDefinitions 分析 ················· 13
2.3.3 parseDefaultElement Spring 原生标签的处理 ················· 14
2.3.4 import 标签解析 ················· 14
2.3.5 alias 标签解析 ················· 18
2.3.6 bean 标签解析 ················· 21
2.3.7 自定义标签解析概述 ················· 28
小结 ················· 28

第 3 章 自定义标签 ················· 29

3.1 创建自定义标签环境搭建 ················· 29
3.1.1 编写 XSD 文件 ················· 29
3.1.2 编写 NamespaceHandler 实现类 ················· 30
3.1.3 编写 BeanDefinitionParser 实现类 ················· 30
3.1.4 编写注册方式 ················· 31
3.1.5 测试用例的编写 ················· 31
3.2 自定义标签解析 ················· 32
3.2.1 NamesapceHandler 和 BeanDefinitionParser 之间的关系 ················· 33

| 3.2.2   获取命名空间地址 ………………………………………………………… 33
| 3.2.3   NamespaceHandler 对象获取 ………………………………………… 33
| 3.2.4   getHandlerMappings 获取命名空间的映射关系 …………………… 34
| 3.2.5   NamespaceHandler 的获取 …………………………………………… 36
| 3.2.6   NamespaceHandler 的 init 方法 ……………………………………… 37
| 3.2.7   NamespaceHandler 缓存的刷新 ……………………………………… 38
| 3.2.8   解析标签 BeanDefinitionParser 对象准备 ………………………… 38
| 3.2.9   解析标签 parse 方法调用 …………………………………………… 39
| 小结 …………………………………………………………………………………… 43

第 4 章 别名注册和 BeanDefinition 注册 …………………………………………… 44

   4.1   别名注册测试环境搭建 ……………………………………………………… 44
   4.2   别名注册接口 ………………………………………………………………… 45
   4.3   SimpleAliasRegistry 中注册别名的实现 …………………………………… 45
   4.4   别名换算真名 ………………………………………………………………… 47
   4.5   BeanDefinition 注册 ………………………………………………………… 48
   4.6   DefaultListableBeanFactory 中存储 BeanDefinition 的容器 ……………… 50
   4.7   DefaultListableBeanFactory 中的注册细节 ………………………………… 50
       4.7.1   BeanDefinition 的验证 ……………………………………………… 51
       4.7.2   容器中存在 BeanName 对应的 BeanDefinition 的处理 …………… 52
       4.7.3   容器中不存在 BeanName 对应的 BeanDefinition 的处理 ………… 54
       4.7.4   BeanDefinition 的刷新处理 ………………………………………… 55
   4.8   BeanDefinition 的获取 ……………………………………………………… 57
   小结 …………………………………………………………………………………… 58

第 5 章 bean 标签解析 ………………………………………………………………… 59

   5.1   创建 bean 标签解析环境 …………………………………………………… 59
       5.1.1   编写 SpringXML 配置文件 ………………………………………… 59
       5.1.2   编写 bean-node 对应的测试用例 …………………………………… 60
   5.2   parseBeanDefinitionElement 方法处理 ……………………………………… 61
       5.2.1   parseBeanDefinitionElement 第一部分处理 ………………………… 61
       5.2.2   parseBeanDefinitionElement 第二部分处理 ………………………… 61
       5.2.3   parseBeanDefinitionElement 第三部分处理 ………………………… 63
   5.3   BeanDefinition 装饰 ………………………………………………………… 96
   5.4   BeanDefinition 细节 ………………………………………………………… 100
       5.4.1   AbstractBeanDefinition 属性 ………………………………………… 100
       5.4.2   RootBeanDefinition 属性 …………………………………………… 101
       5.4.3   ChildBeanDefinition 属性 …………………………………………… 102
       5.4.4   GenericBeanDefinition 属性 ………………………………………… 102

目 录

  5.4.5 AnnotatedGenericBeanDefinition 属性 …………………………… 103
小结 ………………………………………………………………………………… 103

第 6 章 Bean 的生命周期 ……………………………………………………… 105

6.1 Java 对象的生命周期 …………………………………………………… 105
6.2 浅看 Bean 生命周期 ……………………………………………………… 106
6.3 初始化 Bean ……………………………………………………………… 110
  6.3.1 无构造标签 ……………………………………………………… 110
  6.3.2 构造标签中的 index 模式和 name 模式 ………………………… 112
  6.3.3 Spring 中的实例化策略 ………………………………………… 115
6.4 Bean 属性设置 …………………………………………………………… 115
  6.4.1 BeanWrapper 创建 ……………………………………………… 117
  6.4.2 BeanWrapper 属性设置 ………………………………………… 118
  6.4.3 CachedIntrospectionResults 对象介绍 ………………………… 120
  6.4.4 PropertyValue 对象介绍 ………………………………………… 122
  6.4.5 最终的数据设置 ………………………………………………… 123
6.5 Bean 生命周期值 Aware 接口 …………………………………………… 124
6.6 BeanPostProcessor#postProcessBeforeInitialization ………………… 126
6.7 InitializingBean 接口和自定义 init-method 方法 ……………………… 126
6.8 BeanPostProcessor#postProcessAfterInitialization …………………… 129
6.9 Bean 的摧毁 ……………………………………………………………… 129
  6.9.1 DefaultSingletonBeanRegistry 中的摧毁 ……………………… 130
  6.9.2 DefaultListableBeanFactory 中的摧毁 ………………………… 133
小结 ………………………………………………………………………………… 133

第 7 章 Bean 的获取 ……………………………………………………………… 134

7.1 Bean 获取方式配置 ……………………………………………………… 134
7.2 Bean 获取的测试环境搭建 ……………………………………………… 134
7.3 doGetBean 分析 …………………………………………………………… 136
  7.3.1 BeanName 转换 ………………………………………………… 141
  7.3.2 尝试从单例容器中获取 ………………………………………… 141
  7.3.3 从 FactoryBean 接口中获取实例 ……………………………… 142
  7.3.4 尝试从父容器中获取 …………………………………………… 145
  7.3.5 BeanName 标记 ………………………………………………… 146
  7.3.6 非 FactoryBean 的单例对象创建 ……………………………… 146
  7.3.7 非 FactoryBean 的原型对象创建 ……………………………… 160
  7.3.8 既不是单例模式也不是原型模式的非 FactoryBean 创建 …… 161
  7.3.9 类型转换器中获取 Bean ………………………………………… 162
7.4 循环依赖 …………………………………………………………………… 163

7.4.1　Java 中的循环依赖 …… 163
7.4.2　Spring 中的循环依赖处理 …… 165
小结 …… 166

第 8 章　SpringXML 模式下容器的生命周期　167

8.1　SpringXML 模式下容器的生命周期测试环境搭建 …… 167
8.2　XmlBeanFactory 分析 …… 168
8.3　FileSystemXmlApplicationContext 分析 …… 170
　　8.3.1　父上下文处理 …… 170
　　8.3.2　配置文件路径解析 …… 172
　　8.3.3　刷新操作 …… 173
　　8.3.4　关闭方法分析 …… 189
8.4　ClassPathXmlApplicationContext 分析 …… 191
8.5　SpringXML 关键对象附表 …… 191
8.6　初识 LifecycleProcessor …… 194
8.7　LifecycleProcessor 测试环境搭建 …… 195
8.8　start 方法分析 …… 196
8.9　stop 方法分析 …… 198
8.10　LifecycleGroup 相关变量 …… 199
8.11　BeanPostProcessor 注册 …… 200
8.12　BeanFactoryPostProcessor 方法调用 …… 201
小结 …… 205

第 9 章　Spring 注解模式　206

9.1　注解模式测试环境搭建 …… 206
9.2　basePackages 模式启动 …… 207
　　9.2.1　scan 方法分析 …… 208
　　9.2.2　doScan 方法分析 …… 208
　　9.2.3　处理单个 BeanDefinition …… 215
9.3　componentClasses 模式启动 …… 221
小结 …… 224

第 10 章　Spring 配置类解析　225

10.1　parse 方法分析 …… 225
10.2　processConfigurationClass 方法分析 …… 226
10.3　doProcessConfigurationClass 方法分析 …… 229
10.4　处理各类注解 …… 229
　　10.4.1　处理 @Component 注解 …… 229
　　10.4.2　处理 @PropertySource 和 @PropertySources 注解 …… 231

        10.4.3 处理 @ComponentScans 和 @ComponentScan 注解 ············ 238
        10.4.4 处理 @Import 注解 ············ 241
        10.4.5 处理 @ImportResource 注解 ············ 243
        10.4.6 处理 @Bean 注解 ············ 244
    10.5 处理父类配置 ············ 248
    小结 ············ 249

第 11 章 ConfigurationClassPostProcessor 分析 ············ 250

    11.1 初识 ConfigurationClassPostProcessor ············ 250
    11.2 ConfigurationClassPostProcessor 测试用例搭建 ············ 251
    11.3 postProcessBeanDefinitionRegistry 方法分析 ············ 251
        11.3.1 容器内已存在的 Bean 进行候选分类 ············ 255
        11.3.2 候选 BeanDefinition Holder 的排序 ············ 256
        11.3.3 BeanName 生成器的创建 ············ 256
        11.3.4 初始化基本环境信息 ············ 257
        11.3.5 解析候选 Bean ············ 257
        11.3.6 注册 Import Bean 和清理数据 ············ 262
    11.4 postProcessBeanFactory 方法分析 ············ 262
    小结 ············ 265

第 12 章 DeferredImportSelectorHandler 分析 ············ 266

    12.1 初识 DeferredImportSelectorHandler ············ 266
    12.2 DeferredImportSelectorHandler 测试环境搭建 ············ 267
    12.3 handler 方法分析 ············ 269
    12.4 DeferredImportSelectorGroupingHandler 分析 ············ 269
    12.5 processImports 方法分析 ············ 270
    小结 ············ 273

第 13 章 ConfigurationClassBeanDefinitionReader 分析 ············ 274

    13.1 ConfigurationClassBeanDefinitionReader 测试环境搭建 ············ 274
    13.2 ConfigurationClassBeanDefinitionReader 构造函数 ············ 275
    13.3 loadBeanDefinitions 方法分析 ············ 276
    13.4 TrackedConditionEvaluator 分析 ············ 276
    13.5 loadBeanDefinitionsForConfigurationClass 方法分析 ············ 277
    13.6 loadBeanDefinitionsForBeanMethod 方法分析 ············ 278
    13.7 registerBeanDefinitionForImportedConfigurationClass 方法分析 ············ 282
    13.8 loadBeanDefinitionsFromImportedResources 方法分析 ············ 285
    13.9 loadBeanDefinitionsFromRegistrars 方法分析 ············ 287
    小结 ············ 289

第 14 章 Spring 元数据 290

- 14.1 认识 MetadataReaderFactory 290
- 14.2 SimpleMetadataReaderFactory 分析 290
- 14.3 CachingMetadataReaderFactory 分析 292
- 14.4 注解元数据读取工厂总结 293
- 14.5 初识 MetadataReader 293
- 14.6 MetadataReader 接口实现类说明 294
  - 14.6.1 SimpleAnnotationMetadataReadingVisitor 成员变量 294
  - 14.6.2 SimpleAnnotationMetadata 成员变量 294
  - 14.6.3 SimpleMethodMetadata 成员变量 295
  - 14.6.4 MergedAnnotationsCollection 成员变量 295
  - 14.6.5 TypeMappedAnnotation 成员变量 296
  - 14.6.6 AnnotationTypeMappings 成员变量 296
  - 14.6.7 ClassMetadataReadingVisitor 成员变量 297
  - 14.6.8 AnnotationMetadataReadingVisitor 成员变量 297
- 14.7 类元数据接口说明 298
- 14.8 StandardClassMetadata 对象分析 299
- 14.9 注解元数据基础认识 299
- 14.10 Java 中注解数据获取 301
- 14.11 ScopeMetadataResolver 分析 302
  - 14.11.1 ScopeMetadata 分析 303
  - 14.11.2 AnnotationScopeMetadataResolver 分析 303
- 小结 305

第 15 章 Spring 事件 306

- 15.1 Spring 事件测试环境搭建 306
- 15.2 Spring 事件处理器注册 308
  - 15.2.1 事件处理器实例创建后 309
  - 15.2.2 事件处理器实例摧毁前 310
- 15.3 Spring 事件推送和处理 311
- 小结 314

第 16 章 占位符解析 315

- 16.1 基本环节搭建 315
- 16.2 XML 的解析 316
- 16.3 外部配置的读取 318
- 16.4 字符串占位符解析 321
  - 16.4.1 resolveStringValue 分析 326

16.4.2　resolvePlaceholders 分析 ·············· 326
　　　16.4.3　resolveRequiredPlaceholders 分析 ·············· 333
　　　16.4.4　BeanDefinitionVisitor#visitBeanDefinition 分析 ·············· 334
　小结 ·············· 335

第 17 章　Spring 中的转换服务 ·············· 336

　17.1　初识 Spring 转换服务 ·············· 336
　17.2　ConversionServiceFactoryBean 对象的实例化 ·············· 337
　　　17.2.1　afterPropertiesSet 方法分析 ·············· 338
　　　17.2.2　GenericConversionService 对象创建 ·············· 338
　　　17.2.3　注册转换服务 ·············· 339
　　　17.2.4　ConversionServiceFactory.registerConverters 分析 ·············· 344
　17.3　转换过程分析 ·············· 345
　　　17.3.1　ConversionService 分析 ·············· 345
　　　17.3.2　handleResult 分析 ·············· 347
　　　17.3.3　getConverter 分析 ·············· 347
　　　17.3.4　ConversionUtils.invokeConverter 分析 ·············· 350
　　　17.3.5　handleConverterNotFound 分析 ·············· 352
　17.4　脱离 Spring 实现转换服务 ·············· 352
　小结 ·············· 355

第 18 章　MessageSource 源码分析 ·············· 356

　18.1　MessageSource 测试环境搭建 ·············· 356
　18.2　MessageSource 实例化 ·············· 357
　18.3　getMessage 方法分析 ·············· 358
　　　18.3.1　resolveCodeWithoutArguments 方法分析 ·············· 360
　　　18.3.2　resolveCode 方法分析 ·············· 362
　小结 ·············· 363

第 19 章　资源解析器 ·············· 364

　19.1　资源解析器测试环境搭建 ·············· 364
　19.2　ResourcePatternResolver 类图分析 ·············· 365
　19.3　PathMatchingResourcePatternResolver 构造器分析 ·············· 366
　19.4　getResource 方法分析 ·············· 366
　19.5　getResources 方法分析 ·············· 368
　　　19.5.1　findPathMatchingResources 方法分析 ·············· 369
　　　19.5.2　findAllClassPathResources 方法分析 ·············· 380
　小结 ·············· 381

第 20 章 BeanName 生成策略 ·········· 382

20.1 AnnotationBeanNameGenerator 分析 ·········· 382
20.1.1 AnnotatedBeanDefinition 类型的 BeanName 生成策略 ·········· 383
20.1.2 非 AnnotatedBeanDefinition 类型的 BeanName 生成策略 ·········· 384
20.2 FullyQualifiedAnnotationBeanNameGenerator 分析 ·········· 385
20.3 DefaultBeanNameGenerator 分析 ·········· 385
小结 ·········· 386

第 21 章 条件注解 ·········· 387

21.1 条件注解测试环境搭建 ·········· 387
21.2 条件注解分析 ·········· 388
小结 ·········· 391

第 22 章 Spring 排序注解 ·········· 392

22.1 排序注解测试环境搭建 ·········· 392
22.2 OrderComparator.sort 方法分析 ·········· 393
小结 ·········· 396

第1章 Spring容器环境搭建及基本使用

从本章开始将进入SpringXML相关的源码分析,将会介绍Spring框架的简单使用,这个使用是基于SpringXML的一种使用方式。此外,在本章中还会讲述Spring IoC容器的功能和Spring IoC中一些核心对象的作用。

1.1 Spring容器环境搭建

在开始Spring IoC相关分析之前,需要先搭建一个Spring的基本工程,该工程可以从Spring容器中获取一个Bean,不含其他复杂操作。

在IDEA中右击项目顶层,单击New→Module选项,如图1.1所示。

图1.1 选择Module工程

单击Module选项后会弹出如图1.2所示对话框,单击Gradle选项,勾选Java复选框,再单击Next按钮进入下一步。

单击Next按钮后弹出图1.3中的内容,输入Name、GroupId、ArtifactId和Version这四项数据内容。

输入完成后单击Finish按钮完成工程创建。

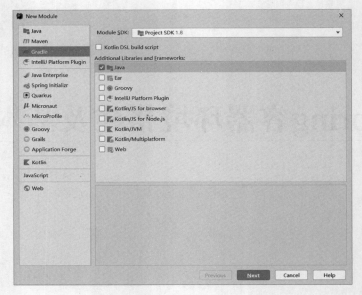

图 1.2 选择 Gradle 和 Java 工程

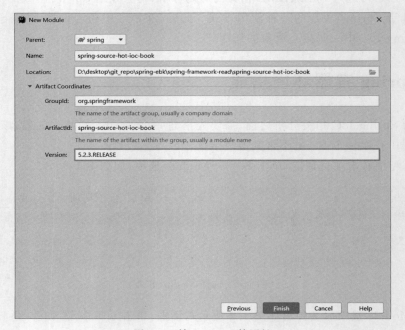

图 1.3 输入 Name 等属性

1.2 Spring 基本代码编辑

基本工程创建完成后,编写几个基本类和配置文件。首先在 build.gradle 文件中添加 Spring 相关的依赖:spring-context 和 spring-core,这两个依赖是 Spring 中的最小依赖单元。

编写后的 build.gradle 文件内容如下。

```
plugins {
  id 'java'
}

group 'org.springframework'
version '5.2.3.RELEASE'

repositories {
  mavenCentral()
}
dependencies {
  compile(project(":spring-context"))
  compile(project(":spring-core"))
  testCompile("org.junit.jupiter:junit-jupiter-api")
  testCompile("org.junit.jupiter:junit-jupiter-params")
}

test {
  useJUnitPlatform()
}
```

注意，上述代码是在 SpringFramework 源码工程中进行创建的，属于 SpringFramework 项目中的一部分，因此不需要编写版本号，如果不是 SpringFramework 环境，就使用下面这种方式进行依赖注入。

compile group：'org.springframework'，name：'spring-core'，version：'5.2.3.RELEASE'

在完成基本的 build.gradle 配置文件编写后，需要创建一个最小的 JavaBean 实体，具体代码如下。

```java
public class PeopleBean {
  private String name;

  public String getName() {
      return name;
  }

  public void setName(String name) {
      this.name = name;
  }
}
```

完成 JavaBean 编写后，需要制作一个 SpringXML 配置文件，该文件用来存储 Spring 的一些相关配置，将其命名为 first-ioc.xml，其中的内容如下。

```
<?xml version="1.0" encoding="UTF-8"?>
<beans xmlns:xsi="http://www.w3.org/2001/XMLSchema-instance"
```

```xml
xmlns="http://www.springframework.org/schema/beans"
xsi:schemaLocation="http://www.springframework.org/schema/beans http://www.springframework.org/schema/beans/spring-beans.xsd">

    <bean id="people" class="com.source.hot.ioc.book.pojo.PeopleBean">
        <property name="name" value="zhangsan"/>
    </bean>
</beans>
```

完成了SpringXML编写后,需要编写测试用例,测试用例类名为FirstIoCDemoTest,具体代码如下。

```java
class FirstIoCDemoTest {
    @Test
    void testIoC() {
        ClassPathXmlApplicationContext context
                = new ClassPathXmlApplicationContext("META-INF/first-ioc.xml");

        PeopleBean people = context.getBean("people", PeopleBean.class);

        String name = people.getName();
        assumeTrue(name.equals("zhangsan"));
    }
}
```

至此,所需要的基本文件都已经编写完成。在这个基本测试用例中,使用 ClassPathXmlApplicationContext 类作为 Spring IoC 容器对象,在 Spring 中还存在另外两种可以用作 Spring IoC 容器的对象:FileSystemXmlApplication 和 XmlBeanFactory。XmlBeanFactory 已经被废弃了,虽然标记了废弃注解,但是还是可以使用这个类。XmlBeanFactory 中的废弃标记信息如下。

```java
@Deprecated
@SuppressWarnings({"serial","all"})
public class XmlBeanFactory extends DefaultListableBeanFactory {}
```

下面进一步增加测试用例。先来编写 XmlBeanFactory 相关的测试用例,具体代码如下。

```java
@Test
void testXmlBeanFactory() {
    XmlBeanFactory beanFactory =
    new XmlBeanFactory(new ClassPathResource("META-INF/first-ioc.xml"));

    PeopleBean people = beanFactory.getBean("people",PeopleBean.class);

    String name = people.getName();
    assumeTrue(name.equals("zhangsan"));
}
```

完成了 XmlBeanFactory 的测试用例编写后,进一步对 FileSystemXmlApplicationContext 类

进行单元测试的编写,在这个测试用例的编写中使用 first-ioc.xml 全路径。具体的测试代码如下。

```
@Test
void testFileSystemXmlApplicationContext() {
  FileSystemXmlApplicationContext context =
new FileSystemXmlApplicationContext("D:\\desktop\\git_repo\\spring-ebk\\spring-framework-read\\spring-source-hot-ioc-book\\src\\test\\resources\\META-INF\\first-ioc.xml");

  PeopleBean people = context.getBean("people",PeopleBean.class);

  String name = people.getName();
  assumeTrue(name.equals("zhangsan"));
}
```

至此,对于 SpringXML 的三种容器处理方式的测试用例都编写完成。

通过前文的测试用例编写,可以发现 Spring IoC 容器至少具备以下两种功能。

(1) 读取 SpringXML 配置文件。
(2) 从容器中获取实例,获取 Bean 实例。

1.3 Spring IoC 核心类

接下来将介绍 Spring IoC 中的核心类(部分),在前文的测试用例中引出了 Spring IoC 容器使用的三个类:ClassPathXmlApplicationContext、XmlBeanFactory 和 FileSystemXmlApplicationContext,这三个类是 Spring IoC 容器入口的核心对象。ClassPathXmlApplicationContext 的类图如图 1.4 所示。

图 1.4　ClassPathXmlApplicationContext 类图

下面是 FileSystemXmlApplicationContext 的类图,如图 1.5 所示。

图 1.5　FileSystemXmlApplicationContext 类图

最后是 XmlBeanFactory 的类图,如图 1.6 所示。

下面将对类图中出现的一些接口做一个简单的说明。
(1) Resource:Spring 中资源的定义接口。
(2) ResourceLoader:该对象提供了资源加载方法。
(3) BeanDefinitionReader:该对象提供了读取资源对象到 Bean 定义的方法。

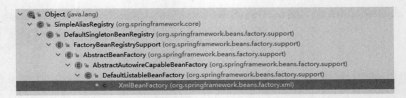

图 1.6　XmlBeanFactory 类图

（4）DocumentLoader：该对象提供了将资源文件转换成 Document 对象的方法。

（5）BeanDefinitionDocumentReader：该对象提供了将 Document 对象读取并注册到容器的方法。

（6）EnvironmentCapable：该对象提供了获取环境配置的方法。

（7）AliasRegistry：该对象提供了关于 Alias 的增、删、改、查方法。

（8）SingletonBeanRegistry：该对象提供了关于单例 Bean 的增删改查方法。

（9）BeanDefinitionRegistry：该对象提供了关于 Bean 定义的增删改查方法。

（10）BeanFactory：该对象提供了获取 Bean 实例的方法。

（11）HierarchicalBeanFactory：该对象在 BeanFactory 基础上提供了关于父 BeanFactory 的支持。

（12）ConfigurableBeanFactory：该对象提供了对 BeanFactory 的设置方法，在其中可以设置关于类加载器、转换服务等配置信息。

（13）AutowireCapableBeanFactory：该对象提供了创建 Bean、注入 Bean、Bean 后置处理器（BeanPostProcessor）、摧毁 Bean 等方法。

（14）ApplicationContext：应用上下文核心接口，各类上下文实现类都是它的实现类。

（15）ConfigurableApplicationContext：该对象提供了配置应用上下文属性的方法。

（16）ListableBeanFactory：该对象提供了搜索 Bean 的方法，从容器中获取关于 Bean 的配置。

（17）Lifecycle：该对象提供了关于容器的生命周期方法。

对于上述 17 个接口可以做出如下 5 种分类。

（1）资源及资源处理类。该分类有 Resource 和 ResourceLoader。

（2）处理注册的类。该分类有 alias（AliasRegistry）、Bean 定义注册（BeanDefinitionRegistry）和单例 Bean 注册（SingletonBeanRegistry）。

（3）关于生命周期相关的分类，该分类还可以继续细分为以下两类。

① 容器的生命周期。容器生命周期的核心接口为 Lifecycle。

② Bean 的生命周期。Bean 生命周期的接口有 InitializingBean 和 DisposableBean 等。

（4）Bean 的拓展。该分类有 BeanPostProcessor 和 Aware 系列接口。

（5）上下文相关分类。该分类以 ApplicationContext 作为主导接口。

小结

本章对 Spring 容器（SpringXML 模式）的使用做了简单说明，介绍了如何创建一个 SpringXML 应用以及如何从容器中获取数据，还对 Spring 中的核心对象进行了简要说明。

第 2 章

IoC 资源读取及注册

在第 1 章中讲述了基于 SpringXML 的 Spring 容器的使用，在 Spring 容器的使用过程中需要编写 SpringXML 配置文件，这个 SpringXML 配置文件就是本章的分析目标：资源。本章将介绍 Spring 是如何对 SpringXML 配置文件进行解析并转换成 BeanDefinition 的。本章将围绕三点进行分析：XML 文档的验证，Document 对象的获取，SpringXML 配置文件转换成 BeanDefinition 并进行注册。

2.1 XML 文档验证

2.1.1 认识 XML 验证模式

现如今的各种编辑器的智能提示功能都十分强大，使得开发者们在编写 XML 文件时可以降低出错的可能，但是还是会出现错误。面对这些错误程序，需要通过一定的规则对 XML 文件进行验证。

对于 XML 文件的验证，在 XML 这一个文件协议提出之时就有了验证方式，各类编程语言也都是围绕这个验证方式进行开发，一般常用的验证方式是 DTD（Document Type Definition）验证。DTD 是一种特殊文档，它规定、约束符合标准通用标示语言（SGML）或 SGML 子集可扩展标示语言（XML）规则的定义和陈述。XSD（XML Schema Definition）是另一种常见的验证方式。它描述了可扩展标记语言文档的结构，可用于替代 DTD。这两种验证方式都需要编写基本文档，它们的验证离不开自身的文档，其中，DTD 文档的拓展名为 *.dtd，XSD 文档的拓展名为 *.xsd。接下来需要在 Spring 框架中寻找到它们的具体路径信息。在 Spring 中，基本的 bean 标签相关的 XML 文件验证方式文件存储在 spring-beans/src/main/resources/org/springframework/beans/factory/xml 路径下，在这个路径中存放了对于 SpringXML 的基础验证文件内容，文件内容如图 2.1 所示。

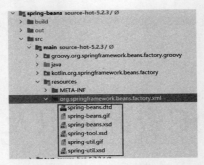

图 2.1 spring-beans 工程下存储的 XML 验证文件

2.1.2 Spring 中 XML 的验证

在 2.1.1 节中介绍了关于 XML 文件的验证方式，下面将详细地讲解 Spring 是如何对 XML 文件进行验证的。XML 有两种验证方式：DTD 和 XSD。Spring 需要对这两种验证方式进行推论，确认是两者中的某一个。下面在 Spring 源码中找到推论 XML 验证方式的方法签名，具体方法签名是 org.springframework.beans.factory.xml.XmlBeanDefinitionReader # getValidationModeForResource，getValidationModeForResource 代码内容如下。

```
protected int getValidationModeForResource(Resource resource) {
    //获取 XML 验证方式
    int validationModeToUse = getValidationMode();
    if (validationModeToUse != VALIDATION_AUTO) {
        return validationModeToUse;
    }
    int detectedMode = detectValidationMode(resource);
    if (detectedMode != VALIDATION_AUTO) {
        return detectedMode;
    }
    return VALIDATION_XSD;
}
```

这段代码对于 XML 验证方式的确认有如下两个操作步骤。

（1）从成员变量 validationMode 中获取验证模式。

（2）验证模式如果是自动推测，那么会交给 detectValidationMode 方法进行处理，处理后得到具体的验证模式。

detectValidationMode 方法的方法签名是 org.springframework.util.xml.XmlValidationModeDetector # detectValidationMode，其中的代码内容如下。

```
public int detectValidationMode(InputStream inputStream) throws IOException {
    BufferedReader reader = new BufferedReader(new InputStreamReader(inputStream));
    try {
        boolean isDtdValidated = false;
        String content;
        while ((content = reader.readLine()) != null) {
```

```
            content = consumeCommentTokens(content);
            if (this.inComment || !StringUtils.hasText(content)) {
                continue;
            }
            if (hasDoctype(content)) {
                isDtdValidated = true;
                break;
            }
            if (hasOpeningTag(content)) {
                //End of meaningful data...
                break;
            }
        }
        return (isDtdValidated ? VALIDATION_DTD: VALIDATION_XSD);
    }
    catch (CharConversionException ex) {
        return VALIDATION_AUTO;
    }
    finally {
        reader.close();
    }
}
```

在这段代码中,需要重点关注以下这段代码。

```
if (hasDoctype(content)) {
    isDtdValidated = true;
    break;
}
```

这段核心代码主要依靠 hasDoctype 方法来进行,hasDoctype 方法的处理行为就是字符串判断,它判断是否存在 DOCTYPE 字符串信息,如果存在 DOCTYPE,则认为是 DTD 验证方式,反之则是 XSD 验证方式。

2.2 Document 对象获取

接下来将进入 Document 对象的获取相关内容分析,注意这里的 Document 对象可以理解为 SpringXML 配置文件,Spring 中将 SpringXML 配置文件的读取并转换成 Document 对象这一个重要任务交给 DocumentLoader 接口负责,下面是 DocumentLoader 接口的定义。

```
public interface DocumentLoader {

    Document loadDocument(
            InputSource inputSource, EntityResolver entityResolver,
            ErrorHandler errorHandler, int validationMode, boolean namespaceAware)
        throws Exception;

}
```

首先需要认识接口的一些数据来源,在这个接口中重点需要关注 InputSource 数据的来源,在 Spring 中对于这个数据的提取是通过 org.springframework.beans.factory.xml.XmlBeanDefinitionReader#loadBeanDefinitions(org.springframework.core.io.support.EncodedResource)方法进行的,具体代码如下。

```
//省略了前后代码
InputStream inputStream = encodedResource.getResource().getInputStream();
try {
  InputSource inputSource = new InputSource(inputStream);
  if (encodedResource.getEncoding() != null) {
    inputSource.setEncoding(encodedResource.getEncoding());
  }
  return doLoadBeanDefinitions(inputSource,encodedResource.getResource());
```

在这段代码中可以看到,它是从 Resource 中获取 InputSource 对象。了解 InputSource 的数据来源后就可以寻找 DocumentLoader 的实现类。在 Spring 中 DocumentLoader 有且仅有一个实现类,这个实现类是 DefaultDocumentLoader。下面是 DefaultDocumentLoader 实现 DocumentLoader 中 loadDocument 方法的细节代码。

```
@Override
public Document loadDocument(InputSource inputSource,EntityResolver entityResolver,
        ErrorHandler errorHandler, int validationMode, boolean namespaceAware) throws Exception {

    //创建 XML Document 构建工具
    DocumentBuilderFactory factory = createDocumentBuilderFactory(validationMode,namespaceAware);
    if (logger.isTraceEnabled()) {
      logger.trace("Using JAXP provider [" + factory.getClass().getName() + "]");
    }

    //DocumentBuilder 类创建
    DocumentBuilder builder = createDocumentBuilder(factory,entityResolver,errorHandler);
    return builder.parse(inputSource);
}
```

在这个方法中对于 Document 对象的创建主要依赖于 javax.xml 和 org.w3c 两个包下的接口或者类,具体创建过程不做具体分析。

2.3 BeanDefinition 注册

通过前文得到了 Document 对象,Spring 在得到 Document 对象后会对它进行解析,将 SpringXML 配置文件中的数据进行解析转换成 BeanDefinition 对象并进行注册,主要负责这项行为的入口方法签名是 org.springframework.beans.factory.xml.XmlBeanDefinitionReader#doLoadBeanDefinitions,该方法的代码信息如下。

```
//去掉了异常处理和日志
protected int doLoadBeanDefinitions(InputSource inputSource,Resource resource)
```

```
    throws BeanDefinitionStoreException {
    //将输入流转换成 Document
    Document doc = doLoadDocument(inputSource,resource);
    //注册 Bean 定义,并获取数量
    int count = registerBeanDefinitions(doc,resource);
    return count;
}
```

下面进一步对 registerBeanDefinitions 方法进行阅读,registerBeanDefinitions 的代码详情如下。

```
public int registerBeanDefinitions(Document doc,Resource resource) throws
BeanDefinitionStoreException {
    //获取基于 Document 的 Bean 定义读取器
    BeanDefinitionDocumentReader documentReader = createBeanDefinitionDocumentReader();
    //历史已有的 Bean 定义数量
    int countBefore = getRegistry().getBeanDefinitionCount();
    //注册 Bean 定义
    documentReader.registerBeanDefinitions(doc,createReaderContext(resource));
    //注册后的数量 - 历史数量
    return getRegistry().getBeanDefinitionCount() - countBefore;
}
```

通过阅读 registerBeanDefinitions 方法已经可以找到重要的类(接口),它是 BeanDefinitionDocumentReader,需要重点分析的方法是 registerBeanDefinitions,BeanDefinitionDocumentReader 接口中对 registerBeanDefinitions 方法的定义信息如下。

```
public interface BeanDefinitionDocumentReader {

    /**
     * 注册 Bean 定义
     */
    void registerBeanDefinitions(Document doc,XmlReaderContext readerContext)
            throws BeanDefinitionStoreException;

}
```

BeanDefinitionDocumentReader#registerBeanDefinitions 方法的作用是 BeanDefinition 的注册。在 Spring 中 BeanDefinitionDocumentReader 的实现类是 DefaultBeanDefinitionDocumentReader,在 DefaultBeanDefinitionDocumentReader 类中找到 registerBeanDefinitions 方法,具体代码如下。

```
@Override
public void registerBeanDefinitions(Document doc,XmlReaderContext readerContext) {
    this.readerContext = readerContext;
    doRegisterBeanDefinitions(doc.getDocumentElement());
}
```

在这段方法中,重点方法是 doRegisterBeanDefinitions,在 doRegisterBeanDefinitions 方法中的处理是进行 Document 对象解析,将解析结果包装成 BeanDefinition 对象并将 BeanDefinition 注册到 Spring IoC 容器中。

2.3.1　doRegisterBeanDefinitions 流程

下面将进入 doRegisterBeanDefinitions 方法的分析，doRegisterBeanDefinitions 的方法签名是 org.springframework.beans.factory.xml.DefaultBeanDefinitionDocumentReader#doRegisterBeanDefinitions，具体代码如下。

```
//删除了注释和日志
protected void doRegisterBeanDefinitions(Element root) {
    //父 BeanDefinitionParserDelegate 一开始为 null
    BeanDefinitionParserDelegate parent = this.delegate;
    //创建 BeanDefinitionParserDelegate
    this.delegate = createDelegate(getReaderContext(),root,parent);

    //判断命名空间是否为默认的命名空间
    //默认命名空间: http://www.springframework.org/schema/beans
    if (this.delegate.isDefaultNamespace(root)) {
        //获取 profile 属性
        String profileSpec = root.getAttribute(PROFILE_ATTRIBUTE);
        //是否存在 profile
        if (StringUtils.hasText(profileSpec)) {
            //profile 切分后的数据
            String[] specifiedProfiles = StringUtils.tokenizeToStringArray(
                profileSpec,
                BeanDefinitionParserDelegate.MULTI_VALUE_ATTRIBUTE_DELIMITERS);
            if (!getReaderContext().getEnvironment().acceptsProfiles(specifiedProfiles)) {
                return;
            }
        }
    }

    //前置处理
    preProcessXml(root);
    //BeanDefinition 处理
    parseBeanDefinitions(root,this.delegate);
    //后置 XML 处理
    postProcessXml(root);

    this.delegate = parent;
}
```

在 doRegisterBeanDefinitions 方法中主要的处理流程如下。

（1）设置父 BeanDefinitionParserDelegate 对象，值得注意的是，这个设置父对象一般情况下是不存在的，即 this.delegate=null。

（2）创建 BeanDefinitionParserDelegate 对象，BeanDefinitionParserDelegate 对象是作为解析的重要方法。

（3）对于 profile 属性的处理。

（4）XML 解析的前置处理。

（5）XML 的解析处理。

（6）XML 解析的后置处理。

（7）设置成员变量。

这里提一个拓展点，在这段处理过程中有 profile 属性，这个属性一般是用来做 Spring 的环境区分，在 SpringBoot 框架中有一个类似的配置 spring.profiles，在 SpringXML 模式中 profile 是 beans 标签的一个属性。

在 doRegisterBeanDefinitions 方法中遇到的 preProcessXml 方法和 postProcessXml 方法目前属于空方法状态，没有任何实现代码。

2.3.2　parseBeanDefinitions 分析

下面进入到 parseBeanDefinitions 方法的分析，parseBeanDefinitions 方法代码如下。

```
protected void parseBeanDefinitions(Element root,BeanDefinitionParserDelegate delegate) {
    //是否是默认的命名空间
    if (delegate.isDefaultNamespace(root)) {
        //子节点列表
        NodeList nl = root.getChildNodes();
        for (int i = 0; i < nl.getLength(); i++) {
            Node node = nl.item(i);
            if (node instanceof Element) {
                Element ele = (Element) node;
                //是否是默认的命名空间
                if (delegate.isDefaultNamespace(ele)) {
                    //处理标签的方法
                    parseDefaultElement(ele,delegate);
                }
                else {
                    //处理自定义标签
                    delegate.parseCustomElement(ele);
                }
            }
        }
    }
    else {
        //处理自定义标签
        delegate.parseCustomElement(root);
    }
}
```

parseBeanDefinitions 方法是对每一个 XML 中的节点（Element）进行处理，节点本身存在多样性，节点是指 XML 文件中的一个标签，节点多样性是指节点的多种可能，在 Spring 框架中对节点有以下两种分类。

（1）Spring 提供的默认标签，即在 DTD 文件或者 XSD 文件中定义的标签。

（2）开发者自行编写（定义）的标签。

节点多样性在 Spring 中提供了两个处理方法，一个是 parseDefaultElement 方法，另一个是 parseCustomElement 方法，前者是为了处理 Spring 原生标签，后者是为了处理开发者自定义的标签。

2.3.3 parseDefaultElement Spring 原生标签的处理

在 Spring 中原生标签包含四种：alias 标签、bean 标签、beans 标签和 import 标签。在 Spring 中对于这四种标签的定义存在一定的层次关系，在 spring-beans.dtd 文件中对四种标签的层次关系定义如下。

```
<!ELEMENT beans (
    description?,
    (import | alias | bean) *
)>
```

四种标签的层次关系为：beans 标签下包含 import 标签、alias 标签和 bean 标签。接下来进入到处理这四种标签的方法分析，parseDefaultElement 方法详细内容如下。

```
private void parseDefaultElement(Element ele,BeanDefinitionParserDelegate delegate) {
    //解析 import 标签
    if (delegate.nodeNameEquals(ele,IMPORT_ELEMENT)) {
        importBeanDefinitionResource(ele);
    }
    //解析 alias 标签
    else if (delegate.nodeNameEquals(ele,ALIAS_ELEMENT)) {
        processAliasRegistration(ele);
    }
    //解析 bean 标签
    else if (delegate.nodeNameEquals(ele,BEAN_ELEMENT)) {
        processBeanDefinition(ele,delegate);
    }
    //解析 beans 标签
    //嵌套的 beans
    else if (delegate.nodeNameEquals(ele,NESTED_BEANS_ELEMENT)) {
        doRegisterBeanDefinitions(ele);
    }
}
```

根据前文所述标签存在的关系，可以将最后一个条件语句当作一个递归，再次进行各类标签的解析。接下来将对 import 标签、alias 标签和 bean 标签分别进行阐述。

2.3.4 import 标签解析

处理 import 标签的方法是 importBeanDefinitionResource，为了更好地分析 import 标签的处理过程，需要先制作一个 import 标签解析的模拟环境，首先创建一个 SpringXML 配置文件，文件名称是 import-beans.xml，文件中的内容如下。

```
<?xml version = "1.0" encoding = "UTF - 8"?>
```

```xml
<beans xmlns="http://www.springframework.org/schema/beans"
       xmlns:xsi="http://www.w3.org/2001/XMLSchema-instance"
       xsi:schemaLocation=" http://www.springframework.org/schema/beans http://www.springframework.org/schema/beans/spring-beans.xsd">
    <import resource="first-ioc.xml"/>
</beans>
```

完成 SpringXML 配置文件编写后再编写一个测试类,类名为 ImportNodeTest,详细代码如下。

```java
class ImportNodeTest {

    @Test
    void testImportNode() {
        ClassPathXmlApplicationContext context =
                new ClassPathXmlApplicationContext("META-INF/import-beans.xml");
        context.close();
    }

}
```

首先,将 import-beans.xml 文件中关于 import 标签相关的内容提取出来,提取后数据如下: `<import resource="first-ioc.xml"/>`。通过阅读提取的数据可以确定在使用中需要设置 resource 属性,进一步追踪源代码来确认 import 标签中的一些属性,spring-beans.dtd 文件中对 import 标签的定义如下。

```
<!ELEMENT import EMPTY>

<!ATTLIST import resource CDATA #REQUIRED>
```

spring-beans.xsd 文件中对于 import 标签的定义信息如下。

```xml
<xsd:element name="import">
   <xsd:annotation>
      <xsd:documentation source="java:org.springframework.core.io.Resource"><![CDATA[
Specifies an XML bean definition resource to import.
      ]]></xsd:documentation>
   </xsd:annotation>
   <xsd:complexType>
      <xsd:complexContent>
         <xsd:restriction base="xsd:anyType">
            <xsd:attribute name="resource" type="xsd:string" use="required">
               <xsd:annotation>
                  <xsd:documentation><![CDATA[
The relative resource location of the XML (bean definition) file to import,
for example "myImport.xml" or "includes/myImport.xml" or "../myImport.xml".
                  ]]></xsd:documentation>
               </xsd:annotation>
            </xsd:attribute>
         </xsd:restriction>
      </xsd:complexContent>
```

```
</xsd: complexType>
</xsd: element>
```

从 spring-beans.dtd 文件和 spring-beans.xsd 文件的描述中可以确定 import 标签只存在一个属性,该属性是 resource。下面先来阅读 import 标签解析方法中的第一部分代码。

```
//获取 resource 属性
String location = ele.getAttribute(RESOURCE_ATTRIBUTE);
//是否存在地址
if (!StringUtils.hasText(location)) {
    getReaderContext().error("Resource location must not be empty",ele);
    return;
}

//处理配置文件占位符
location = getReaderContext().getEnvironment().resolveRequiredPlaceholders(location);

//资源集合
Set<Resource> actualResources = new LinkedHashSet<>(4);

//是不是绝对地址
boolean absoluteLocation = false;
try {
    //判断是否为 URL
    //通过转换成 URI 判断是否是绝对地址
    absoluteLocation = ResourcePatternUtils.isUrl(location) ||
ResourceUtils.toURI(location).isAbsolute();
}
catch (URISyntaxException ex) {
}
```

在这段代码中必不可少的处理是提取 resource 属性,通过调试及对比前文的 SpringXML 配置文件观察变量 location 的数据情况,location 变量存储的是 import 标签中的 resource 属性,location 的数据信息如图 2.2 所示。

图 2.2 location 变量信息

在得到 location 变量信息后,Spring 会对其进行占位符解析操作,常用占位符是 ${},在后续的解析过程中会将占位符 ${}解析成为一个具体的地址,占位符解析相关的接口是 PropertyResolver。

获取 location 变量后对它做了一个是否是绝对路径的判断,判断逻辑是否为 URL 或者转换成 URI 后判断是否是绝对地址。在得到是否是绝对路径的判断结果后第一部分的处理结束。下面继续进行 import 标签的解析,关于 import 标签解析第二部分和第三部分相关代码如下。

```
//删除了异常处理和日志
//第二部分
//是不是绝对地址
if (absoluteLocation) {
    //获取 import 的数量(Bean 定义的数量)
    int importCount =
getReaderContext().getReader().loadBeanDefinitions(location,actualResources);
}
//第三部分
else {
    //import 的数量
    int importCount;
    //资源信息
    Resource relativeResource = getReaderContext().getResource().createRelative(location);
    //资源是否存在
    if (relativeResource.exists()) {
        //确定加载的 Bean 定义数量
        importCount = getReaderContext().getReader().loadBeanDefinitions(relativeResource);
        //加入资源集合
        actualResources.add(relativeResource);
    }
    //资源不存在处理方案
    else {
        //获取资源 URL 的数据
        String baseLocation = getReaderContext().getResource().getURL().toString();
        //获取 import 数量
        importCount = getReaderContext().getReader().loadBeanDefinitions(
            StringUtils.applyRelativePath(baseLocation,location),actualResources);
    }
}
```

在第二部分代码和第三部分代码处理过程中，主要进行的是 loadBeanDefinitions 方法，这个方法本质上就是在做 Bean 的加载，它和 doLoadBeanDefinitions 方法有着相同的能力，在这个方法中其主要目的是解析 beans 标签，这种处理是一种嵌套的处理。前文讲述了四大标签（beans、bean、alias 和 import）的关系，四个标签之间的嵌套关系如图 2.3 所示。

从图 2.3 中可以知道，import 标签下会间接包含 import 标签，这个标签的数据从 beans 标签引入，在 beans 标签下直接包含 bean 标签、alias 标签和 import 标签。

对于 import 标签的解析其实就是对 SpringXML 配置文件的解析，一个 SpringXML 配置文件中有什么标签就会被引入什么标签。完成了 import 标签的解析后会进行 import 事件处理，import 事件处理方法如下。

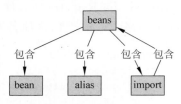

图 2.3 标签之间的嵌套关系

```
//第四部分
Resource[] actResArray = actualResources.toArray(new Resource[0]);
//唤醒 import 处理事件
getReaderContext().fireImportProcessed(location,actResArray,extractSource(ele));
```

在这个方法中执行了 import 解析完成的事件发布,在这个事件发布中需要依赖 ReaderEventListener 对象,在这里对于 import 事件处理的 fireImportProcessed 方法,在 Spring 中关于 ReaderEventListener 的实现有两个:CollectingReaderEventListener 和 EmptyReaderEventListener。CollectingReaderEventListener 位于测试工程中,并不是常规使用中的 ReaderEventListener,通常,Spring 会使用 EmptyReaderEventListener,下面是 EmptyReaderEventListener 的代码信息。

```
public class EmptyReaderEventListener implements ReaderEventListener {

    @Override
    public void defaultsRegistered(DefaultsDefinition defaultsDefinition) {
    }

    @Override
    public void componentRegistered(ComponentDefinition componentDefinition) {
    }

    @Override
    public void aliasRegistered(AliasDefinition aliasDefinition) {
    }

    @Override
    public void importProcessed(ImportDefinition importDefinition) {
    }

}
```

可以看到,EmptyReaderEventListener 类什么都不处理。

2.3.5 alias 标签解析

在开始分析源码之前,需要先找到分析目标,同时也需要搭建一个测试环境,需要分析的目标方法是 org.springframework.beans.factory.xml.DefaultBeanDefinitionDocumentReader#processAliasRegistration,确认需要分析的目标后进行测试用例的编写。编写一个 SpringXML 配置文件,文件名为 alias-node.xml,文件中的详细代码如下。

```
<?xml version = "1.0" encoding = "UTF-8"?>
<beans xmlns:xsi = "http://www.w3.org/2001/XMLSchema-instance"
    xmlns = "http://www.springframework.org/schema/beans"
    xsi:schemaLocation = " http://www.springframework.org/schema/beans http://www.springframework.org/schema/beans/spring-beans.xsd">

    <alias name = "people" alias = "p1"/>
    <bean id = "people" class = "com.source.hot.ioc.book.pojo.PeopleBean">
        <property name = "name" value = "zhangsan"/>
    </bean>
</beans>
```

完成 SpringXML 配置文件编写后,进行测试类的编写,测试类是 AliasNodeTest,代码信息如下。

```java
class AliasNodeTest {

    @Test
    void testAlias(){
        ClassPathXmlApplicationContext context
                = new ClassPathXmlApplicationContext("META-INF/alias-node.xml");

        Object people = context.getBean("people");
        Object p1 = context.getBean("p1");

        assert people.equals(p1);
    }
}
```

进入源码分析之前需要先了解 DTD 文件和 XSD 文件中对 alias 标签的定义,spring-beans.dtd 文件中对 alias 标签的定义如下。

```
<!ELEMENT alias EMPTY>

<!ATTLIST alias name CDATA #REQUIRED>

<!ATTLIST alias alias CDATA #REQUIRED>
```

Spring-beans.xsd 文件中对 alias 标签的定义如下。

```xml
<xsd:element name="alias">
    <xsd:annotation>
        <xsd:documentation><![CDATA[
Defines an alias for a bean (which can reside in a different definition resource).
        ]]></xsd:documentation>
    </xsd:annotation>
    <xsd:complexType>
        <xsd:complexContent>
            <xsd:restriction base="xsd:anyType">
                <xsd:attribute name="name" type="xsd:string" use="required">
                    <xsd:annotation>
                        <xsd:documentation><![CDATA[
The name of the bean to define an alias for.
                        ]]></xsd:documentation>
                    </xsd:annotation>
                </xsd:attribute>
                <xsd:attribute name="alias" type="xsd:string" use="required">
                    <xsd:annotation>
                        <xsd:documentation><![CDATA[
The alias name to define for the bean.
                        ]]></xsd:documentation>
                    </xsd:annotation>
```

```
            </xsd:attribute>
        </xsd:restriction>
      </xsd:complexContent>
    </xsd:complexType>
</xsd:element>
```

从 DTD 和 XSD 文件中可以确认 alias 标签存在两个属性：name 和 alias。name 属性是原始名称，alias 属性是别名。对于在测试用例中所编写的<alias name="people" alias="p1"/>，可以理解为 people 又被称为 p1。

下面进行 processAliasRegistration 方法分析，首先来看 processAliasRegistration 方法中的前两行代码。

```
//获取 name 属性
String name = ele.getAttribute(NAME_ATTRIBUTE);
//获取 alias 属性
String alias = ele.getAttribute(ALIAS_ATTRIBUTE);
```

这两行代码的主要目的是提取 alias 标签的 name 属性和 alias 属性，在获取完成属性信息后 Spring 会对这两个属性进行验证，验证代码如下。

```
boolean valid = true;
//name 属性验证
if (!StringUtils.hasText(name)) {
    getReaderContext().error("Name must not be empty",ele);
    valid = false;
}
//alias 属性验证
if (!StringUtils.hasText(alias)) {
    getReaderContext().error("Alias must not be empty",ele);
    valid = false;
}
```

在这段验证代码中，Spring 对 name 和 alias 属性都做了非空判断，当开发者在编写 alias 标签时遗漏了两个属性中的任何一个时都会抛出异常，如果需要进行异常模拟，可以将<alias name="people" alias="p1"/>修改为<alias name="people" alias=""/>，此时启动 Spring 会出现下面的异常信息。

```
Configuration problem: Alias must not be empty
Offending resource: class path resource [META-INF/alias-node.xml]
    org.springframework.beans.factory.parsing.BeanDefinitionParsingException:
Configuration problem: Alias must not be empty
    Offending resource: class path resource [META-INF/alias-node.xml]
```

继续向下阅读 processAliasRegistration 方法，在完成数据验证后 Spring 进行了下面两个操作。

（1）别名注册。
（2）发布别名注册事件。

这两个操作的代码如下。

```
//删除了异常处理
if (valid) {
    //注册
    getReaderContext().getRegistry().registerAlias(name,alias);

    //alias 注册事件触发
    getReaderContext().fireAliasRegistered(name,alias,extractSource(ele));
}
```

接下来将介绍别名注册相关的处理,在 registerAlias(name,alias);方法中可以看到参数是 name 和 alias,这两个属性必然和别名注册有关,在这里主要解决 name 和 alias 的关系绑定。举个例子,有一个人叫张三,在使用 QQ 时使用的昵称为王五,在使用微信时的昵称为李四,此时李四和王五这两个昵称都是张三的别名。在这个例子中可以设计两种关系模型,第一种是真名一对多别名,第二种是别名一对一真名,数据结构如下。

(1) 真名一对多别名:Map < String,List < String >>。
(2) 别名一对一真名:Map < String,String >。

在 Spring 中对别名和真名的关系绑定选择的处理形式是别名一对一真名,下面是 Spring 中存储别名和真名的容器。

```
private final Map < String,String > aliasMap = new ConcurrentHashMap <>(16);
```

根据数据结构的定义可以知道,在这个容器中 key 和 value 分别存储了什么,key 存储别名,value 存储真名。下面来看测试用例中进行别名注册时的 aliasMap 信息,如图 2.4 所示。

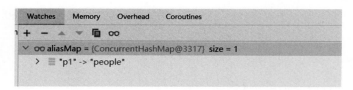

图 2.4　别名存储容器信息

最后在别名容器数据保存后需要执行别名注册事件的发布,在前文关于 import 标签的处理中已经明确 eventListener 的类型是 EmptyReaderEventListener,该方法中对于别名注册事件的处理代码是没有实现的,因此在这里不会执行任何处理。

2.3.6　bean 标签解析

下面将进入 bean 标签解析相关源码分析,在开始分析之前需要先进行测试用例的环境搭建,首先编写 SpringXML 配置文件,文件名为 bean-node.xml,文件内容如下。

```
<?xml version = "1.0" encoding = "UTF - 8"?>
< beans xmlns: xsi = "http://www.w3.org/2001/XMLSchema - instance"
    xmlns = "http://www.springframework.org/schema/beans"
    xsi: schemaLocation = " http://www.springframework.org/schema/beans http://www.springframework.org/schema/beans/spring - beans.xsd">
```

```
        <bean id="people" class="com.source.hot.ioc.book.pojo.PeopleBean">
        </bean>
</beans>
```

完成 SpringXML 配置编写后再编写测试类,测试类名称为 BeanNodeTest,相关代码如下。

```
class BeanNodeTest {
    @Test
    void testBean() {
        ClassPathXmlApplicationContext context
                = new ClassPathXmlApplicationContext("META-INF/bean-node.xml");

        Object people = context.getBean("people");
        context.close();
    }
}
```

接下来介绍 bean 标签的定义,在测试用例中所编写的 bean 标签如下。

```
<bean id="people" class="com.source.hot.ioc.book.pojo.PeopleBean"></bean>
```

在这个例子中只设置了 id 和 class 两个属性,在 Spring 中对于 bean 标签的定义远不止这两个属性,下面通过阅读 DTD 和 XSD 文件对 bean 标签的信息做补充认识。首先查看 spring-beans.dtd 文件中的 bean 标签定义。

```
<!ELEMENT bean (
    description?,
    (meta | constructor-arg | property | lookup-method | replaced-method) *
)>

<!ATTLIST bean id ID #IMPLIED>

<!ATTLIST bean name CDATA #IMPLIED>

<!ATTLIST bean class CDATA #IMPLIED>

<!ATTLIST bean parent CDATA #IMPLIED>

<!ATTLIST bean scope CDATA #IMPLIED>

<!ATTLIST bean abstract (true | false) #IMPLIED>

<!ATTLIST bean lazy-init (true | false | default) "default">

<!ATTLIST bean autowire (no | byName | byType | constructor | autodetect | default) "default">

<!ATTLIST bean depends-on CDATA #IMPLIED>

<!ATTLIST bean autowire-candidate (true | false) #IMPLIED>

<!ATTLIST bean init-method CDATA #IMPLIED>
```

```
<!ATTLIST bean destroy-method CDATA #IMPLIED>

<!ATTLIST bean factory-method CDATA #IMPLIED>

<!ATTLIST bean factory-bean CDATA #IMPLIED>
```

在上述代码中可以先关注下面这段代码。

```
<!ELEMENT bean (
    description?,
    (meta | constructor-arg | property | lookup-method | replaced-method) *
)>
```

在这段代码中可以看到 bean 标签下的各类标签情况。

（1）meta 标签。

（2）constructor-arg 标签。

（3）property 标签。

（4）lookup-method 标签。

（5）replaced-method 标签。

在 bean 标签中存在的属性如下。

（1）id：Bean 的 ID。

（2）name：Bean 的名称。

（3）class：Bean 的类全路径。

（4）parent：父类。

（5）scope：作用域。

（6）lazy-init：是否懒加载。

（7）abstract：是否抽象类。

（8）autowire：注入方式。

（9）depends-on：依赖。

（10）autowire-candidate：注入候选者。

（11）init-method：实例化后执行的方法。

（12）destroy-method：摧毁时调用的方法。

（13）factory-method：工厂方法的方法名。

（14）factory-bean：工厂 Bean 的名称。

上述下级标签和属性组合起来就是 bean 标签的完整内容，接下来介绍 spring-beans.xsd 文件中对于 bean 标签的定义，具体代码如下。

```
<xsd:group name="beanElements">
    <xsd:sequence>
        <xsd:element ref="description" minOccurs="0"/>
        <xsd:choice minOccurs="0" maxOccurs="unbounded">
            <xsd:element ref="meta"/>
            <xsd:element ref="constructor-arg"/>
            <xsd:element ref="property"/>
```

```xml
            <xsd:element ref="qualifier"/>
            <xsd:element ref="lookup-method"/>
            <xsd:element ref="replaced-method"/>
            <xsd:any namespace="##other" processContents="strict" minOccurs="0" maxOccurs="unbounded"/>
        </xsd:choice>
    </xsd:sequence>
</xsd:group>

<xsd:attributeGroup name="beanAttributes">
    <xsd:attribute name="name" type="xsd:string">
        <xsd:annotation>
            <xsd:documentation></xsd:documentation>
        </xsd:annotation>
    </xsd:attribute>
    <xsd:attribute name="class" type="xsd:string">
        <xsd:annotation>
            <xsd:documentation source="java:java.lang.Class"></xsd:documentation>
        </xsd:annotation>
    </xsd:attribute>
    <xsd:attribute name="parent" type="xsd:string">
        <xsd:annotation>
            <xsd:documentation></xsd:documentation>
        </xsd:annotation>
    </xsd:attribute>
    <xsd:attribute name="scope" type="xsd:string">
        <xsd:annotation>
            <xsd:documentation></xsd:documentation>
        </xsd:annotation>
    </xsd:attribute>
    <xsd:attribute name="abstract" type="xsd:boolean">
        <xsd:annotation>
            <xsd:documentation></xsd:documentation>
        </xsd:annotation>
    </xsd:attribute>
    <xsd:attribute name="lazy-init" default="default" type="defaultable-boolean">
        <xsd:annotation>
            <xsd:documentation></xsd:documentation>
        </xsd:annotation>
    </xsd:attribute>
    <xsd:attribute name="autowire" default="default">
        <xsd:annotation>
            <xsd:documentation></xsd:documentation>
        </xsd:annotation>
        <xsd:simpleType>
            <xsd:restriction base="xsd:NMTOKEN">
                <xsd:enumeration value="default"/>
                <xsd:enumeration value="no"/>
                <xsd:enumeration value="byName"/>
                <xsd:enumeration value="byType"/>
                <xsd:enumeration value="constructor"/>
```

```
                </xsd:restriction>
            </xsd:simpleType>
        </xsd:attribute>
        <xsd:attribute name="depends-on" type="xsd:string">
            <xsd:annotation>
                <xsd:documentation></xsd:documentation>
            </xsd:annotation>
        </xsd:attribute>
        <xsd:attribute name="autowire-candidate" default="default" type="defaultable-boolean">
            <xsd:annotation>
                <xsd:documentation></xsd:documentation>
            </xsd:annotation>
        </xsd:attribute>
        <xsd:attribute name="primary" type="xsd:boolean">
            <xsd:annotation>
                <xsd:documentation></xsd:documentation>
            </xsd:annotation>
        </xsd:attribute>
        <xsd:attribute name="init-method" type="xsd:string">
            <xsd:annotation>
                <xsd:documentation></xsd:documentation>
            </xsd:annotation>
        </xsd:attribute>
        <xsd:attribute name="destroy-method" type="xsd:string">
            <xsd:annotation>
                <xsd:documentation></xsd:documentation>
            </xsd:annotation>
        </xsd:attribute>
        <xsd:attribute name="factory-method" type="xsd:string">
            <xsd:annotation>
                <xsd:documentation></xsd:documentation>
            </xsd:annotation>
        </xsd:attribute>
        <xsd:attribute name="factory-bean" type="xsd:string">
            <xsd:annotation>
                <xsd:documentation></xsd:documentation>
            </xsd:annotation>
        </xsd:attribute>
        <xsd:anyAttribute namespace="##other" processContents="lax"/>
    </xsd:attributeGroup>
```

完成 bean 标签的基础认识后将进入 bean 标签的解析方法 processBeanDefinition 的分析，processBeanDefinition 方法代码如下。

```
//删除了异常处理
protected void processBeanDefinition(Element ele, BeanDefinitionParserDelegate delegate) {
    //创建 BeanDefinition
    BeanDefinitionHolder bdHolder = delegate.parseBeanDefinitionElement(ele);
    if (bdHolder != null) {
        //BeanDefinition 装饰
```

```
        bdHolder = delegate.decorateBeanDefinitionIfRequired(ele,bdHolder);
        //注册 BeanDefinition
        BeanDefinitionReaderUtils.registerBeanDefinition(bdHolder,getReaderContext().getRegistry());
        //Component 注册事件触发
        getReaderContext().fireComponentRegistered(new BeanComponentDefinition(bdHolder));
    }
}
```

在这段代码中很难直接看到关于 bean 标签的解析过程,只能在其中模糊地看到一些关于 BeanDefinition 的 处 理 操 作:registerBeanDefinition 和 fireComponentRegistered,在 Spring IoC 对 bean 标签的处理中 Spring 将其完全托管给 BeanDefinitionParserDelegate 类进行处理。对于 BeanDefinitionParserDelegate 的分析将在后续章节中进行,在这段代码中只需要知道整体的处理流程即可,处理流程如下。

(1) 交由 BeanDefinitionParserDelegate 对象进行 bean 标签的处理得到 BeanDefinitionHolder 对象。

(2) 将得到的 BeanDefinitionHolder 对象进行 BeanDefinition 注册。

(3) 发布 Bean 注册事件。

接下来对 BeanDefinition 对象的注册进行说明,下面是 registerBeanDefinition 方法的源代码。

```
public static void registerBeanDefinition(
        BeanDefinitionHolder definitionHolder,BeanDefinitionRegistry registry)
        throws BeanDefinitionStoreException {

    //获取 BeanName
    String beanName = definitionHolder.getBeanName();
    //注册 BeanDefinition
    registry.registerBeanDefinition(beanName,definitionHolder.getBeanDefinition());

    //别名列表
    String[] aliases = definitionHolder.getAliases();
    //注册别名列表
    if (aliases != null) {
        for (String alias: aliases) {
            registry.registerAlias(beanName,alias);
        }
    }
}
```

在这段代码中总共有以下两个处理行为。

(1) 处理 BeanName 和 BeanDefinition 之间的关系。

(2) 处理 BeanName 和 alias 之间的关系。

在上述两个处理行为中需要重点关注 BeanName 和 BeanDefinition 之间的关系处理,对于关系处理的方法是 org.springframework.beans.factory.support.DefaultListableBeanFactory#registerBeanDefinition,在 Spring 中对于 BeanName 和 BeanDefinition 之间的关系存储结构是一个 Map 结构,具体结构如下。

```
private final Map<String,BeanDefinition> beanDefinitionMap = new ConcurrentHashMap<>(256);
```

beanDefinitionMap 的数据信息：key 表示 BeanName，value 表示 BeanDefinition 对象。测试用例中的 BeanName 和 BeanDefinition 之间的关系信息如图 2.5 所示。

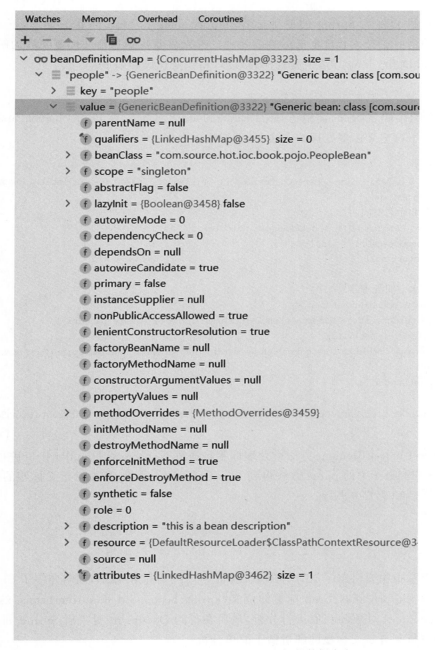

图 2.5　BeanName 和 BeanDefinition 之间的数据绑定

接下来介绍关于 Bean 注册事件的推送，在前文关于 import 标签处理和 alias 标签处理是已经确认处理事件的对象中是一个空方法，不做任何行为，同样地，在 bean 标签处理完成后的 Bean 注册事件推送也是一个空方法，没有执行任何行为。

2.3.7 自定义标签解析概述

前文介绍了关于 Spring 的原生标签 import、alias 和 bean 标签的处理，下面将介绍 Spring 中的另一个技术：自定义标签的处理。首先需要找到自定义标签处理的解析入口，在 parseBeanDefinitions 方法中有这样的处理，相关处理代码如下。

```
delegate.parseCustomElement(root);
```

上述代码中的 delegate 类型是 BeanDefinitionParserDelegate，在 BeanDefinitionParserDelegate 中提供了关于自定义标签处理的方法，处理方法内容如下。

```
@Nullable
public BeanDefinition parseCustomElement（Element ele,@Nullable BeanDefinition containingBd）{
    //获取命名空间的 URL
    String namespaceUri = getNamespaceURI(ele);
    if (namespaceUri == null) {
        return null;
    }
    //命名空间处理器
    NamespaceHandler handler =
this.readerContext.getNamespaceHandlerResolver().resolve(namespaceUri);
    if (handler == null) {
        error("Unable to locate Spring NamespaceHandler for XML schema namespace [" + namespaceUri +
"]",ele);
        return null;
    }
    return handler.parse(ele,new ParserContext(this.readerContext,this,containingBd));
}
```

在 parseCustomElement 方法中处理标签是通过 NamespaceHandler 接口进行的，通过 NamespaceHandler 中的 parse 方法得到 BeanDefinition 对象，有关自定义标签的使用及源码分析将在后续章节中进行。

小结

在这一章中主要讲解了 XML 的两种验证方式：DTD 和 XSD，同时介绍了 Spring 中对于原生标签 import、alias、bean 标签的定义（spring-beans.dtd 和 spring-beans.xsd 文件），接着对 Document 对象的获取进行分析，最后在得到 Document 对象后 Spring 如何进行原生标签的解析和自定义标签的解析做了简单分析。

第3章

自定义标签

在第 1 章中对 Spring 的简单使用进行了说明,简单介绍了 bean 标签的使用。bean 标签属于 Spring 的原生标签,在 Spring 中除了原生标签以外还能够支持自定义标签,本章将介绍 SpringXML 配置文件中的自定义标签如何进行自定义、如何使用自定义标签,并对 SpringXML 的自定义标签相关的内容进行源码分析。

3.1 创建自定义标签环境搭建

在开始自定义标签分析之前,需要先编写自定义标签解析相关的测试用例,编写自定义标签需要执行下面四个步骤。

(1) 编写 XSD 文件或者 DTD 文件。
(2) 编写 NamespaceHandler 实现类。
(3) 编写 BeanDefinitionParser 实现类。
(4) 编写注册方式,向 Spring 中注册。

接下来对上述四个步骤做详细说明。

3.1.1 编写 XSD 文件

首先编写一个 Java 对象用来存储自定义标签解析后的数据,编写 UserXsdJava 对象,代码信息如下。

```
//省略 getter&setter
public class UserXsd {
    private String name;
    private String idCard;
}
```

完成 XSD 文件解析结果的存储对象后进一步编写 XSD 文件，该 XSD 文件名为 user.xsd，文件内容如下。

```xml
<?xml version = "1.0" encoding = "UTF - 8"?>
< schema xmlns = "http: //www.w3.org/2001/XMLSchema"
         targetNamespace = "http: //www.huifer.com/schema/user"
         elementFormDefault = "qualified">

    < element name = "user_xsd">
        < complexType >
            < attribute name = "id" type = "string"/>
            < attribute name = "name" type = "string"/>
            < attribute name = "idCard" type = "string"/>
        </complexType >
    </element >
</schema >
```

3.1.2　编写 NamespaceHandler 实现类

完成 XSD 文件编写后进一步编写 NamespaceHandler 接口的实现类，Spring 提供了 NamespaceHandlerSupport 对象让开发者更加简单地使用，开发者只需要重写 init 方法即可向 Spring 注册标签和标签的解析对象，编写 UserXsdNamespaceHandler 类，详细代码如下。

```java
public class UserXsdNamespaceHandler extends NamespaceHandlerSupport {

    @Override
    public void init() {
        registerBeanDefinitionParser("user_xsd",new UserXsdParser());
    }

}
```

3.1.3　编写 BeanDefinitionParser 实现类

在编写 NamespaceHandler 实现类的时候引入了一个新的 Java 对象 UserXsdParser，该对象是 BeanDefinitionParser 接口的实现类，在 Spring 中可以通过继承 AbstractSingleBeanDefinitionParser 类重写 getBeanClass 和 doParse 方法即可完成 BeanDefinitionParser 的实现，下面是 UserXsdParser 的代码内容。

```java
public class UserXsdParser extends AbstractSingleBeanDefinitionParser {
    @Override
    protected Class<?> getBeanClass(Element element) {
        return UserXsd.class;
    }
```

```
@Override
protected void doParse(Element element,BeanDefinitionBuilder builder) {
    String name = element.getAttribute("name");
    String idCard = element.getAttribute("idCard");
    builder.addPropertyValue("name",name);
    builder.addPropertyValue("idCard",idCard);
}
```

在这段代码中通过提取 Element 对象的 name 和 idCard 属性将其设置到 BeanDefinitionBuilder 对象的属性表中。

3.1.4 编写注册方式

下面编写注册方式。注册自定义标签解析能力需要编写两个文件，一个是 spring.handlers 文件，另一个是 spring.schemas 文件。在这个测试用例中需要向 spring.handlers 文件中填写下面这段内容。

```
http\://www.huifer.com/schema/user=com.source.hot.ioc.book.namespace.handler.UserXsdNamespaceHandler
```

对于 spring.handlers 文件可以分成两部分来进行理解，第一部分是等号前面的内容，等号前的内容是指命名空间和 XSD 文件中 schema 中的 targetNamespace 属性之间的关系；第二部分是等号后面的内容，它是指接口 NamespaceHandler 实现类的完整类路径。

完成 spring.handlers 编写后进一步编写 spring.schemas 文件，向 spring.schemas 文件中添加下面这段代码。

```
http\://www.huifer.com/schema/user.xsd=META-INF/user.xsd
```

对于 spring.schemas 文件可以分成两部分来进行理解，第一部分是等号前面的内容，它是指 schemaLocation 的一个链接地址；第二部分是等号后面的内容，它是指 schemaLocation 对应的 XSD 文件描述路径。

3.1.5 测试用例的编写

完成了各项基本准备工作后进一步编写一个自定义标签处理的 Java 程序，首先需要编写 SpringXML 配置文件，文件名为 custom-xml.xml，向 custom-xml.xml 文件中填写下面的内容。

```xml
<?xml version="1.0" encoding="UTF-8"?>
<beans xmlns="http://www.springframework.org/schema/beans"
       xmlns:xsi="http://www.w3.org/2001/XMLSchema-instance"
       xmlns:myname="http://www.huifer.com/schema/user"
       xsi:schemaLocation=" http://www.springframework.org/schema/beans http://www.springframework.org/schema/beans/spring-beans.xsd
```

```
    http://www.huifer.com/schema/user http://www.huifer.com/schema/user.xsd
">

    < myname: user_xsd id = "testUserBean" name = "huifer" idCard = "123"/>

</beans >
```

完成 SpringXML 配置文件的编写后再编写一个测试类,测试类名称为 CustomXmlTest,CustomXmlTest 中代码如下。

```
class CustomXmlTest {

    @Test
    void testXmlCustom() {
        ClassPathXmlApplicationContext context =
new ClassPathXmlApplicationContext("META - INF/custom - xml.xml");
        UserXsd testUserBean = context.getBean("testUserBean",UserXsd.class);
        assert testUserBean.getName().equals("huifer");
        assert testUserBean.getIdCard().equals("123");
        context.close();
    }
}
```

完成测试类及测试方法的编写后自定义标签测试环境即搭建完成。

3.2 自定义标签解析

下面将进入自定义标签解析相关源代码分析,首先需要找到自定义标签解析的源码入口,该入口的方法签名为 org. springframework. beans. factory. xml. BeanDefinitionParserDelegate#parseCustomElement(org. w3c. dom. Element,org. springframework. beans. factory. config. BeanDefinition),详细代码如下。

```
@Nullable
public BeanDefinition parseCustomElement(Element ele,@Nullable BeanDefinition containingBd) {
    //获取命名空间的 URL
    String namespaceUri = getNamespaceURI(ele);
    if (namespaceUri == null) {
        return null;
    }
    //命名空间处理器
    NamespaceHandler handler
 = this.readerContext.getNamespaceHandlerResolver().resolve(namespaceUri);
    if (handler == null) {
        error(" Unable to locate Spring NamespaceHandler for XML schema namespace [ " +
namespaceUri + "]",ele);
        return null;
    }
    return handler.parse(ele,new ParserContext(this.readerContext,this,containingBd));
}
```

3.2.1 NamesapceHandler 和 BeanDefinitionParser 之间的关系

parseCustomElement 方法中体现了 namespaceUri 的一些关系，在测试用例中 namespaceUri 和 spring.handlers 中的文件存在关系，命名空间对应命名空间处理器，通过这个关系可以确认 NamespaceHandler 是 UserXsdNamespaceHandler 对象，在 UserXsdNamespaceHandler 类中有 init 方法来注册标签和标签的解析能力提供类的关系。具体关系如图 3.1 所示。

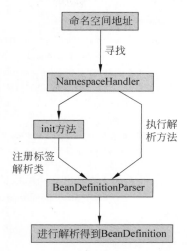

图 3.1 NamespaceHandler 关系图

3.2.2 获取命名空间地址

接下来将介绍命名空间地址的获取，命名空间地址及 namespaceUri 属性，获取该属性的方法是由 org.w3c.dom 提供，具体细节不做展开，获取命名空间地址后的数据是 http://www.huifer.com/schema/user。

3.2.3 NamespaceHandler 对象获取

通过前文获得了 namespaceUri 数据信息后会通过该信息寻找到对应的 NamespaceHandler 对象，具体处理方法如下。

```
NamespaceHandler handler
 = this.readerContext.getNamespaceHandlerResolver().resolve(namespaceUri);
```

在这段方法中负责命名空间解析的对象是 NamespaceHandlerResolver 接口，NamespaceHandlerResolver 接口定义如下。

```
@FunctionalInterface
public interface NamespaceHandlerResolver {

    /**
     * 解析命名空间的 URL 获得命名空间处理器
     */
    @Nullable
    NamespaceHandler resolve(String namespaceUri);

}
```

在 Spring 中它只有一个实现类 DefaultNamespaceHandlerResolver，在 DefaultNamespaceHandlerResolver 中有一个成员变量 DEFAULT_HANDLER_MAPPINGS_LOCATION，成员变量详细信息如下。

```
public static final String DEFAULT_HANDLER_MAPPINGS_LOCATION =
"META-INF/spring.handlers";
```

在 DEFAULT_HANDLER_MAPPINGS_LOCATION 成员变量中定义了默认的命名空间处理器存储路径，spring.handlers 中存储的内容是(key,value)结构，key 是命名空间，value 是命名空间处理器的类全路径，接下来查看 resolve 方法，详细代码如下。

```
//删除异常处理和日志
public NamespaceHandler resolve(String namespaceUri) {
    //获取 Namespace Handler 映射表
    Map < String,Object > handlerMappings = getHandlerMappings();
    //从映射表中获取 URI 对应的 Handler
    //字符串(名称)
    //实例
    Object handlerOrClassName = handlerMappings.get(namespaceUri);
    if(handlerOrClassName == null) {
        return null;
    } else if(handlerOrClassName instanceof NamespaceHandler) {
        return(NamespaceHandler) handlerOrClassName;
    }
    //其他情况都做字符串处理
    else {
        String className = (String) handlerOrClassName;
        Class <?> handlerClass = ClassUtils.forName(className,this.classLoader);
        //通过反射构造 namespaceHandler 实例
        NamespaceHandler namespaceHandler =
(NamespaceHandler) BeanUtils.instantiateClass(handlerClass);
        //初始化
        namespaceHandler.init();
        //重写缓存
        handlerMappings.put(namespaceUri,namespaceHandler);
        return namespaceHandler;
    }
}
```

在 resolve 方法中整体处理流程如下。

（1）获取命名空间映射关系，即命名空间地址（namespaceUri）对应命名空间解析对象类全路径，这个关系是一个 map 集合。

（2）从 map 集合中获取命名空间地址（namespaceUri）对应的数据。

（3）根据提取数据的不同情况进行处理。

① 通过命名空间地址获取的对象是 NamespaceHandler 类型，直接返回使用。

② 通过命名空间地址获取的对象不是 NamespaceHandler 类型。当类型不是 NamespaceHandler 对象时，它的类型只可能是字符串类型，字符串是没有办法直接调用 NamespaceHandler 所提供的方法的，因此需要将字符串转换成 NamespaceHandler 对象。

3.2.4　getHandlerMappings 获取命名空间的映射关系

在前文讲到字符串转换为 NamespaceHandler 的内容在本节会对其做补充，负责这部分处理的方法是 getHandlerMappings，详细代码如下。

```
//删除异常处理和日志
private Map < String,Object > getHandlerMappings() {
    //设置容器
    Map < String,Object > handlerMappings = this.handlerMappings;
    if(handlerMappings == null) {
        synchronized(this) {
            handlerMappings = this.handlerMappings;
            if(handlerMappings == null) {
                //读取资源文件地址
                Properties mappings =
PropertiesLoaderUtils.loadAllProperties(this.handlerMappingsLocation,this.classLoader);
                handlerMappings = new ConcurrentHashMap < > (mappings.size());
                //数据合并,将 mappings 数据复制给 handlerMappings
                CollectionUtils.mergePropertiesIntoMap(mappings,handlerMappings);
                this.handlerMappings = handlerMappings;
            }
        }
    }
    return handlerMappings;
}
```

在这段方法中首先需要关注的是下面这段代码。

```
Properties mappings =
    PropertiesLoaderUtils.loadAllProperties(this.handlerMappingsLocation,this.classLoader);
```

这段代码是用来提取 META-INF/spring.handlers 文件中的数据内容,Spring 将 META-INF/spring.handlers 文件当作拓展名为 properties 类型的文件进行处理,处理之后得到的是 Map 对象,在完成 META-INF/spring.handlers 文件的读取后进行了合并操作,分别将历史的 handlerMappings 和本次读取得到的 handlerMappings 进行合并。图 3.2 为本次读取得到的 handlerMappings 数据。

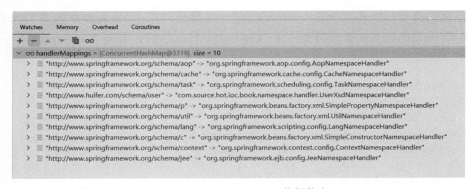

图 3.2 handlerMappings 数据信息

通过图 3.2 可以看到一条这样的信息:http://www.huifer.com/schema/user -> com.source.hot.ioc.book.namespace.handler.UserXsdNamespaceHandler,这段信息的来源是前文在测试用例中文件 spring.handlers 的内容,除此之外,其他的数据是 Spring 项目中存放的,文件依旧还是在 META-INF/spring.handlers 中。

spring-beans 工程下的 spring.handlers 数据信息如下。

```
http\://www.springframework.org/schema/c = org.springframework.beans.factory.xml.SimpleConstructorNamespaceHandler
http\://www.springframework.org/schema/p = org.springframework.beans.factory.xml.SimplePropertyNamespaceHandler
http\://www.springframework.org/schema/util = org.springframework.beans.factory.xml.UtilNamespaceHandler
```

spring-aop 工程下的 spring.handlers 数据信息如下。

```
http\://www.springframework.org/schema/aop = org.springframework.aop.config.AopNamespaceHandler
```

spring-context 工程下的 spring.handlers 数据信息如下。

```
http\://www.springframework.org/schema/context = org.springframework.context.config.ContextNamespaceHandler
http\://www.springframework.org/schema/jee = org.springframework.ejb.config.JeeNamespaceHandler
http\://www.springframework.org/schema/lang = org.springframework.scripting.config.LangNamespaceHandler
http\://www.springframework.org/schema/task = org.springframework.scheduling.config.TaskNamespaceHandler
http\://www.springframework.org/schema/cache = org.springframework.cache.config.CacheNamespaceHandler
```

上述这些文件是 Spring 容器中提供的 NamespaceHandler 数据，除此之外，还有一些其他的 spring.handlers 文件本文不做赘述。

3.2.5 NamespaceHandler 的获取

在前面的操作中已经准备好 handlerMappings 数据对象，下面需要将数据对象进行初始化（实例化，实例化的前提是类型不是 NamespaceHandler）。有关 NamespaceHandler 的获取代码如下。

```
Object handlerOrClassName = handlerMappings.get(namespaceUri);
if(handlerOrClassName == null) {
    return null;
} else if(handlerOrClassName instanceof NamespaceHandler) {
    return(NamespaceHandler) handlerOrClassName;
}
//其他情况都做字符串处理
else {
    String className = (String) handlerOrClassName;
    Class <?> handlerClass = ClassUtils.forName(className,this.classLoader);
    //通过反射构造 namespaceHandler 实例
    NamespaceHandler namespaceHandler =
(NamespaceHandler) BeanUtils.instantiateClass(handlerClass);
    //初始化
```

```
        namespaceHandler.init();
        //重写缓存
        handlerMappings.put(namespaceUri,namespaceHandler);
        return namespaceHandler;
}
```

上述代码中有以下三种情况需要处理。

（1）容器中没有当前 namespaceUri 的值。

（2）容器中有当前 namespaceUri 的值，并且类型是 NamespaceHandler。

（3）容器中有当前 namespaceUri 的值，但类型不是 NamespaceHandler。

对于上述三种情况需要重点分析的是第三种。在第三种情况中，value 的数据类型是 String，Spring 需要将 String 类型转换为最终的 Java 对象，通过类全路径转换成 Java 对象需要执行以下三个步骤。

（1）通过 Class.forName 得到 Class 对象。

（2）通过 Class 对象提取构造函数。

（3）通过构造函数创建实例。

上述三个操作步骤可以分别对应下面这些代码，第一步对应"Class<?> handlerClass = ClassUtils.forName(className,this.classLoader);"，第二步和第三步对应 Spring 项目中的一个工具类 BeanUtils。通过 BeanUtils.instantiateClass 方法即可得到 Java 对象。

3.2.6　NamespaceHandler 的 init 方法

接下来将介绍 NamespaceHandler 的 init 方法，在测试用例中，UserXsdNamespaceHandler 对象继承 NamespaceHandlerSupport 类，接下来需要分析的重点内容都在 NamespaceHandlerSupport 类中。首先需要指出一点，在 Spring 中的 spring.handlers 文件中的实现类都有继承 NamespaceHandlerSupport 类，可见它的重要性。在测试用例中使用了 registerBeanDefinitionParser 方法进行了标签和解析类的注册，下面对 registerBeanDefinitionParser 方法进行分析，先看 registerBeanDefinitionParser 代码。

```
protected final void registerBeanDefinitionParser(String elementName,BeanDefinitionParser parser) {
    this.parsers.put(elementName,parser);
}
```

在这段代码中需要了解以下两个参数的含义。

（1）elementName：XML 标签的名称。

（2）parser：提供标签解析能力的实际对象，是 BeanDefinitionParser 接口的实现类。

下面介绍 parsers 的数据结构，在 Spring 中对 parsers 的定义如下。

```
private final Map<String,BeanDefinitionParser> parsers = new HashMap<>();
```

key 表示 XML 标签的名称，value 表示 BeanDefinitionParser 接口的实现类。parsers 的数据存储情况如图 3.3 所示。

图 3.3　parsers 数据信息

3.2.7　NamespaceHandler 缓存的刷新

在 NamespaceHandler 的生命周期方法 init 执行完成后会刷新 namespaceUri 对应的 value，此时会产生 handlerMappings 中 value 的多种情况。

（1）value 不存在。
（2）value 存在，且是 NamespaceHandler 实例。
（3）value 存在，且是字符串。

刷新 handlerMappings 变量的操作代码如下。

```
handlerMappings.put(namespaceUri,namespaceHandler)
```

这段代码的执行就是将 namespaceUri 和 namespaceHandler 的关系重新绑定，此时放入的 value 就会变成 NamespaceHandler 实例，后续在需要使用时就会进入下面这段代码。

```
else if (handlerOrClassName instanceof NamespaceHandler) {
    return (NamespaceHandler) handlerOrClassName;
}
```

在这里 Spring 通过了一个 Map 对象的刷新操作来提高了性能，避免了每次从字符串出发进行反射获取实例，获取实例之后再做其他操作。图 3.4 为经过刷新操作后的 handlerMappings 数据情况。

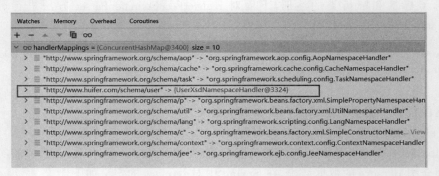

图 3.4　刷新后的 handlerMappings

3.2.8　解析标签 BeanDefinitionParser 对象准备

接下来将进入自定义标签解析的重要环节——BeanDefinitionParser 实现类的准备阶

段。首先阅读 parse 方法：

```
NamespaceHandler handler
  = this.readerContext.getNamespaceHandlerResolver().resolve(namespaceUri);
return handler.parse(ele,new ParserContext(this.readerContext,this,containingBd));
```

在这段代码中得到了 NamespaceHandler 对象，这个对象的实际类型是 UserXsdNamespaceHandler，确认实际类型后对于 parse 方法入口的查询有了方向，parse 方法位于 UserXsdNamespaceHandler 的父类 NamespaceHandlerSupport 中，具体代码如下。

```
public BeanDefinition parse(Element element,ParserContext parserContext) {
    //搜索 element 对应的 BeanDefinitionParser
    BeanDefinitionParser parser = findParserForElement(element,parserContext);
    //解析
    return (parser != null ? parser.parse(element,parserContext) : null);
}
```

从这段操作代码中可以发现一个重点类 BeanDefinitionParser，在测试用例中编写的内容中有一个与之存在关系，这个类是 UserXsdParser。在测试用例中和 UserXsdParser 对象产生关系的内容是一个注册方法 registerBeanDefinitionParser("user_xsd", new UserXsdParser())，这个方法表示 user_xsd 标签需要通过 UserXsdParser 对象进行解析。在 parse 方法中出现的 findParserForElement 方法目的就是通过标签名称找到对应的标签处理对象。findParserForElement 方法代码如下。

```
private BeanDefinitionParser findParserForElement(Element element,ParserContext parserContext) {
    //获取 element 的名称
    String localName = parserContext.getDelegate().getLocalName(element);
    //从容器中获取
    BeanDefinitionParser parser = this.parsers.get(localName);
    if (parser == null) {
        parserContext.getReaderContext().fatal(
            "Cannot locate BeanDefinitionParser for element [" + localName + "]",element);
    }
    return parser;
}
```

在这段方法中可以看到以下两个处理步骤。
（1）提取 Element 的名称数据。
（2）从 parsers 中获取解析 BeanDefinitionParser 对象。
此时得到的 BeanDefinitionParser 是 UserXsdParser 对象。

3.2.9 解析标签 parse 方法调用

在 Spring 中 parse 方法提供者是 AbstractBeanDefinitionParser 对象，详细代码如下。

```
//删除异常处理和日志
```

```java
public final BeanDefinition parse(Element element,ParserContext parserContext) {
    //解析 Element 得到 BeanDefinition 对象
    AbstractBeanDefinition definition = parseInternal(element,parserContext);
    if(definition != null && !parserContext.isNested()) {
        //解析标签的 id
        String id = resolveId(element,definition,parserContext);
        if(!StringUtils.hasText(id)) {
            parserContext.getReaderContext().error("Id is required for element '" + parserContext.getDelegate().getLocalName(element) + "' when used as a top-level tag",element);
        }
        String[] aliases = null;
        if(shouldParseNameAsAliases()) {
            //标签的 name 属性
            String name = element.getAttribute(NAME_ATTRIBUTE);
            if(StringUtils.hasLength(name)) {
                //别名处理,根据逗号进行字符串切割
                aliases = StringUtils.trimArrayElements(StringUtils.commaDelimitedListToStringArray(name));
            }
        }
        //创建 BeanDefinitionHolder
        BeanDefinitionHolder holder = new BeanDefinitionHolder(definition,id,aliases);
        //注册 BeanDefinition
        registerBeanDefinition(holder,parserContext.getRegistry());
        //是否需要触发事件
        if(shouldFireEvents()) {
            //组件注册事件
            BeanComponentDefinition componentDefinition = new BeanComponentDefinition(holder);
            postProcessComponentDefinition(componentDefinition);
            parserContext.registerComponent(componentDefinition);
        }
    }
    return definition;
}
```

在这段代码中比较难以查看到 UserXsdParser 对象的踪迹,UserXsdParser 类图信息如图 3.5 所示。

接下来看第一段代码。

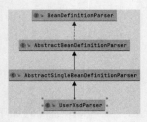

图 3.5 UserXsdParser 类图

AbstractBeanDefinition definition = parseInternal(element,parserContext);

parseInternal 方法在 AbstractBeanDefinitionParser 是一个抽象方法,真正的实现在 AbstractSingleBeanDefinitionParser 中,下面是 AbstractSingleBeanDefinitionParser#parseInternal 方法代码。

```java
@Override
protected final AbstractBeanDefinition parseInternal(Element element,ParserContext parserContext) {
    //BeanDefinition 构造器
    BeanDefinitionBuilder builder = BeanDefinitionBuilder.genericBeanDefinition();
    //获取 parent 属性
    String parentName = getParentName(element);
    if (parentName != null) {
        builder.getRawBeanDefinition().setParentName(parentName);
    }

    //获取 class 属性
    Class<?> beanClass = getBeanClass(element);
    if (beanClass != null) {
        builder.getRawBeanDefinition().setBeanClass(beanClass);
    }
    else {
        String beanClassName = getBeanClassName(element);
        if (beanClassName != null) {
            builder.getRawBeanDefinition().setBeanClassName(beanClassName);
        }
    }
    //设置源
    builder.getRawBeanDefinition().setSource(parserContext.extractSource(element));

    //获取已存在的 BeanDefinition,该对象仅用来设置 scope 属性
    BeanDefinition containingBd = parserContext.getContainingBeanDefinition();
    if (containingBd != null) {
        builder.setScope(containingBd.getScope());
    }
    if (parserContext.isDefaultLazyInit()) {
        builder.setLazyInit(true);
    }
    //真正的调用
    doParse(element,parserContext,builder);
    //BeanDefinition 构造器中获取 BeanDefinition
    return builder.getBeanDefinition();
}
```

在 UserXsdParser 类中重写了 getBeanClass 和 doParse 方法,测试用例中所编写的 doParse 方法的调用需要通过一层外部调用才可以抵达测试用例中 UserXsdParser 类的 doParse 方法,具体调用过程的代码如下。

```java
//AbstractSingleBeanDefinitionParser#parseInternal 中调用
protected void doParse(Element element,ParserContext parserContext,
BeanDefinitionBuilder builder) {
    doParse(element,builder);
}
//需要重写的方法
protected void doParse(Element element,BeanDefinitionBuilder builder) {
}
```

AbstractSingleBeanDefinitionParser#parseInternal 方法处理的细节如下。

（1）准备基本数据，基本数据包含 parentName、beanClass、source 和 scope。
（2）执行自定义的 doParse 方法。
（3）通过 BeanDefinitionBuilder 来获取 BeanDefinition。

在 parse 方法中首先需要进行 id 的处理，代码如下。

```
String id = resolveId(element,definition,parserContext);
```

在这段代码中对于 id 属性的获取其本质是提取 XML 标签中的 id 属性。完成 id 数据获取后需要执行的事项是针对别名的处理，相关代码如下。

```
String[] aliases = null;
if (shouldParseNameAsAliases()) {
    //标签的 name 属性
    String name = element.getAttribute(NAME_ATTRIBUTE);
    if (StringUtils.hasLength(name)) {
        //别名处理,根据逗号进行字符串切割
        aliases =
StringUtils.trimArrayElements(StringUtils.commaDelimitedListToStringArray(name));
    }
}
```

在别名处理阶段会判断是否需要处理别名，默认是都需要处理的。别名处理方式是将 alias 标签中的 name 属性根据分隔符（逗号）切分，关于切分其本质为 name.split(",")，具体的处理方法是 StringUtils.commaDelimitedListToStringArray。在数据信息准备完成之后需要进行 BeanDefinition 对象的注册和事件发布，相关代码如下。

```
BeanDefinitionHolder holder =
new BeanDefinitionHolder(definition,id,aliases);
//注册 BeanDefinition
registerBeanDefinition(holder,parserContext.getRegistry());
//是否需要出发事件
if (shouldFireEvents()) {
    //组件注册事件
    BeanComponentDefinition componentDefinition =
new BeanComponentDefinition(holder);
    postProcessComponentDefinition(componentDefinition);
    parserContext.registerComponent(componentDefinition);
}
```

在这部分代码处理中，关于事件发布相关内容是一个预留方法，暂时是一个空处理，对于 BeanDefinition 对象的注册会在后续章节中进行详细分析，在这仅需要了解 BeanDefinition 对象的存储容器：

```
private final Map<String,BeanDefinition> beanDefinitionMap =
new ConcurrentHashMap<>(256);
```

当 BeanDefinition 对象被放置到容器后这段方法的处理就完成了。

小结

通过本章的阐述，相信读者对于自定义标签的处理流程有了更加详细的认知。在本章了解了 NamespaceHandler、NamespaceHandlerSupport、AbstractSingleBeanDefinitionParser、AbstractBeanDefinitionParser 和 BeanDefinitionParser 之间的关系，这些内容在 Spring 中是一个十分重要的技术点，Spring 后续的一些内容都强依赖于这一门技术（自定义标签解析），常见的有 SpringMVC、Spring 事务、SpringAOP 等，它们都是基于此作为一个拓展实现了更强大的功能。

第4章

别名注册和BeanDefinition注册

在第3章中介绍了SpringXML的自定义标签,本章将回归到SpringXML的原生标签中。本章分为两部分,第一部分是别名注册,第二部分是BeanDefinition注册。在别名注册中会使用到SpringXML原生标签中的alias标签,在分析别名注册时会从别名注册的测试环境搭建开始到别名注册分析结束,涵盖alias标签的使用和注册流程的分析。在别名注册相关内容分析后进行BeanDefinition注册的分析,对BeanDefinition的注册细节进行分析。

4.1 别名注册测试环境搭建

开始进行别名注册分析之前,需要先进行别名注册测试环境的搭建,这里所使用到的测试用例和第3章中的用例相同,首先需要编写SpringXML配置文件,文件名为alias-node.xml,文件内代码如下。

```
<?xml version="1.0" encoding="UTF-8"?>
<beans xmlns:xsi="http://www.w3.org/2001/XMLSchema-instance"
    xmlns="http://www.springframework.org/schema/beans"
    xsi:schemaLocation=" http://www.springframework.org/schema/beans http://www.springframework.org/schema/beans/spring-beans.xsd">

    <alias name="people" alias="p1"/>
    <bean id="people" class="com.source.hot.ioc.book.pojo.PeopleBean">
        <property name="name" value="zhangsan"/>
    </bean>
</beans>
```

完成SpringXML配置文件编写后进行测试类编写,测试类是AliasNodeTest,具体代码如下。

```java
class AliasNodeTest {

    @Test
    void testAlias() {
        ClassPathXmlApplicationContext context
                = new ClassPathXmlApplicationContext("META-INF/alias-node.xml");

        Object people = context.getBean("people");
        Object p1 = context.getBean("p1");

        assert people.equals(p1);
    }
}
```

测试用例准备完毕下面将进入别名注册相关源码分析。

4.2 别名注册接口

首先需要确认分析目标,本节的分析目标是下面这段代码:

`getReaderContext().getRegistry().registerAlias(name,alias)`

在这段代码中需要使用到一个接口,该接口是别名注册接口。

4.3 SimpleAliasRegistry 中注册别名的实现

接下来将分析 SimpleAliasRegistry 对象中关于别名注册的实现方法,具体处理代码如下。

```java
//删除日志和验证相关代码
public void registerAlias(String name, String alias) {
    synchronized(this.aliasMap) {
        //别名和真名是否相同
        if(alias.equals(name)) {
            //移除
            this.aliasMap.remove(alias);
        } else {
            //通过别名获取真名
            String registeredName = this.aliasMap.get(alias);
            //真名不为空
            if(registeredName != null) {
                //真名等于参数的真名
                if(registeredName.equals(name)) {
                    return;
                }
                //是否覆盖别名
                if(!allowAliasOverriding()) {
                    throw new IllegalStateException("Cannot define alias '" + alias + "' for name '" + name + "': It is already registered for name '" + registeredName + "'.");
```

```
            }
        }
        //别名是否循环使用
        checkForAliasCircle(name,alias);
        //设置别名对应真名
        this.aliasMap.put(alias,name);
    }
  }
}
```

在这段代码中需要关注以下两个参数。

（1）name：这个参数是 alias 标签中属性 name 的数据值，一般情况下将其称为真名、BeanName。

（2）alias：这个参数是 alias 标签中属性 alias 的数据值，一般情况下将其称为别名。

接下来对这段代码的处理流程进行分析，整个处理方法中围绕一个判断做各类处理，该判断是判断别名是否和真名相同，相等时的处理很简单，在别名容器中删除这个别名的键值信息即可。不相同的时候处理逻辑有以下三步。

（1）从别名容器中用别名去获取容器中可能存在的真名。

（2）当容器中可能存在的真名存在的情况下会和参数真名进行比较是否相同，相同就直接返回了，不做其他处理。

（3）当容器中可能存在的真名不存在的情况下，Spring 会进行别名是否循环使用，当检测通过时加入别名容器中。

整个流程中比较复杂的方法在 checkForAliasCircle 中，其他的步骤相对来说是一个比较容易理解的流程。接下来对别名检查做一个分析，首先查看处理代码。

```
protected void checkForAliasCircle(String name,String alias) {
    //是否存在别名
    if (hasAlias(alias,name)) {
        throw new IllegalStateException("Cannot register alias '" + alias +
            "' for name '" + name + "': Circular reference - '" +
            name + "' is a direct or indirect alias for '" + alias + "' already");
    }
}
```

在这段方法中还有 hasAlias 方法重点都在这个方法中了，hasAlias 方法的代码内容如下。

```
public boolean hasAlias(String name,String alias) {
    //从别名 map 中获取已注册的真名
    String registeredName = this.aliasMap.get(alias);
    //注册的真名和参数真名是否相同
    //递归判断是否存在别名
    return ObjectUtils.nullSafeEquals(registeredName,name) || (registeredName != null
        && hasAlias(name,registeredName));
}
```

在这段方法中处理逻辑：从别名容器中获取当前参数 alias 对应的容器内真名

(registeredName),将 registeredName 和 name 做比较确认是否是相等的。

4.4 别名换算真名

接下来需要解决使用上的一个问题：别名去获取 Bean 对象的时候，Spring 是如何找到真正的名字的？在 Spring 中有一个叫作 transformedBeanName 的方法，提供者是 AbstractBeanFactory，这个方法就是专门用来求解 BeanName 的方法。具体处理方法如下。

```
protected String transformedBeanName(String name) {
    //转换 BeanName
    //通过 BeanFactoryUtils.transformedBeanName 求 BeanName
    //如果是有别名的(方法参数是别名),会从别名列表中获取对应的 BeanName
    return canonicalName(BeanFactoryUtils.transformedBeanName(name));
}
```

在 transformedBeanName 中得分为以下两步看。
(1) 传入参数的处理。
(2) 别名容器中获取对应的 BeanName。
首先查看第一步中 BeanFactoryUtils.transformedBeanName(name)的处理代码。

```
public static String transformedBeanName(String name) {
    Assert.notNull(name,"'name' must not be null");
    //名字不是 & 开头直接返回
    if (!name.startsWith(BeanFactory.FACTORY_BEAN_PREFIX)) {
        return name;
    }
    //截取字符串,再返回
    return transformedBeanNameCache.computeIfAbsent(name,beanName -> {
        do {
            beanName = beanName.substring(BeanFactory.FACTORY_BEAN_PREFIX.length());
        }
        while (beanName.startsWith(BeanFactory.FACTORY_BEAN_PREFIX));
        return beanName;
    });
}
```

在这个方法中根据参数是否以"&"字符开头进行两种处理。
(1) 不是以"&"开头：直接返回参数。
(2) 是以"&"开头：切割字符串后返回。
在这个方法中,得到的是一个字符串,可能是别名,可能是 BeanName。
回到测试用例,p1 是别名,这个别名不以"&"开头,就直接返回 p1 了。
这样就理解了第一步 BeanFactoryUtils.transformedBeanName(name) 的作用,下面就是对 canonicalName 方法的分析了。
首先找到方法的提供者 SimpleAliasRegistry#canonicalName,具体代码如下。

```
public String canonicalName(String name) {
```

```
            String canonicalName = name;
            String resolvedName;
            do {
                //别名的获取
                resolvedName = this.aliasMap.get(canonicalName);
                if (resolvedName != null) {
                    canonicalName = resolvedName;
                }
            }
            while (resolvedName != null);
            return canonicalName;
}
```

上述方法就是一个简单的处理,直接从容器中获取对应的名称返回即可。

4.5 BeanDefinition 注册

在 Spring 中对于 BeanDefinition 对象的注册是通过 registerBeanDefinition 方法进行的,具体代码如下。

```
public static void registerBeanDefinition(
        BeanDefinitionHolder definitionHolder,BeanDefinitionRegistry registry)
        throws BeanDefinitionStoreException {

    //获取 BeanName
    String beanName = definitionHolder.getBeanName();
    //注册 BeanDefinition
    registry.registerBeanDefinition(beanName,definitionHolder.getBeanDefinition());

    //别名列表
    String[] aliases = definitionHolder.getAliases();
    //注册别名列表
    if (aliases != null) {
        for (String alias: aliases) {
            registry.registerAlias(beanName,alias);
        }
    }
}
```

通过阅读 registerBeanDefinition 方法可以确定该方法具体处理了以下两个事项。
(1) 将 BeanName 和 BeanDefinition 进行关系绑定。
(2) 将 BeanName 和 alias 进行关系绑定。

在 registerBeanDefinition 方法中可以进一步确定定义 BeanDefinition 注册方法的是 BeanDefinitionRegistry 接口,图 4.1 为 BeanDefinitionRegistry 的类图。

在 Spring 中关于 BeanDefinition 的注册操作,频繁使用的实现类是 DefaultListableBeanFactory,该对象会经常出现在各个需要进行 BeanDefinition 注册的地方。BeanDefinitionRegistry 的一些成员变量属性如图 4.2 所示。

第4章　别名注册和BeanDefinition注册

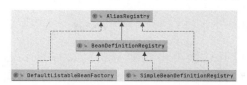

图 4.1　BeanDefinitionRegistry 类图

图 4.2　BeanDefinitionRegistry 成员变量信息

4.6 DefaultListableBeanFactory 中存储 BeanDefinition 的容器

通过前文的分析可以确认接下来需要分析的目标是 org.springframework.beans.factory.support.DefaultListableBeanFactory#registerBeanDefinition 方法。首先进行 Java 中常用数据存储集合的对比，在 Java 中对于数据存储常用的对象有 List、Map 等结构（接口），首先分析 List 作为 BeanDefinition 的存储结构，对于存储 BeanDefinition 而言 List 结构足以满足，但是光进行数据存储还不够，还需要考虑使用时的情况，当使用者想要通过 BeanName 来得到 BeanDefinition 时，程序需要循环 List 再判断 BeanDefinition 的名称是否和需要的 BeanName 相同，相同就返回。但是在 Spring 中，BeanDefinition 接口并没有提供获取 BeanName 的方法，如果需要根据这个模式进行处理，那么需要对 BeanDefinition 接口进行额外的定义（在 BeanDefinition 接口中定义一个获取 BeanName 的方法）。分析完成 List 作为存储结构后再来分析 Map 作为 BeanDefinition 的存储结构。对于存储 BeanDefinition 而言，Map 也可以满足，但是 Map 结构需要存储为键值对的形式，根据前文所述，在使用时需要通过 BeanName 获取 BeanDefinition，那么对于 Map 的键值对设计就可以是 key 存储 BeanName，value 存储 BeanDefinition。关于 List 和 Map 两种存储结构的选择，还可以从时间复杂度上进行考虑，对于获取而言，List 的时间复杂度为 $O(n)$，Map 的时间复杂度为 $O(1)$，相比较之下 Map 作为存储结构更有优势。下面的代码是 Spring 中用来存储 BeanDefinition 的数据结构。

```
private final Map<String,BeanDefinition> beanDefinitionMap =
    new ConcurrentHashMap<>(256);
```

在这个存储结构中和前文所述的 Map 存储如出一辙，key 表示 BeanName，value 表示 BeanDefinition 对象。

4.7 DefaultListableBeanFactory 中的注册细节

本节将介绍 BeanDefinition 对象注册时的处理细节，在 DefaultListableBeanFactory 类中所提供的注册方法具体流程如下。

（1）对 BeanName 进行验证。
（2）BeanName 在 BeanDefinition 容器中的不同情况处理。
① BeanName 已经在容器中存在，并且拥有对应的 BeanDefinition 对象。
② BeanName 在容器中搜索 BeanDefinition 失败，容器中没有对应的 BeanDefinition 对象。
（3）根据 BeanName 进行 BeanDefinition 的刷新操作。
接下来将对上述三个操作步骤进行详细分析。

4.7.1 BeanDefinition 的验证

在方法最开始需要做的是对 BeanDefinition 的数据验证,下面是关于 BeanDefinition 验证的代码。

```
if (beanDefinition instanceof AbstractBeanDefinition) {
    try {
        //Bean 定义验证
        ((AbstractBeanDefinition) beanDefinition).validate();
    }
    catch (BeanDefinitionValidationException ex) {
        throw new BeanDefinitionStoreException(beanDefinition.getResourceDescription(),
beanName,"Validation of bean definition failed",ex);
    }
}
```

在 Spring 中通过 processBeanDefinition 方法得到的 BeanDefinition,具体类型是 GenericBeanDefinition。GenericBeanDefinition 类图如图 4.3 所示。

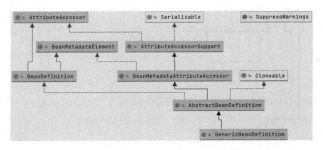

图 4.3 GenericBeanDefinition 类图

从 GenericBeanDefinition 类图中可以发现 GenericBeanDefinition 类是 AbstractBeanDefinition 的子类,有关 validate 方法在 AbstractBeanDefinition 类中有具体的处理代码。值得注意的是,validate 方法存在两个实现,第一个是 AbstractBeanDefinition,第二个是 ChildBeanDefinition。AbstractBeanDefinition 中的 validate 方法详情如下。

```
public void validate() throws BeanDefinitionValidationException {
    //是否存在重写方法,factory_method_name 是否为空
    if (hasMethodOverrides() && getFactoryMethodName() != null) {
        throw new BeanDefinitionValidationException(
                "Cannot combine factory method with container-generated method overrides: " +
                    "the factory method must create the concrete bean instance.");
    }
    //bean class 是否等于 Class
    if (hasBeanClass()) {
        //方法重写 + 验证
        prepareMethodOverrides();
    }
}
```

在这段代码中有很多封装的方法,下面对整个方法的主体流程及封装方法的作用进行说明。

(1) 判断 MethodOverrides 数据是否为空,处理方法为 hasMethodOverrides()。

(2) 判断是否存在工厂函数,注意此时判断的是字符串并非 Method 对象,处理方法为 getFactoryMethodName()。

(3) 判断 BeanClass 属性是否是 Class 类型,BeanClass 的数据类型有两种情况,第一种是 Class 类型,第二种是 String 类型,处理方法为 hasBeanClass()。

(4) 进行方法重写标记处理和验证,处理方法为 prepareMethodOverrides,具体处理逻辑如下。

① 判断是否存在需要重写的方法,处理方法为 hasMethodOverrides。

② 如果需要重新循环处理 MethodOverride 对象,将 overloaded 属性标记为 false。

下面介绍另一个 validate 方法的提供者中的处理逻辑,ChildBeanDefinition 类中关于 validate 的代码信息如下。

```
@Override
public void validate() throws BeanDefinitionValidationException {
    super.validate();
    if (this.parentName == null) {
        throw new BeanDefinitionValidationException ( " ' parentName ' must be set in ChildBeanDefinition");
    }
}
```

在 ChildBeanDefinition#validate 方法中首先会调用父类 AbstractBeanDefinition 中的 validate 方法,这段方法的分析在前文提及,在处理完成父类提供的方法之后在 ChildBeanDefinition 中对 parentName 做验证,判断 parentName 是否存在,不存在则抛出异常,异常信息为 'parentName' must be set in ChildBeanDefinition。

4.7.2 容器中存在 BeanName 对应的 BeanDefinition 的处理

接下来分析容器中 BeanName 存在对应 BeanDefinition 对象的操作相关源码。下面是处理代码。

```
//第二部分 BeanName 存在 BeanDefinition 的情况
//从 map 中根据 BeanName 获取 BeanDefinition
BeanDefinition existingDefinition = this.beanDefinitionMap.get(beanName);
if (existingDefinition != null) {
    //BeanName 是否允许重复注册
    if (!isAllowBeanDefinitionOverriding()) {
        throw new BeanDefinitionOverrideException(beanName,beanDefinition,existingDefinition);
    }
    //role 值比较
    else if (existingDefinition.getRole() < beanDefinition.getRole()) {
        if (logger.isInfoEnabled()) {
            logger.info("Overriding user-defined bean definition for bean '" + beanName +
```

```
                    "' with a framework - generated bean definition: replacing [" +
                    existingDefinition + "] with [" + beanDefinition + "]");
        }
    }
    //map 中存储的 BeanDefinition 是否和参数相同
    else if (!beanDefinition.equals(existingDefinition)) {
        if (logger.isDebugEnabled()) {
            logger.debug("Overriding bean definition for bean '" + beanName +
                    "' with a different definition: replacing [" + existingDefinition +
                    "] with [" + beanDefinition + "]");
        }
    }
    else {
        if (logger.isTraceEnabled()) {
            logger.trace("Overriding bean definition for bean '" + beanName +
                    "' with an equivalent definition: replacing [" + existingDefinition +
                    "] with [" + beanDefinition + "]");
        }
    }
    //设置 BeanName 和 BeanDefinition 关系
    this.beanDefinitionMap.put(beanName,beanDefinition);
}
```

在这段方法中可以关注下面两个内容。

（1）existingDefinition 变量是指通过 BeanName 在容器中获取的历史 BeanDefinition。

（2）isAllowBeanDefinitionOverriding 方法是判断是否允许覆盖 BeanDefinition。

在整个流程中上面两个变量会直接影响处理方式。首先，existingDefinition 的存在与否决定了后续的操作，在本节主要分析的是存在的处理情况，4.7.3 节会对不存在的情况进行分析。其次是 isAllowBeanDefinitionOverriding 方法，当 isAllowBeanDefinitionOverriding 方法推论结果是不允许覆盖时，也就是下面这段代码成立，抛出异常 BeanDefinitionOverrideException。关于异常抛出的详细代码如下。

```
if (!isAllowBeanDefinitionOverriding()) {
    throw new
BeanDefinitionOverrideException(beanName,beanDefinition,existingDefinition);
}
```

在处理 BeanName 存在 BeanDefinition 的情况下上面就是核心内容，那么剩下的还有一些小的 else if，如 Role 的处理。当前 BeanDefinition 和 历史 BeanDefinition 的对比，具体的处理行为如下。

（1）在对比 Role 和两个 BeanDefinition 的时候都是一个简单的日志输出，未做复杂处理。

（2）在方法最后，如果是允许覆盖 BeanDefinition 的，那么 Spring 就会将数据覆盖。即调用 this.beanDefinitionMap.put(beanName,beanDefinition)代码。调用后 BeanDefinition 容器会更新 BeanName 和 BeanDefinition 之间的关系。

4.7.3 容器中不存在 BeanName 对应的 BeanDefinition 的处理

接下来分析容器中 BeanName 不存在对应 BeanDefinition 的处理相关源码，对于这部分的处理源代码如下。

```
else {
    //检查 Bean 是否已经开始创建
    if (hasBeanCreationStarted()) {
        synchronized (this.beanDefinitionMap) {
            //设置 BeanName 和 BeanDefinition 关系
            this.beanDefinitionMap.put(beanName,beanDefinition);
            //BeanDefinition 的名称列表
            List < String > updatedDefinitions =
    new ArrayList <>(this.beanDefinitionNames.size() + 1);
            //加入内存数据
            updatedDefinitions.addAll(this.beanDefinitionNames);
            //加入当前的 BeanName
            updatedDefinitions.add(beanName);
            //对象替换
            this.beanDefinitionNames = updatedDefinitions;
            //移除当前的 BeanName
            removeManualSingletonName(beanName);
        }
    }
    else {
        //设置容器数据
        this.beanDefinitionMap.put(beanName,beanDefinition);
        this.beanDefinitionNames.add(beanName);
        //移除当前的 BeanName
        removeManualSingletonName(beanName);
    }
    this.frozenBeanDefinitionNames = null;
}
```

在这段代码中首先需要强调一个方法 hasBeanCreationStarted，该方法能够确认 Bean 是否开始创建，在 hasBeanCreationStarted 方法中进行推论的主要依据是 alreadyCreated 对象，该对象存储了正在创建的 Bean，具体处理代码如下。

```
protected boolean hasBeanCreationStarted() {
    return !this.alreadyCreated.isEmpty();
}
```

在这个方法中执行的判断就是判断 alreadyCreated 对象是否不为空，在这段方法使用时还需要了解 alreadyCreated 对象是什么时候进行数据设置。在 Spring 中对于 alreadyCreated 数据赋值的方法是 markBeanAsCreated，该方法是用来标记 Bean 正在被创建，markBeanAsCreated 方法的调用是在 getBean 时，即在创建 Bean 实例时会被调用。

关于 alreadyCreated 的分析到此结束，下面来看其他一些代码。在整个处理操作中大部分都是针对Java集合的一些操作，在处理流程最后有一个removeManualSingletonName方法，该方法执行的内容和手动注册的单例Bean有关，具体处理代码如下。

```
private void removeManualSingletonName(String beanName) {
    updateManualSingletonNames(
    set -> set.remove(beanName), set -> set.contains(beanName));
}
```

在这段代码中进行了另一个方法updateManualSingletonNames的调用，详细代码如下。

```
private void updateManualSingletonNames(Consumer<Set<String>> action, Predicate<Set<String>> condition) {
    if (hasBeanCreationStarted()) {
        synchronized (this.beanDefinitionMap) {
            //输入的BeanName是否在manualSingletonNames中存在
            if (condition.test(this.manualSingletonNames)) {
                //数据重写
                Set<String> updatedSingletons = new LinkedHashSet<>(this.manualSingletonNames);
                //删除BeanName
                action.accept(updatedSingletons);
                //数据重写
                this.manualSingletonNames = updatedSingletons;
            }
        }
    }
    else {
        //输入的BeanName是否在manualSingletonNames中存在
        if (condition.test(this.manualSingletonNames)) {
            //删除BeanName
            action.accept(this.manualSingletonNames);
        }
    }
}
```

在这段方法中需要关注action变量，整个Spring IoC中一般有下面这些可能。

（1）environment。

（2）systemProperties。

（3）systemEnvironment。

（4）messageSource。

（5）applicationEventMulticaster。

（6）lifecycleProcessor。

4.7.4　BeanDefinition的刷新处理

最后分析BeanDefinition的刷新操作，刷新BeanDefinition数据信息相关代码如下。

```
if (existingDefinition != null || containsSingleton(beanName)) {
    //刷新 BeanDefinition
    resetBeanDefinition(beanName);
}
```

在这段代码中可以清晰地知道,只要满足下面两个条件的其中一个就会进行 BeanDefinition 刷新操作。

(1) 通过 BeanName 搜索 BeanDefinition 成功。

(2) 单例对象容器中包含当前 BeanName 对应的 Bean 实例。

下面介绍刷新 BeanDefinition 的方法细节,在刷新操作中分为四个主要步骤。

(1) 将合并的 BeanDefinition(mergedBeanDefinitions)中可能存在 BeanDefinition 中的 stale 属性设置为 true。

(2) 摧毁 BeanName 对应的单例对象。

(3) MergedBeanDefinitionPostProcessor 的后置方法执行。

(4) 处理 BeanDefinition 名称列表中名称和当前 BeanName 相同的数据。

在上述四个步骤中需要关注 mergedBeanDefinitions 对象的获取过程,通过 AbstractBeanFactory 提供的 getMergedBeanDefinition 方法可以获取。注意 mergedBeanDefinition 的本质还是 BeanDefinition,在这里主要以 RootBeanDefinition 类型的形式出现,对于名称中 merged 的定义是指合并父 BeanDefinition 对象和当前 BeanDefinition,细节处理代码如下。

```
//复制父 BeanDefinition 对象
mbd = new RootBeanDefinition(pbd);
//覆盖 BeanDefinition
mbd.overrideFrom(bd);
```

在这段代码中通过父 BeanDefinition(pbd:ParentBeanDefinition)进行创建 mergedBeanDefinition(mbd),在 RootBeanDefinition 中通过数据重载方法 overrideFrom 将当前 BeanDefinition 中的数据和父 BeanDefinition 的数据进行覆盖,覆盖数据后得到的对象就是 mergedBeanDefinition。

在后续操作中(步骤(2))需要对单例 Bean 进行摧毁,有关 Bean 摧毁的相关分析将会在后续章节中进行分析,现在只需要知道单例对象的摧毁就是从单例对象容器中剔除当前 BeanName 对应的 Bean 实例。

下面进入步骤(3)的分析。这部分是关于 MergedBeanDefinitionPostProcessor 的处理, MergedBeanDefinitionPostProcessor 对象是 BeanPostProcessor 的子类,这是 Spring 开发中经常会说到的 Bean 后置处理器,在 Spring 中对于 MergedBeanDefinitionPostProcessor 接口的实现只有 AutowiredAnnotationBeanPostProcessor,具体代码如下。

```
public void resetBeanDefinition(String beanName) {
    this.lookupMethodsChecked.remove(beanName);
    this.injectionMetadataCache.remove(beanName);
}
```

在这段代码中通过 BeanName 移除了两个容器中的相关数据,第一个容器是

lookupMethodsChecked，第二个容器是 injectionMetadataCache。

最后进入步骤（4）的分析。步骤四其实是一个递归处理，处理的目标是在 BeanDefinition 的名称列表中寻找和当前 BeanName 相同的数据，如果符合就会重复步骤一到步骤四。

下面这段代码是 resetBeanDefinition 完整内容，可以对照处理流程进行阅读。

```
protected void resetBeanDefinition(String beanName) {
    //清空合并的 BeanDefinition
    clearMergedBeanDefinition(beanName);

    //摧毁单例 Bean
    destroySingleton(beanName);

    //后置处理器执行 resetBeanDefinition 方法
    for (BeanPostProcessor processor: getBeanPostProcessors()) {
        if (processor instanceof MergedBeanDefinitionPostProcessor) {
            //执行后置方法的 resetBeanDefinition
            ((MergedBeanDefinitionPostProcessor) processor).resetBeanDefinition(beanName);
        }
    }

    //处理其他的 BeanName
    for (String bdName: this.beanDefinitionNames) {
        if (!beanName.equals(bdName)) {
            BeanDefinition bd = this.beanDefinitionMap.get(bdName);
            //BeanName 等于父名称
            if (bd != null && beanName.equals(bd.getParentName())) {
                //递归刷新 beanName 对应的 beanDefinition
                resetBeanDefinition(bdName);
            }
        }
    }
}
```

4.8　BeanDefinition 的获取

在前文通过种种操作将 BeanName 和 BeanDefinition 的关系已经存储在容器中，主要操作是进行 BeanDefinition 对象注册，下面将介绍如何通过 BeanName 获取 BeanDefinition 对象，在 Spring 中负责处理这个行为的方法是 DefaultListableBeanFactory#getBeanDefinition，具体代码如下。

```
@Override
public BeanDefinition getBeanDefinition(String beanName)
throws NoSuchBeanDefinitionException {
    BeanDefinition bd = this.beanDefinitionMap.get(beanName);
    if (bd == null) {
        if (logger.isTraceEnabled()) {
```

```
                logger.trace("No bean named '" + beanName + "' found in " + this);
            }
            throw new NoSuchBeanDefinitionException(beanName);
        }
        return bd;
    }
```

在这段代码中对于 BeanDefinition 的获取可以看到十分直接，直接在 BeanDefinition 容器中通过名称获取即调用 Map.get 方法获取，如果在容器中获取不到 BeanName 对应的 BeanDefinition 对象，会抛出 NoSuchBeanDefinitionException 异常。

小结

本章介绍了别名注册和别名使用的源码分析，它们分别位于 AliasRegistry、SimpleAliasRegistry 和 BeanFactoryUtils 之中。从使用角度上看是关于 alias 标签在 SpringXML 文件中的使用。在 BeanDefinition 注册分析中重点关注的是 BeanDefinition 的存储过程和 BeanDefinition 对象获取的过程，在分析 BeanDefinition 的存储过程时需要关注存储方法中的各类数据验证，在获取时同样需要关注获取失败的异常处理。需要注意的是，现在 BeanName 对应的还只是 BeanDefinition，并不是一个 Bean 实例，它还没有办法直接进行使用。

第5章 bean标签解析

在第 4 章中对 BeanDefinition 的注册流程进行了相关分析,在 SpringXML 开发模式下 BeanDefinition 的数据来源是 SpringXML 原始标签中的 bean 标签,该标签会定义 BeanDefinition 对象中的各个属性。本章将围绕 bean 标签进行测试环境搭建和 bean 标签解析流程的分析。在 Spring 中对于 SpringXML 文件中 bean 标签的解析过程如下。

(1) 将 Element 交给 BeanDefinitionParserDelegate 解析。
(2) 注册 BeanDefinition。
(3) 发布组件注册事件。

5.1 创建 bean 标签解析环境

在进行 bean 标签的源码分析之前,还需要做一些准备工作:搭建一个 bean 标签解析的环境。

5.1.1 编写 SpringXML 配置文件

下面先创建一个 SpringXML 配置文件,并将其命名为 bean-node.xml,向该 bean-node.xml 文件中填充下面这段代码。

```
<?xml version = "1.0" encoding = "UTF-8"?>
<beans xmlns:xsi = "http://www.w3.org/2001/XMLSchema-instance"
    xmlns = "http://www.springframework.org/schema/beans"
    xsi:schemaLocation = " http://www.springframework.org/schema/beans http://www.springframework.org/schema/beans/spring-beans.xsd">

    <bean id = "people" class = "com.source.hot.ioc.book.pojo.PeopleBean">
```

```
        </bean>
</beans>
```

5.1.2 编写 bean-node 对应的测试用例

创建一个名为 BeanNodeTest 的 Java 对象，代码如下。

```
class BeanNodeTest {
    @Test
    void testBean() {
        ClassPathXmlApplicationContext context
                = new ClassPathXmlApplicationContext("META-INF/bean-node.xml");

        Object people = context.getBean("people");
        context.close();
    }
}
```

通过上述操作测试用例准备完毕，下面请回忆一下 Spring 中对于 bean 标签的解析方法。Spring 中对于 bean 标签解析的方法签名是 org.springframework.beans.factory.xml.DefaultBeanDefinitionDocumentReader#processBeanDefinition。processBeanDefinition 方法的代码如下。

```
protected void processBeanDefinition(Element ele, BeanDefinitionParserDelegate delegate) {
    //创建 BeanDefinition
    BeanDefinitionHolder bdHolder = delegate.parseBeanDefinitionElement(ele);
    if (bdHolder != null) {
        //BeanDefinition 装饰
        bdHolder = delegate.decorateBeanDefinitionIfRequired(ele, bdHolder);
        try {
            //注册 BeanDefinition
            BeanDefinitionReaderUtils.registerBeanDefinition(bdHolder, getReaderContext().getRegistry());
        }
        catch (BeanDefinitionStoreException ex) {
            getReaderContext().error("Failed to register bean definition with name '" +
                    bdHolder.getBeanName() + "'", ele, ex);
        }
        //component 注册事件触发
        getReaderContext().fireComponentRegistered(new BeanComponentDefinition(bdHolder));
    }
}
```

Spring 中通过 BeanDefinitionParserDelegate#parseBeanDefinitionElement 方法来对 Element 对象进行解析，映射到 XML 文件中 Element 对象就是一个 XML 标签，下面将进入到该方法的分析阶段。

5.2 parseBeanDefinitionElement 方法处理

5.2.1 parseBeanDefinitionElement 第一部分处理

通过前文找到了需要分析的方法,Spring 中对于 bean 标签的 id 属性和 name 属性的处理代码如下。

```
String id = ele.getAttribute(ID_ATTRIBUTE);
//获取 name
String nameAttr = ele.getAttribute(NAME_ATTRIBUTE);

//别名列表
List<String> aliases = new ArrayList<>();
//是否有 name 属性
if (StringUtils.hasLength(nameAttr)) {
   //获取名称列表,根据 ,; 进行分隔
   String[] nameArr =
StringUtils.tokenizeToStringArray(nameAttr,MULTI_VALUE_ATTRIBUTE_DELIMITERS);
   //添加所有
   aliases.addAll(Arrays.asList(nameArr));
}
```

第一部分的处理很简单,获取 bean 标签中的 id 和 name 属性,对 name 属性做分隔符切分后将切分结果作为别名。在第一部分中提到了分隔符这一信息,在这段代码中的分隔符总共有以下 3 种。

(1) 逗号。
(2) 分号。
(3) 空格。

5.2.2 parseBeanDefinitionElement 第二部分处理

接下来将进入第二部分的分析,Spring 中对于 BeanName 的处理代码如下。

```
String beanName = id;
if (!StringUtils.hasText(beanName) && !aliases.isEmpty()) {
 //别名的第一个设置为 beanName
 beanName = aliases.remove(0);
 if (logger.isTraceEnabled()) {
    logger.trace("No XML 'id' specified - using '" + beanName +
        "' as bean name and " + aliases + " as aliases");
 }
}

//BeanDefinition 为空
if (containingBean == null) {
```

```
//判断beanName是否被使用,Bean别名是否被使用
checkNameUniqueness(beanName,aliases,ele);
}
```

在第二部分的处理中可以再将其分为以下两小段内容。

(1) 关于 BeanName 的推论。

(2) 关于 BeanName 是否被使用,别名是否被使用的验证。

Spring 中对于 BeanName 的推论规则如下。

(1) BeanName 可以是 bean 标签中的 id 属性。

(2) BeanName 可以是 bean 标签中的 name 属性,注意 name 属性可以存在多个,多个 name 使用分隔符分隔,当存在多个 name 属性时会取第一个值作为 BeanName。

(3) 当 bean 标签中同时出现 id 属性和 name 属性时会用 id 作为 BeanName。

在了解了 BeanName 的生成规则后来编写相关测试用例,首先要变写的是关于 id 的测试用例。

```
<bean id="people" class="com.source.hot.ioc.book.pojo.PeopleBean">
</bean>
```

根据前文的分析此时可以推论出 BeanName 应该是 people,调试信息如图 5.1 所示。

接下来编写关于 name 的测试用例。

```
<bean name="peopleBean,people" class="com.source.hot.ioc.book.pojo.PeopleBean">
</bean>
```

根据前文的分析此时可以推论出 BeanName 应该是 peopleBean,调试信息如图 5.2 所示。

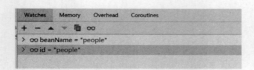
图 5.1 设置 id 时的 BeanName

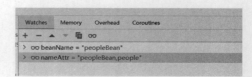
图 5.2 设置 name 时的 BeanName

最后来编写 id 和 name 同时设置的测试用例。

```
<bean id="p1" name="peopleBean,people"
class="com.source.hot.ioc.book.pojo.PeopleBean">
</bean>
```

根据前文的分析此时可以推论出 BeanName 应该是 p1,调试信息如图 5.3 所示。

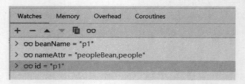
图 5.3 同时设置 id 和 name 时的 BeanName

关于 BeanName 的推论相关的分析到此结束，下面进入到第二部分关于 BeanName 是否被使用，别名是否被使用的验证的分析。Spring 中 checkNameUniqueness 的完整代码内容如下。

```java
protected void checkNameUniqueness(String beanName,List<String> aliases,Element beanElement) {
    //当前寻找的 name
    String foundName = null;

    //是否有 BeanName
    //使用过的 name 中是否存在
    if (StringUtils.hasText(beanName) && this.usedNames.contains(beanName)) {
        foundName = beanName;
    }
    if (foundName == null) {
        //寻找匹配的第一个
        foundName = CollectionUtils.findFirstMatch(this.usedNames,aliases);
    }
    //抛出异常
    if (foundName != null) {
        error("Bean name '" + foundName + "' is already used in this <beans> element",
beanElement);
    }

    //加入使用队列
    this.usedNames.add(beanName);
    this.usedNames.addAll(aliases);
}
```

在 checkNameUniqueness 方法中需要重点关注的是 usedNames 变量，usedNames 变量是指已经使用过的名称包含 BeanName 和 Alias。整个验证过程就是判断传递的参数是否在 usedNames 容器中存在，一旦存在就会抛出异常。如果不存在则将 BeanName 和 Alias 加入 usedNames 容器等待后续使用。

至此，parseBeanDefinitionElement 第二部分的分析告一段落，下面将进入第三部分的分析。

5.2.3　parseBeanDefinitionElement 第三部分处理

Spring 中 parseBeanDefinitionElement 方法中的第三部分代码如下。

```java
AbstractBeanDefinition beanDefinition = parseBeanDefinitionElement(ele,beanName,containingBean);
```

这段代码其实是一个方法的调用，真正应该阅读的代码应该是下面这一段。

```java
public AbstractBeanDefinition parseBeanDefinitionElement(
    Element ele,String beanName,@Nullable BeanDefinition containingBean) {

    //第一部分
    //设置阶段 Bean 定义解析阶段
```

```java
        this.parseState.push(new BeanEntry(beanName));

        String className = null;
        //是否包含属性 class
        if (ele.hasAttribute(CLASS_ATTRIBUTE)) {
            className = ele.getAttribute(CLASS_ATTRIBUTE).trim();
        }
        String parent = null;
        //是否包含属性 parent
        if (ele.hasAttribute(PARENT_ATTRIBUTE)) {
            parent = ele.getAttribute(PARENT_ATTRIBUTE);
        }

        //第二部分

        //创建 BeanDefinition
        AbstractBeanDefinition bd = createBeanDefinition(className,parent);

        //BeanDefinition 属性设置
        parseBeanDefinitionAttributes(ele,beanName,containingBean,bd);
        //设置描述
        bd.setDescription(DomUtils.getChildElementValueByTagName(ele,DESCRIPTION_ELEMENT));
        //元信息设置 meta 标签解析
        parseMetaElements(ele,bd);
        //lookup-override 标签解析
        parseLookupOverrideSubElements(ele,bd.getMethodOverrides());
        //replaced-method 标签解析
        parseReplacedMethodSubElements(ele,bd.getMethodOverrides());

        //constructor-arg 标签解析
        parseConstructorArgElements(ele,bd);
        //property 标签解析
        parsePropertyElements(ele,bd);
        //qualifier 标签解析
        parseQualifierElements(ele,bd);
        //资源设置
        bd.setResource(this.readerContext.getResource());
        //source 设置
        bd.setSource(extractSource(ele));

        return bd;

}
```

在这个方法中 Spring 做了 11 步操作,下面将围绕这 11 步依次解析它们的处理行为。操作细节如下。

(1) 处理 className 和 parent 属性。

(2) 创建基本的 BeanDefinition 对象,具体类:AbstractBeanDefinition、GenericBeanDefinition。

(3) 读取 bean 标签的属性,为 BeanDefinition 对象进行赋值。

（4）处理描述标签 description。
（5）处理 meta 标签。
（6）处理 lookup-override 标签。
（7）处理 replaced-method 标签。
（8）处理 constructor-arg 标签。
（9）处理 property 标签。
（10）处理 qualifier 标签。
（11）设置资源对象和 source 属性。

注意：在这段方法中出现的 parseState 对象操作不属于本节的分析内容，可以简单被理解成这是一个阶段标记。

1. 处理 class name 和 parent 属性

Spring 中对于 bean 标签的 class name 和 parent 两个属性的处理代码如下。

```
String className = null;
//是否包含属性 class
if (ele.hasAttribute(CLASS_ATTRIBUTE)) {
    className = ele.getAttribute(CLASS_ATTRIBUTE).trim();
}
String parent = null;
//是否包含属性 parent
if (ele.hasAttribute(PARENT_ATTRIBUTE)) {
    parent = ele.getAttribute(PARENT_ATTRIBUTE);
}
```

在这段代码中对于 className 和 parent 的处理十分简单，直接从 Element 对象中提取，提取完成数据后为下面创建 AbstractBeanDefinition 对象提供了基础数据支持。下面来看创建 AbstractBeanDefinition 对象的细节。

2. 创建 AbstractBeanDefinition 对象

Spring 中对于创建 AbstractBeanDefinition 对象的代码如下。

```
//parseBeanDefinitionElement 方法内调用
AbstractBeanDefinition bd = createBeanDefinition(className,parent);

//createBeanDefinition 详情
protected AbstractBeanDefinition
createBeanDefinition(@Nullable String className,@Nullable String parentName)
    throws ClassNotFoundException {


    return BeanDefinitionReaderUtils.createBeanDefinition(
        parentName,className,this.readerContext.getBeanClassLoader());
}
```

在这段方法中存在一层引用方法，这段引用方法的代码如下。

```
public static AbstractBeanDefinition createBeanDefinition(
```

```
        @Nullable String parentName, @Nullable String className, @Nullable ClassLoader
classLoader) throws ClassNotFoundException {

    GenericBeanDefinition bd = new GenericBeanDefinition();
    //设置父 BeanName
    bd.setParentName(parentName);
    if (className != null) {
        if (classLoader != null) {
            //设置 class
            //内部是通过反射创建 class
            bd.setBeanClass(ClassUtils.forName(className,classLoader));
        }
        else {
            //设置 className
            bd.setBeanClassName(className);
        }
    }
    return bd;
}
```

在这段方法中,createBeanDefinition 方法是一个静态方法,其完整的方法签名是 org.springframework.beans.factory.support.BeanDefinitionReaderUtils#createBeanDefinition,在这个方法中 Spring 对其做出了如下操作。

(1) 创建类型为 GenericBeanDefinition 的 BeanDefinition 对象。

(2) 为创建的 GenericBeanDefinition 对象进行数据设置,设置的属性是 parentName 和 beanClass。

注意:beanClass 的数据类型存在以下两种情况。

(1) beanClass 的类型是 String。

(2) beanClass 的类型是 Class。

下面进入调试阶段,首先编写一段 SpringXML 配置:

```
<bean id="people" class="com.source.hot.ioc.book.pojo.PeopleBean">
</bean>
```

在完成配置文件编写后启动项目观察 createBeanDefinition 执行后的结果对象 bd(BeanDefinition:bean 定义信息),如图 5.4 所示。

可以看到,beanClass 还是字符串类型。现在 AbstractBeanDefinition 基础对象已经准备完毕,接下来就是补充这个对象的其他属性。

3. 设置 BeanDefinition 的基本信息

接下来进行 parseBeanDefinitionAttributes 方法的解析,Spring 中 parseBeanDefinitionAttribute 的细节代码如下。

```
public AbstractBeanDefinition parseBeanDefinitionAttributes(Element ele,
        String beanName,
        @Nullable BeanDefinition containingBean,AbstractBeanDefinition bd) {

    //是否存在 singleton 属性
```

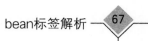

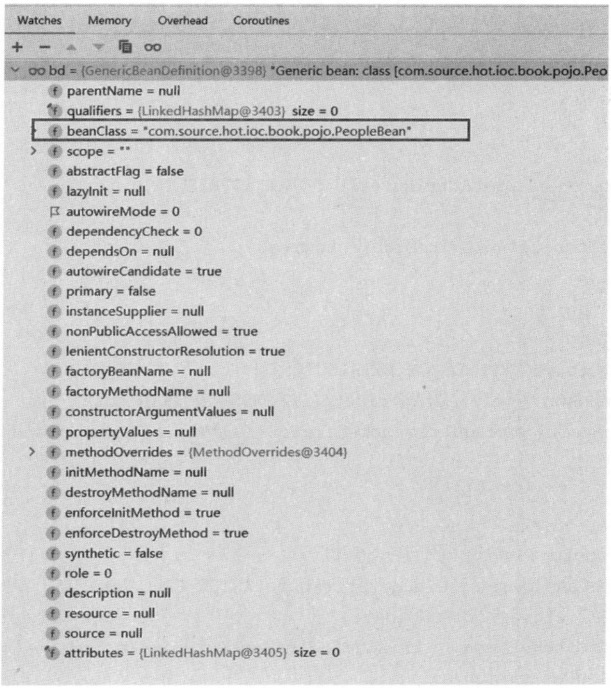

图 5.4 createBeanDefinition 执行后结果

```
if (ele.hasAttribute(SINGLETON_ATTRIBUTE)) {
    error("Old 1.x 'singleton' attribute in use - upgrade to 'scope' declaration",ele);
}
//是否存在 scope 属性
else if (ele.hasAttribute(SCOPE_ATTRIBUTE)) {
    //设置 scope 属性
    bd.setScope(ele.getAttribute(SCOPE_ATTRIBUTE));
}
//Bean 定义是否为空
else if (containingBean != null) {
    //设置 BeanDefinition 中的 scope
    bd.setScope(containingBean.getScope());
}

//是否存在 abstract 属性
if (ele.hasAttribute(ABSTRACT_ATTRIBUTE)) {
    //设置 abstract 属性

    bd.setAbstract(TRUE_VALUE.equals(ele.getAttribute(ABSTRACT_ATTRIBUTE)));
}

//获取 lazy-init 属性
String lazyInit = ele.getAttribute(LAZY_INIT_ATTRIBUTE);
//是否是默认的 lazy-init 属性
if (isDefaultValue(lazyInit)) {
    //获取默认值
    lazyInit = this.defaults.getLazyInit();
```

```java
        }
        //设置 lazy-init 属性
        bd.setLazyInit(TRUE_VALUE.equals(lazyInit));

        //获取注入方式
        //autowire 属性
        String autowire = ele.getAttribute(AUTOWIRE_ATTRIBUTE);
        //设置注入方式
        bd.setAutowireMode(getAutowireMode(autowire));

        //依赖的 Bean
        //depends-on 属性
        if (ele.hasAttribute(DEPENDS_ON_ATTRIBUTE)) {
            String dependsOn = ele.getAttribute(DEPENDS_ON_ATTRIBUTE);
            bd.setDependsOn(StringUtils.tokenizeToStringArray(dependsOn, MULTI_VALUE_ATTRIBUTE_DELIMITERS));
        }

        //autowire-candidate 是否自动注入判断
        String autowireCandidate = ele.getAttribute(AUTOWIRE_CANDIDATE_ATTRIBUTE);
        if (isDefaultValue(autowireCandidate)) {
            String candidatePattern = this.defaults.getAutowireCandidates();
            if (candidatePattern != null) {
                String[] patterns = StringUtils.commaDelimitedListToStringArray(candidatePattern);
                // * 匹配设置数据
                bd.setAutowireCandidate(PatternMatchUtils.simpleMatch(patterns, beanName));
            }
        }
        else {
            bd.setAutowireCandidate(TRUE_VALUE.equals(autowireCandidate));
        }

        //获取 primary 属性
        if (ele.hasAttribute(PRIMARY_ATTRIBUTE)) {
            bd.setPrimary(TRUE_VALUE.equals(ele.getAttribute(PRIMARY_ATTRIBUTE)));
        }

        //获取 init-method 属性
        if (ele.hasAttribute(INIT_METHOD_ATTRIBUTE)) {
            String initMethodName = ele.getAttribute(INIT_METHOD_ATTRIBUTE);
            bd.setInitMethodName(initMethodName);
        }
        //没有 init-method 的情况处理
        else if (this.defaults.getInitMethod() != null) {
            bd.setInitMethodName(this.defaults.getInitMethod());
            bd.setEnforceInitMethod(false);
        }

        //获取 destroy-method 属性
        if (ele.hasAttribute(DESTROY_METHOD_ATTRIBUTE)) {
```

```
            String destroyMethodName =
ele.getAttribute(DESTROY_METHOD_ATTRIBUTE);
        bd.setDestroyMethodName(destroyMethodName);
    }
    //没有 destroy-method 的情况处理
    else if (this.defaults.getDestroyMethod() != null) {
        bd.setDestroyMethodName(this.defaults.getDestroyMethod());
        bd.setEnforceDestroyMethod(false);
    }

    //获取 factory-method 属性
    if (ele.hasAttribute(FACTORY_METHOD_ATTRIBUTE)) {
        bd.setFactoryMethodName(ele.getAttribute(FACTORY_METHOD_ATTRIBUTE));
    }
    //获取 factory-bean 属性
    if (ele.hasAttribute(FACTORY_BEAN_ATTRIBUTE)) {
        bd.setFactoryBeanName(ele.getAttribute(FACTORY_BEAN_ATTRIBUTE));
    }

    return bd;
}
```

这段代码的处理过程与 className 和 parent 的处理过程基本类似，它们的处理思路都是从 bean 标签中提取属性对应的属性值，将属性值设置给 BeanDefinition 对象（BeanDefinition 对象是通过上一步创建出来），在这个处理过程中有一个值得关注的变量 defaults，先来看 defaults 变量的定义代码。

```
private final DocumentDefaultsDefinition defaults = new DocumentDefaultsDefinition();
```

这段代码中 defaults 指定了一个具体的数据类型 DocumentDefaultsDefinition，在 DocumentDefaultsDefinition 类型中存放了 Spring 对一些属性的默认值，DocumentDefaultsDefinition 中的定义的属性及属性默认值如表 5.1 所示。

表 5.1　DocumentDefaultsDefinition 属性及属性默认值

默认值属性名称	默 认 值
lazyInit	false
merge	false
autowire	no
autowireCandidates	null
initMethod	null
destroyMethod	null
source	null

在了解 defaults 对象中存有的数据后，来进一步观察 BeanDefinition 经过 parseBeanDefinitionAttributes 方法处理后得到的信息，如图 5.5 所示。

4. 设置 BeanDefinition 描述信息

接下来进行 BeanDefintion 描述信息设置方法的解析，首先需要编写一段 SpringXML

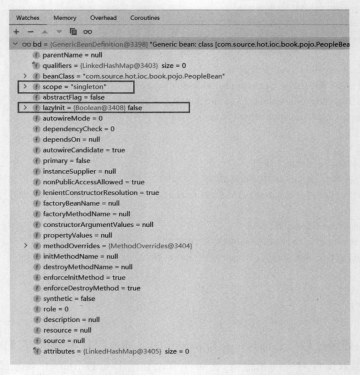

图 5.5　parseBeanDefinitionAttributes 方法执行后结果

配置，在这段配置中需要设置 bean 标签的属性值，具体代码如下。

```
<bean id="people" class="com.source.hot.ioc.book.pojo.PeopleBean">
    <description>this is a bean description</description>
</bean>
```

下面是设置 bean 描述信息的代码内容。

```
bd.setDescription(DomUtils.getChildElementValueByTagName(ele,DESCRIPTION_ELEMENT));
```

这段代码的处理比较明了，提取当前节点下 description 节点中的数据，将这个数据赋值给 BeanDefinition 中的 description 属性，处理结果如图 5.6 所示。

5. 设置 meta 属性

接下来进行 parseBeanDefinitionElement 方法的解析。Spring 中 parseBeanDefinitionElement 方法的详细信息如下。

```
public void parseMetaElements(Element ele,
BeanMetadataAttributeAccessor attributeAccessor) {
    //获取下级标签
    NodeList nl = ele.getChildNodes();
    //循环子标签
    for (int i = 0; i < nl.getLength(); i++) {
        Node node = nl.item(i);
        //设置数据
```

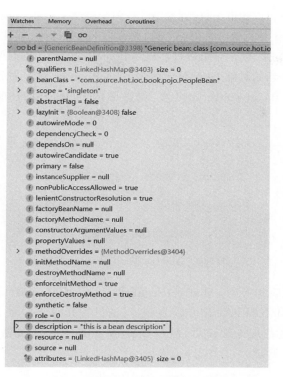

图 5.6 设置 bean 描述信息后结果

```
//是不是 meta 标签
if (isCandidateElement(node) && nodeNameEquals(node,META_ELEMENT)) {
    Element metaElement = (Element) node;
    //获取 key 属性
    String key = metaElement.getAttribute(KEY_ATTRIBUTE);
    //获取 value 属性
    String value = metaElement.getAttribute(VALUE_ATTRIBUTE);
    //元数据对象设置
    BeanMetadataAttribute attribute = new BeanMetadataAttribute(key,value);
    //设置 source
    attribute.setSource(extractSource(metaElement));
    //信息添加
    attributeAccessor.addMetadataAttribute(attribute);
  }
 }
}
```

下面先来编写一段测试用例再进行源代码分析，这段代码处理的是 meta 标签的数据，改进 SpringXML 配置文件。

```
<bean id="people" class="com.source.hot.ioc.book.pojo.PeopleBean">
    <description>this is a bean description</description>
    <meta key="key" value="value"/>
</bean>
```

通过这段配置文件可以知道，meta 标签中存在两个属性 key 和 value。在 parseMetaElements 方法中对 bean 标签的下级标签 meta 的处理就是提取 key 和 value 属性的属性值，在提取后创建 BeanMetadataAttribute 对象，创建对象后将其放到 BeanMetadataAttributeAccessor 集合中。在这段处理过程中出现了一个新的对象 BeanMetadataAttributeAccessor，这个对象是什么呢？回答这个问题需要从 AbstractBeanDefinition 出发，观察 AbstractBeanDefinition 的类图，如图 5.7 所示。

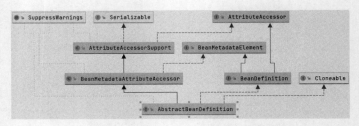

图 5.7　AbstractBeanDefinition 类图

在类图中可以直观地看到，AbstractBeanDefinition 是 BeanMetadataAttributeAccessor 的子类。对于 BeanMetadataAttributeAccessor 的设置其实就是为 AbstractBeanDefinition 添加属性。了解了这些理论知识后，经过 parseMetaElements 方法处理后 BeanDefinition 的对象信息如图 5.8 所示。

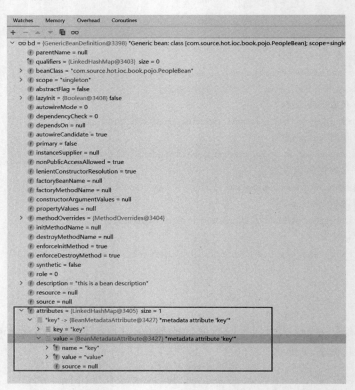

图 5.8　parseMetaElements 执行后 BeanDefinition 的信息

6. lookup-override 标签处理

在处理完 meta 标签后将处理 lookup-override 标签。前文所编写的一些测试代码在这个方法中是不足以支持断点调试的,需要对测试代码进行补充。下面编写测试用例,假设现在有一个商店在出售水果,水果可以是苹果、香蕉等,此时客户需要通过不同的商店获取各个商店所售卖的内容,下面先定义几个 Java 对象。

水果对象详细代码如下。

```
public class Fruits {
    private String name;

    public String getName() {
        return name;
    }

    public void setName(String name) {
        this.name = name;
    }
}
```

苹果对象详细代码如下。

```
public class Apple extends Fruits {
    public Apple() {
        this.setName("apple");
    }

    public void hello() {
        System.out.println("hello");
    }

}
```

商店对象详细代码如下。

```
public abstract class Shop {
    public abstract Fruits getFruits();
}
```

编写 SpringXML 配置文件,文件名称为 spring-lookup-method.xml,详细代码如下。

```
<?xml version="1.0" encoding="UTF-8"?>
<beans xmlns:xsi="http://www.w3.org/2001/XMLSchema-instance"
    xmlns="http://www.springframework.org/schema/beans"
    xsi:schemaLocation="http://www.springframework.org/schema/beans http://www.springframework.org/schema/beans/spring-beans.xsd">
    <bean id="apple" class="com.source.hot.ioc.book.pojo.lookup.Apple">
    </bean>

    <bean id="shop" class="com.source.hot.ioc.book.pojo.lookup.Shop">
        <lookup-method name="getFruits" bean="apple"/>
```

```
    </bean>

</beans>
```

编写测试用例,详细代码如下。

```
@Test
void testLookupMethodBean() {
   ClassPathXmlApplicationContext context =
new ClassPathXmlApplicationContext("META-INF/spring-lookup-method.xml");

   Shop shop = context.getBean("shop",Shop.class);
   System.out.println(shop.getFruits().getName());
   assert context.getBean("apple").equals(shop.getFruits());
}
```

测试方法 testLookupMethodBean 的执行结果输出 apple 并且测试通过。

先来整理测试用例中的执行流程,在 SpringXML 配置文件中配置 bean 标签的 lookup-method 属性,在调用方法时会根据 lookup-method 中的 bean 属性在 Spring 容器中寻找 bean 属性值对应的 Bean 实例,当调用 lookup-method 中 name 属性值的方法时将 bean 属性对应的 Bean 实例作为返回结果返回。在测试用例中调用 shop#getFruits 方法时会在容器中找到 apple 这个 Bean 实例将其作为返回值。

简单了解使用逻辑后,现在来看 Spring 是如何处理 lookup-method 标签的。先阅读处理方法 parseLookupOverrideSubElements。

```
public void parseLookupOverrideSubElements(Element beanEle,MethodOverrides overrides) {
   //获取子标签
   NodeList nl = beanEle.getChildNodes();
   for (int i = 0; i < nl.getLength(); i++) {
      Node node = nl.item(i);
      //是否有 lookup-method 标签
      if (isCandidateElement(node) && nodeNameEquals(node,LOOKUP_METHOD_ELEMENT)) {
         Element ele = (Element) node;
         //获取 name 属性
         String methodName = ele.getAttribute(NAME_ATTRIBUTE);
         //获取 bean 属性
         String beanRef = ele.getAttribute(BEAN_ELEMENT);
         //创建覆盖依赖
         LookupOverride override = new LookupOverride(methodName,beanRef);
         //设置 source
         override.setSource(extractSource(ele));
         overrides.addOverride(override);
      }
   }
}
```

这段处理模式和 meta 标签的处理模式基本相同。处理方式:获取所有子节点,如果子节点是 lookup-method 就提取 name 和 bean 两个属性,在属性获取后创建对应的 Java 对象 LookupOverride 并设置给 MethodOverrides，MethodOverrides 是 BeanDefinition 的一个成员变量。了解理论内容后再看处理结果，如图 5.9 所示。

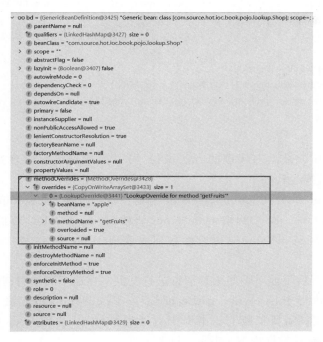

图 5.9　parseLookupOverrideSubElements 执行后 BeanDefinition 的信息

现在对于 lookup-method 标签的解析已经完成，下面思考一个问题：在执行时发生了什么？即当调用 shop.getFruits() 方法时发生了什么？回答这个问题需要了解另一个知识点对象代理，本章不会对代理过程做一个完善的分析，仅做调用方法的分析。

先看 shop 对象，shop 对象并不是一个普通的 Java 对象，它是一个增强对象，shop 对象信息如图 5.10 所示。

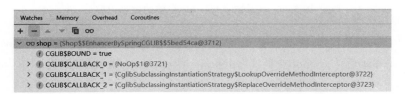

图 5.10　shop 对象

通过图 5.10 可以确认，shop 对象中存在 LookupOverrideMethodInterceptor 对象，接下来需要找到 LookupOverrideMethodInterceptor 对象，在这个对象中存有需要分析的方法：

```
@Override
public Object intercept(Object obj,Method method,Object[] args,MethodProxy mp) throws Throwable {
    LookupOverride lo =
(LookupOverride) getBeanDefinition().getMethodOverrides().getOverride(method);
    Assert.state(lo != null,"LookupOverride not found");
```

```
            Object[] argsToUse = (args.length > 0 ? args : null);
            if (StringUtils.hasText(lo.getBeanName())) {
                return (argsToUse != null ? this.owner.getBean(lo.getBeanName(),argsToUse) :
                        this.owner.getBean(lo.getBeanName()));
            }
            else {
                return (argsToUse != null ?
            this.owner.getBean(method.getReturnType(),argsToUse) :
                        this.owner.getBean(method.getReturnType()));
            }
        }
```

这段方法的本质就是动态代理的方法增强,可以简单理解为原有方法的执行结果被 intercept 替换了,替换过程如下。

(1) 通过参数 method 在 lookupOverride 容器中找到替换的 LookupOverride 对象。

(2) 从 LookupOverride 中提取 beanName 属性,在 Spring IoC 容器(BeanFactory)中通过 beanName 或者 beanName+构造参数列表获得 Bean 实例,将 Bean 实例作为返回结果。

了解了执行流程后下面对一些关键信息进行复盘。在这个测试用例中不存在构造参数列表数据,代码会执行 this.owner.getBean(lo.getBeanName())方法,注意 owner 就是 BeanFactory,提取实例需要依赖它。下面来看通过 method 找到 LookupOverride 对象中 LookupOverride 存储的数据内容,如图 5.11 所示。

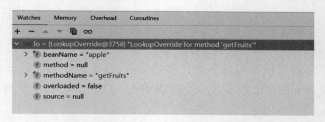

图 5.11 LookupOverride 存储的数据内容

在图 5.11 中可以看到,需要加载的 beanName 是 apple,现在具备获取 Bean 实例的工具 owner,接下来就是从 owner 中将 beanName 为 apple 的 Bean 实例提取作为返回值返回即可。

7. replaced-method 标签处理

前文对 lookup-override 做了充分的分析,在这个基础上阅读 replaced-method 方法将事半功倍,它们的底层处理方式可以说是大同小异。首先来编写 replaced-method 的测试用例。

第一步需要编写 MethodReplacer 接口实现类,实现类代码如下。

```
public class MethodReplacerApple implements MethodReplacer {
    @Override
    public Object reimplement(Object obj,Method method,Object[] args) throws Throwable {
        System.out.println("方法替换");
```

```
        return obj;
    }
}
```

编写 SpringXML 配置文件,文件名称为 spring-replaced-method.xml,文件内容如下。

```xml
<?xml version="1.0" encoding="UTF-8"?>
<beans xmlns:xsi="http://www.w3.org/2001/XMLSchema-instance"
       xmlns="http://www.springframework.org/schema/beans"
       xsi:schemaLocation="http://www.springframework.org/schema/beans http://www.springframework.org/schema/beans/spring-beans.xsd">
    <bean id="apple" class="com.source.hot.ioc.book.pojo.lookup.Apple">
        <replaced-method replacer="methodReplacerApple" name="hello">
            <arg-type>String</arg-type>
        </replaced-method>

    </bean>

    <bean id="methodReplacerApple" class="com.source.hot.ioc.book.pojo.replacer.MethodReplacerApple">
    </bean>

</beans>
```

编写测试方法,具体代码如下。

```
@Test
void testReplacedMethod(){
    ClassPathXmlApplicationContext context =
       new ClassPathXmlApplicationContext("META-INF/spring-replaced-method.xml");
    Apple apple = context.getBean("apple",Apple.class);
    apple.hello();
}
```

测试方法执行结果:控制台输出方法替换。

接下来进行 replaced-method 标签解析的讲解,先阅读处理代码。

```
public void parseReplacedMethodSubElements(Element beanEle,
MethodOverrides overrides) {
    //子节点获取
    NodeList nl = beanEle.getChildNodes();
    for (int i = 0; i < nl.getLength(); i++) {
        Node node = nl.item(i);
        //是否包含 replaced-method 标签
        if (isCandidateElement(node)
&& nodeNameEquals(node, REPLACED_METHOD_ELEMENT)) {
            Element replacedMethodEle = (Element) node;
            //获取 name 属性
            String name = replacedMethodEle.getAttribute(NAME_ATTRIBUTE);
            //获取 replacer
            String callback = replacedMethodEle.getAttribute(REPLACER_ATTRIBUTE);
```

```
            //对象组装
            ReplaceOverride replaceOverride = new ReplaceOverride(name,callback);
            //Look for arg-type match elements.
            //子节点属性
            //处理 arg-type 标签
            List<Element> argTypeEles
    = DomUtils.getChildElementsByTagName(replacedMethodEle,ARG_TYPE_ELEMENT);

            for (Element argTypeEle : argTypeEles) {
                //获取 match 数据值
                String match =
argTypeEle.getAttribute(ARG_TYPE_MATCH_ATTRIBUTE);
                //match 信息设置
                match = (StringUtils.hasText(match) ? match :
DomUtils.getTextValue(argTypeEle));
                if (StringUtils.hasText(match)) {
                    //添加类型标识
                    replaceOverride.addTypeIdentifier(match);
                }
            }
            //设置 source
            replaceOverride.setSource(extractSource(replacedMethodEle));
            //重载列表添加
            overrides.addOverride(replaceOverride);
        }
    }
}
```

在这一段代码中对于 replaced-method 标签的处理分为以下三个步骤。

（1）提取 replaced-method 标签中的 name 和 replacer 属性。

（2）提取子标签 arg-type 的属性。

（3）将提取得到的数据进行组装并赋值给 BeanDefinition 对象。

在了解执行流程后再看看 parseReplacedMethodSubElements 方法执行后 BeanDefinition 的数据信息，如图 5.12 所示。

下面来看 apple.hello 方法执行时发生了什么？首先明确一点，apple 对象是一个代理对象，并不是一个原始的 Java 对象，数据信息如图 5.13 所示。

从图 5.13 中可以看到一个叫作 ReplaceOverrideMethodInterceptor 的类名，这个类就是真正需要分析的目标，在 ReplaceOverrideMethodInterceptor 对象中关于 intercept 方法的处理池代码如下。

```
@Override
public Object intercept(Object obj,Method method,Object[] args,MethodProxy mp) throws Throwable {
    ReplaceOverride ro =
(ReplaceOverride) getBeanDefinition().getMethodOverrides().getOverride(method);
    Assert.state(ro != null,"ReplaceOverride not found");
    MethodReplacer mr = this.owner.getBean(
ro.getMethodReplacerBeanName(),MethodReplacer.class);
```

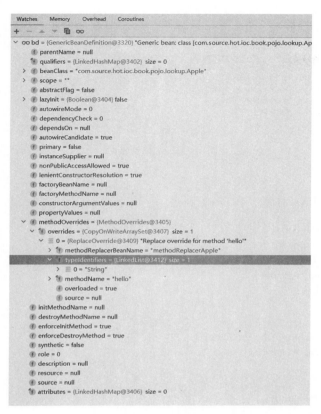

图 5.12　parseReplacedMethodSubElements 执行后的 BeanDefinition 信息

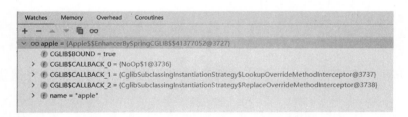

图 5.13　replaced-method 中的 apple 对象

```
return mr.reimplement(obj,method,args);
}
```

在这个方法中 Spring 会根据 method 进行 ReplaceOverride 的搜索，在这里 ReplaceOverride 就是通过 parseReplacedMethodSubElements 方法解析得到的对象，ro 数据信息如图 5.14 所示。

在图 5.14 中可以看到，在 SpringXML 配置中配置过的一些数据信息，Spring 会通过 methodReplacerBeanName＋MethodReplace.class 在 Spring IoC 容器中找到最终的实现类并调用其方法，将方法的处理结果作为返回值返回，从而达到方法替换的作用。

8. constructor-arg 标签处理

接下来将进入 constructor-arg 标签的源码分析阶段，进入方法分析之前需要制作一些

图 5.14 ro 数据信息

测试用例。

首先创建一个 Java 对象。

```
public class PeopleBean {
    private String name;

    public String getName() {
        return name;
    }

    public void setName(String name) {
        this.name = name;
    }

    public PeopleBean() {
    }

    public PeopleBean(String name) {
        this.name = name;
    }
}
```

在这个 Java 对象中定义了一个无参构造和有参构造，下面的分析将围绕有参构造进行。

在完成 Java 对象创建后需要编写 SpringXML 配置文件，文件名称为 spring-constructor-arg.xml，具体代码如下。

```
<?xml version="1.0" encoding="UTF-8"?>
<beans xmlns="http://www.springframework.org/schema/beans"
       xmlns:xsi="http://www.w3.org/2001/XMLSchema-instance"
       xsi:schemaLocation=" http://www.springframework.org/schema/beans http://www.springframework.org/schema/beans/spring-beans.xsd">

    <bean id="people" class="com.source.hot.ioc.book.pojo.PeopleBean">
        <constructor-arg index="0" type="java.lang.String" value="zhangsan"/>
    </bean>

</beans>
```

最后来编写测试方法。

```
@Test
void testConstructArg() {
    ClassPathXmlApplicationContext context =
new ClassPathXmlApplicationContext("META-INF/spring-constructor-arg.xml");
    PeopleBean people = context.getBean("people",PeopleBean.class);
    assert people.getName().equals("zhangsan");
}
```

测试用例准备完成接下来进入源代码的分析。首先找到需要分析的方法签名 org.springframework.beans.factory.xml.BeanDefinitionParserDelegate#parseConstructorArgElements，找到方法签名后进入方法内部阅读内部的代码。

```
public void parseConstructorArgElements(Element beanEle,BeanDefinition bd) {
    //获取
    NodeList nl = beanEle.getChildNodes();
    for (int i = 0; i < nl.getLength(); i++) {
        Node node = nl.item(i);
        if (isCandidateElement(node) &&
nodeNameEquals(node,CONSTRUCTOR_ARG_ELEMENT)) {
            //解析 constructor-arg 下级标签
            parseConstructorArgElement((Element) node,bd);
        }
    }
}
```

这段方法中进行了另一个方法 parseConstructorArgElement 的引用，下面请阅读在 parseConstructorArgElement 中的第一部分代码。

```
//获取 index 属性
String indexAttr = ele.getAttribute(INDEX_ATTRIBUTE);
//获取 type 属性
String typeAttr = ele.getAttribute(TYPE_ATTRIBUTE);
//获取 name 属性
String nameAttr = ele.getAttribute(NAME_ATTRIBUTE);
```

在第一部分代码中的处理逻辑很好理解，其主要目的是获取 constructor-arg 标签中的 index、type 和 name 三个属性，当属性获取完成 Spring 会根据 index 属性是否存在做出两种不同的处理，在前文提到的测试用例中 index 属性是存在的情况，下面先进行 index 属性存在的分析。先阅读处理代码。

```
try {
    //构造参数的索引位置
    int index = Integer.parseInt(indexAttr);
    if (index < 0) {
        error("'index' cannot be lower than 0",ele);
    }
    else {
        try {
            //设置阶段,构造函数处理阶段
            this.parseState.push(new ConstructorArgumentEntry(index));
```

```
            //解析 property 标签
            Object value = parsePropertyValue(ele,bd,null);
            //创建构造函数的属性控制类
            ConstructorArgumentValues.ValueHolder valueHolder =
    new ConstructorArgumentValues.ValueHolder(value);
            if (StringUtils.hasLength(typeAttr)) {
                //类型设置
                valueHolder.setType(typeAttr);
            }
            if (StringUtils.hasLength(nameAttr)) {
                //名称设置
                valueHolder.setName(nameAttr);
            }
            //源设置
            valueHolder.setSource(extractSource(ele));
            if (bd.getConstructorArgumentValues().hasIndexedArgumentValue(index)) {
                error("Ambiguous constructor-arg entries for index " + index,ele);
            }
            else {
                //添加构造函数信息
                bd.getConstructorArgumentValues().addIndexedArgumentValue(index,valueHolder);
            }
        }
        finally {
            //移除当前阶段
            this.parseState.pop();
        }
    }
}
catch (NumberFormatException ex) {
    error("Attribute 'index' of tag 'constructor-arg' must be an integer",ele);
}
```

在这段代码中需要重点关注的对象是 ConstructorArgumentValues，注意在这段方法的分析中对于标签 constructor-arg 的解析会需要进行 property 标签的解析，这一标签的分析会在 5.4.9 节中进行，本节对于 property 标签处理暂时跳过。

通过阅读源代码可以确认在标签中提取的 index、value 和 type 三个属性最终放到了 ConstructorArgumentValues.ValueHolder 对象和 ConstructorArgumentValues 中，进入调试阶段观察具体的数据存储，首先观察 type 和 value 的存储，信息如图 5.15 所示。

进一步观察 index、type 和 value 的存储，信息如图 5.16 所示。

现在对于 index 数据存在的情况相关源码分析结束，下面进入 index 不存在的情况，来看 Spring 是如何对这种情况进行处理的。首先需要编写一个 index 不填写的测试用例，然后再编写一个 SpringXML 配置文件，文件为 spring-constructor-arg.xml，具体代码如下。

```
<?xml version = "1.0" encoding = "UTF-8"?>
< beans xmlns = "http://www.springframework.org/schema/beans"
        xmlns: xsi = "http://www.w3.org/2001/XMLSchema-instance"
        xsi: schemaLocation = "http://www.springframework.org/schema/beans http://www.
```

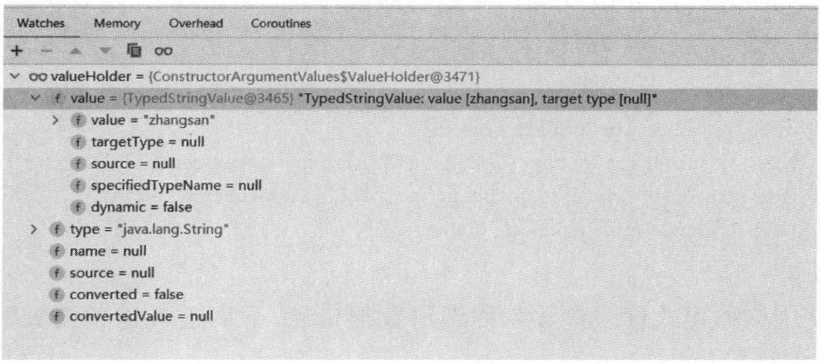

第5章　bean标签解析

图 5.15　type 和 value 的存储信息

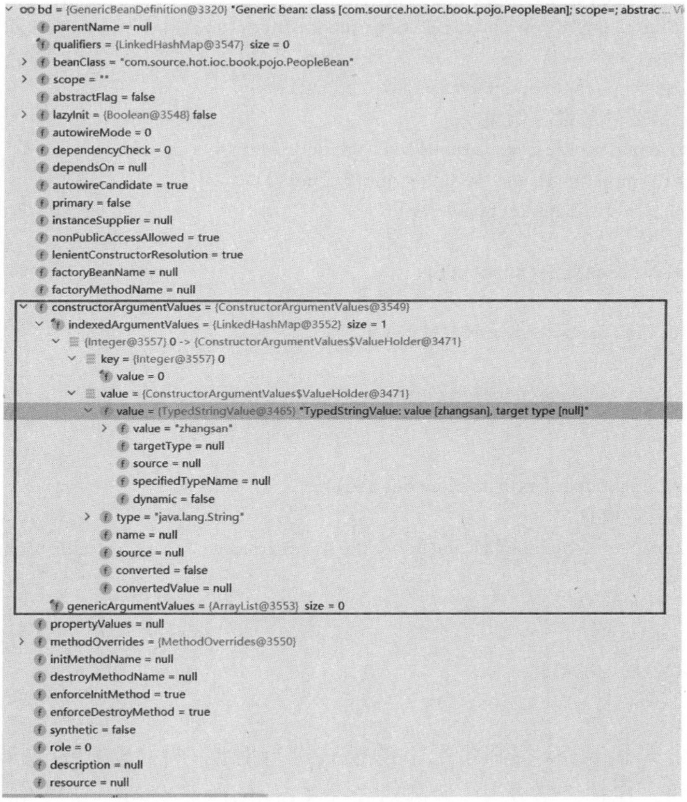

图 5.16　index、type 和 value 的存储信息

springframework.org/schema/beans/spring-beans.xsd">

< bean id = "people2" class = "com. source. hot. ioc. book. pojo. PeopleBean">
　　< constructor - arg name = "name" type = "java. lang. String" value = "zhangsan">
　　</constructor - arg >
</bean >

</beans >

编写完成 SpringXML 配置文件后来编写测试方法。

```
@Test
void testConstructArgForName() {
    ClassPathXmlApplicationContext context =
    new ClassPathXmlApplicationContext("META-INF/spring-constructor-arg.xml");
    PeopleBean people = context.getBean("people2",PeopleBean.class);
    assert people.getName().equals("zhangsan");
}
```

在 Spring 中这种模式称为参数名称绑定数据信息，下面进入这种模式的源代码分析，首先请阅读下面这段代码。

```
try {
    //设置阶段,构造函数处理阶段
    this.parseState.push(new ConstructorArgumentEntry());
    //解析 property 标签
    Object value = parsePropertyValue(ele,bd,null);
    //创建构造函数的属性控制类
    ConstructorArgumentValues.ValueHolder valueHolder =
    new ConstructorArgumentValues.ValueHolder(value);
    if (StringUtils.hasLength(typeAttr)) {
        //类型设置
        valueHolder.setType(typeAttr);
    }
    if (StringUtils.hasLength(nameAttr)) {
        //名称设置
        valueHolder.setName(nameAttr);
    }
    //源设置
    valueHolder.setSource(extractSource(ele));
    //添加构造函数信息
    bd.getConstructorArgumentValues().addGenericArgumentValue(valueHolder);
}
finally {
    //移除当前阶段
    this.parseState.pop();
}
```

将这段处理形式和 index 存在情况下的模式进行对比，两种模式的差异是 index 的控制，其他信息处理相同。在这个方法中，关键对象是 ConstructorArgumentValues.ValueHolder，处理模式相同的情况下分析内容不多直接进入调试阶段来看经过处理后的 valueHolder 对象数据，信息如图 5.17 所示。

执行这段代码之后再进一步观察 BeanDefinition 的对象信息，如图 5.18 所示。

现在构造器的数据信息已经准备完毕，请思考一个问题：构造函数的信息都准备完毕如何使用？在 Spring 中对于这种情况的处理是交给一个方法进行处理的，方法的具体签名是 org.springframework.beans.BeanUtils#instantiateClass(java.lang.reflect.Constructor<T>, java.lang.Object…)，可以根据需求对该方法进行阅读，本节仅该方法的调用链路整理出来，如图 5.19 所示。

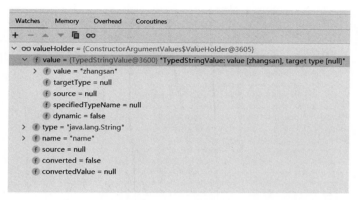

图 5.17　参数名称绑定模式下 valueHolder 对象信息

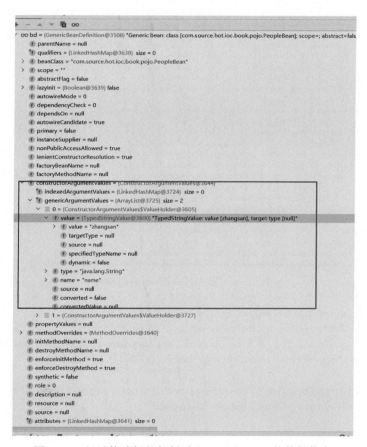

图 5.18　经过构造标签解析后 BeanDefinition 的数据信息

9. property 标签处理

接下来将进入 property 标签处理的源码分析。首先需要编写一个测试用例，第一步编写 SpringXML 配置文件，文件名为 spring-property.xml，该文件的代码内容如下。

```
<?xml version = "1.0" encoding = "UTF-8"?>
```

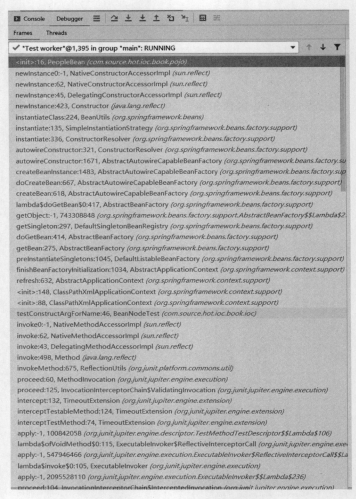

图 5.19 instantiateClass 调用堆栈

```
< beans xmlns: xsi = "http://www.w3.org/2001/XMLSchema-instance"
    xmlns = "http://www.springframework.org/schema/beans"
     xsi: schemaLocation = " http://www. springframework. org/schema/beans  http://www.
springframework.org/schema/beans/spring-beans.xsd">

    < bean id = "people" class = "com. source. hot. ioc. book. pojo. PeopleBean">
        < property name = "name" value = "zhangsan"/>
    </bean>

</beans>
```

完成 SpringXML 配置文件的编写后需要进行单元测试的编写,具体单元测试代码如下。

```
@Test
void testProperty(){
    ClassPathXmlApplicationContext context =
```

```java
new ClassPathXmlApplicationContext("META-INF/spring-property.xml");
    PeopleBean people = context.getBean("people",PeopleBean.class);
    assert people.getName().equals("zhangsan");
}
```

测试用例准备完毕，找到需要分析的方法 parsePropertyElements，该方法的方法签名是 org.springframework.beans.factory.xml.BeanDefinitionParserDelegate#parsePropertyElements，具体代码内容如下。

```java
public void parsePropertyElements(Element beanEle,BeanDefinition bd) {
    NodeList nl = beanEle.getChildNodes();
    for (int i = 0; i < nl.getLength(); i++) {
        Node node = nl.item(i);
        //是否存在 property 标签
        if (isCandidateElement(node) && nodeNameEquals(node,PROPERTY_ELEMENT)) {
            //解析单个标签
            parsePropertyElement((Element) node,bd);
        }
    }
}
```

在这段方法中没有太多的处理细节，最终的能力提供者是 parsePropertyElement 方法，该方法是最终的分析目标，先阅读方法的内容。

```java
public void parsePropertyElement(Element ele,BeanDefinition bd) {
    String propertyName = ele.getAttribute(NAME_ATTRIBUTE);
    if (!StringUtils.hasLength(propertyName)) {
        error("Tag 'property' must have a 'name' attribute",ele);
        return;
    }
    this.parseState.push(new PropertyEntry(propertyName));
    try {
        if (bd.getPropertyValues().contains(propertyName)) {
            error("Multiple 'property' definitions for property '" + propertyName + "'",ele);
            return;
        }
        //解析 property 标签
        Object val = parsePropertyValue(ele,bd,propertyName);
        //构造 PropertyValue 对象
        PropertyValue pv = new PropertyValue(propertyName,val);
        //解析元信息
        parseMetaElements(ele,pv);
        pv.setSource(extractSource(ele));
        //添加 pv 结构
        bd.getPropertyValues().addPropertyValue(pv);
    }
    finally {
        this.parseState.pop();
    }
}
```

在 parsePropertyElement 方法中总共有以下四步处理。

（1）提取 property 标签的 name 属性值。
（2）提取 property 标签的 value 属性值。
（3）解析 property 标签中可能存在的 meta 标签。
（4）将数据解析封装后设置给 BeanDefinition。

在整个处理流程中需要关注 property 标签的数据存储对象 PropertyValue。在上述四个处理步骤中第一步和第三步是一个比较简单的处理，比较复杂的内容是第二步操作，下面将对第二步操作进行详细分析。首先阅读 spring-beans.dtd 文件中对于 property 标签的定义。

```
<!ELEMENT property (
    description?,meta*,
    (bean | ref | idref | value | null | list | set | map | props)?
)>

<!ATTLIST property name CDATA #REQUIRED>

<!ATTLIST property ref CDATA #IMPLIED>

<!ATTLIST property value CDATA #IMPLIED>
```

进一步阅读 Spring 中对 property 标签的另一种定义方式，该方式的数据在 spring-beans.xsd 中存放，具体内容如下。

```
<xsd:element name="property" type="propertyType">
  <xsd:annotation>
    <xsd:documentation><![CDATA[
Bean definitions can have zero or more properties.
Property elements correspond to JavaBean setter methods exposed
by the bean classes. Spring supports primitives,references to other
beans in the same or related factories,lists,maps and properties.
    ]]></xsd:documentation>
  </xsd:annotation>
</xsd:element>


<xsd:complexType name="propertyType">
  <xsd:sequence>
    <xsd:element ref="description" minOccurs="0"/>
    <xsd:choice minOccurs="0" maxOccurs="1">
        <xsd:element ref="meta"/>
        <xsd:element ref="bean"/>
        <xsd:element ref="ref"/>
        <xsd:element ref="idref"/>
        <xsd:element ref="value"/>
        <xsd:element ref="null"/>
        <xsd:element ref="array"/>
        <xsd:element ref="list"/>
        <xsd:element ref="set"/>
```

```xml
<xsd:element ref="map"/>
<xsd:element ref="props"/>
<xsd:any namespace="##other" processContents="strict"/>
            </xsd:choice>
        </xsd:sequence>
        <xsd:attribute name="name" type="xsd:string" use="required">
            <xsd:annotation>
                <xsd:documentation><![CDATA[
The name of the property, following JavaBean naming conventions.
                ]]></xsd:documentation>
            </xsd:annotation>
        </xsd:attribute>
        <xsd:attribute name="ref" type="xsd:string">
            <xsd:annotation>
                <xsd:documentation><![CDATA[
A short-cut alternative to a nested "<ref bean='...'/>".
                ]]></xsd:documentation>
            </xsd:annotation>
        </xsd:attribute>
        <xsd:attribute name="value" type="xsd:string">
            <xsd:annotation>
                <xsd:documentation><![CDATA[
A short-cut alternative to a nested "<value>...</value>" element.
                ]]></xsd:documentation>
            </xsd:annotation>
        </xsd:attribute>
</xsd:complexType>
```

在了解 property 的两种定义内容后下面对 parsePropertyValue 方法的处理流程进行分析,先阅读处理方法。

```java
@Nullable
public Object parsePropertyValue(Element ele, BeanDefinition bd, @Nullable String propertyName) {
    String elementName = (propertyName != null ?
        "<property> element for property '" + propertyName + "'" :
        "<constructor-arg> element");

    //计算子节点
    //Should only have one child element: ref, value, list, etc.
    NodeList nl = ele.getChildNodes();
    Element subElement = null;
    for (int i = 0; i < nl.getLength(); i++) {
        Node node = nl.item(i);
        if (node instanceof Element &&
            !nodeNameEquals(node, DESCRIPTION_ELEMENT) &&
            !nodeNameEquals(node, META_ELEMENT)) {
            if (subElement != null) {
                error(elementName + " must not contain more than one sub-element", ele);
            }
            else {
                subElement = (Element) node;
```

```java
            }
        }
    }

    //ref 属性是否存在
    boolean hasRefAttribute = ele.hasAttribute(REF_ATTRIBUTE);
    //value 属性是否存在
    boolean hasValueAttribute = ele.hasAttribute(VALUE_ATTRIBUTE);
    if ((hasRefAttribute && hasValueAttribute) ||
            ((hasRefAttribute || hasValueAttribute) && subElement != null)) {
        error(elementName +
                " is only allowed to contain either 'ref' attribute OR 'value' attribute OR sub-element", ele);
    }

    if (hasRefAttribute) {
        //获取 ref 属性值
        String refName = ele.getAttribute(REF_ATTRIBUTE);
        if (!StringUtils.hasText(refName)) {
            error(elementName + " contains empty 'ref' attribute", ele);
        }
        //创建连接对象
        RuntimeBeanReference ref = new RuntimeBeanReference(refName);

        ref.setSource(extractSource(ele));
        return ref;
    }
    else if (hasValueAttribute) {
        //获取 value
        TypedStringValue valueHolder =
    new TypedStringValue(ele.getAttribute(VALUE_ATTRIBUTE));
        valueHolder.setSource(extractSource(ele));
        return valueHolder;
    }
    else if (subElement != null) {
        return parsePropertySubElement(subElement, bd);
    }
    else {
        error(elementName + " must specify a ref or value", ele);
        return null;
    }
}
```

在 parsePropertyValue 方法中可以将其分为下面四个步骤进行阅读。

(1) 提取 property 下的子节点标签中非 description 和 meta 标签。

(2) 提取 property 中的 ref 属性,存在的情况下创建 RuntimeBeanReference 对象并返回。

(3) 提取 property 中的 value 属性,存在的情况下创建 TypedStringValue 对象并返回。

（4）解析第一步中得到的子标签信息。

通过前文对 property 标签的两个定义文件中的内容可以得出 property 的子标签有 11 个，分别如下。

（1）meta。

（2）bean。

（3）ref。

（4）idref。

（5）value。

（6）null。

（7）array。

（8）list。

（9）set。

（10）map。

（11）props。

在这 11 个下级标签中，除了 meta 标签以外都是第一步中会提取得到的数据内容。第二步中获取 ref 属性值操作很简单，直接使用 getAttribute 即可获取，创建 RuntimeBeanReference 对象的过程也是一个简单的 new 操作。第三步和第二步的操作理论上相同，都是获取标签中的数据内容再进行特定对象的创建。第四步处理子标签的行为其实就是对上述标签的处理（除 meta 标签外），处理这些标签时所使用的方法是 parsePropertySubElement，下面将对这个方法做一个分析。

开始分析之前需要准备测试用例中需要的代码，首先需要的就是 JavaBean 对象，该 JavaBean 内容如下。

```
public class PeopleBean {
    private String name;

    private List<String> list;

    private Map<String,String> map;
    //省略构造函数,getter,setter
}
```

完成 JavaBean 的改造后进一步修改 SpringXML 配置文件的内容，配置文件名称为 spring-property.xml，文件内容如下。

```
<bean id="people" class="com.source.hot.ioc.book.pojo.PeopleBean">
    <property name="name" value="zhangsan"/>
    <property name="list">
        <list value-type="java.lang.String" merge="default">
            <value>a</value>
            <value>b</value>
        </list>

    </property>
```

```xml
<property name = "map">
    <map key-type = "java.lang.String" value-type = "java.lang.String">
        <entry key = "a" value = "1"/>
    </map>
</property>
</bean>
```

编写完测试代码后先对 list 标签进行分析，首先找到 list 标签的处理方法 parseListElement，这里需要注意对于 list 标签、array 标签、set 标签最终都是交给 parseCollectionElements 方法来进行处理，在这个处理过程中就是将 value 标签的数据提取转换为对应的 Java 对象，下面来看经过 parseListElement 处理过后的数据内容，如图 5.20 所示。

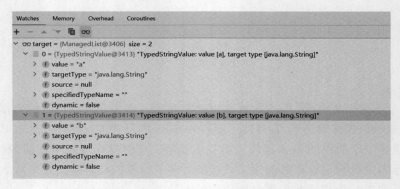

图 5.20　parseListElement 执行后的数据

完成了 list 标签的处理流程分析后接下来需要分析的就是 map 标签。在 SpringXML 配置文件中编写了 map 标签的 key-type 属性和 value-type 属性，以及子标签 entry 和 entry 标签的 key 属性和 value 属性。在当前的测试用例中并没有采用 ref 属性，仅以字符串作为直接字面量作为分析目标，处理 map 标签的方法是 parseMapElement。在处理 list 标签时采取的是字面量，最终得到的对象类型是 TypedStringValue。在 map 标签处理中用到的也是这个类型，只不过是从 list 存储转换成 map 存储。下面来看处理后的结果，如图 5.21 所示。

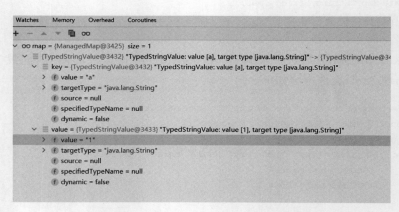

图 5.21　parseMapElement 方法执行后处理结果

现在完成了 map 标签的处理内容，剩下还有一些标签的处理，读者可以对这些具体标签的处理方法再做进一步的阅读。

10. qualifier 标签处理

接下来将进入 qualifier 标签的解析分析。首先需要编写一个测试用例，第一步创建一个 JavaBean 对象，该对象名称是 PeopleBeanTwo，详细代码如下。

```java
public class PeopleBeanTwo {
    @Autowired
    @Qualifier("p1")
    private PeopleBean peopleBean;
    //省略 getter 和 setter
}
```

完成 JavaBean 编写后进一步编写 SpringXML 配置文件内容，该文件名称为 spring-qualifier.xml，详细代码如下。

```xml
<?xml version="1.0" encoding="UTF-8"?>
<beans xmlns:xsi="http://www.w3.org/2001/XMLSchema-instance"
    xmlns:context="http://www.springframework.org/schema/context"
    xmlns="http://www.springframework.org/schema/beans"
    xsi:schemaLocation=" http://www.springframework.org/schema/beans http://www.springframework.org/schema/beans/spring-beans.xsd http://www.springframework.org/schema/context https://www.springframework.org/schema/context/spring-context.xsd">

    <context:annotation-config/>

    <bean id="p2" class="com.source.hot.ioc.book.pojo.PeopleBeanTwo">

    </bean>

    <bean id="peopleBean" class="com.source.hot.ioc.book.pojo.PeopleBean">
        <property name="name" value="zhangsan"/>
        <qualifier value="p1"/>
    </bean>
</beans>
```

最后编写单元测试。

```java
@Test
void testQualifier() {
    ClassPathXmlApplicationContext context =
        new ClassPathXmlApplicationContext("META-INF/spring-qualifier.xml");
    PeopleBeanTwo peopleTwo = context.getBean("p2", PeopleBeanTwo.class);
    assert peopleTwo.getPeopleBean().equals(
        context.getBean("peopleBean", PeopleBean.class));
}
```

现在测试用例准备完毕，接下来找到需要分析的方法 parseQualifierElements，先阅读其中的代码内容。

```java
public void parseQualifierElements(Element beanEle,AbstractBeanDefinition bd) {
    NodeList nl = beanEle.getChildNodes();
    for (int i = 0; i < nl.getLength(); i++) {
        Node node = nl.item(i);
        if (isCandidateElement(node) && nodeNameEquals(node,QUALIFIER_ELEMENT)) {
            //单个解析
            parseQualifierElement((Element) node,bd);
        }
    }
}
```

由于 Spring 是允许 qualifier 标签在 bean 标签下存在多个的，在 Spring 中对单个 qualifier 标签的处理是交给 parseQualifierElement 方法进行，下面先阅读 parseQualifierElement 的方法内容。

```java
public void parseQualifierElement(Element ele,AbstractBeanDefinition bd) {
    //获取 type 属性
    String typeName = ele.getAttribute(TYPE_ATTRIBUTE);
    if (!StringUtils.hasLength(typeName)) {
        error("Tag 'qualifier' must have a 'type' attribute",ele);
        return;
    }
    //设置阶段,处理 qualifier 阶段
    this.parseState.push(new QualifierEntry(typeName));
    try {
        //自动注入对象创建
        AutowireCandidateQualifier qualifier = new AutowireCandidateQualifier(typeName);
        //设置源
        qualifier.setSource(extractSource(ele));
        //获取 value 属性
        String value = ele.getAttribute(VALUE_ATTRIBUTE);
        if (StringUtils.hasLength(value)) {
            //设置属性
            qualifier.setAttribute(AutowireCandidateQualifier.VALUE_KEY,value);
        }
        NodeList nl = ele.getChildNodes();
        for (int i = 0; i < nl.getLength(); i++) {
            Node node = nl.item(i);
            if (isCandidateElement(node) && nodeNameEquals(node,QUALIFIER_ATTRIBUTE_ELEMENT)) {
                Element attributeEle = (Element) node;
                //获取 key 属性
                String attributeName = attributeEle.getAttribute(KEY_ATTRIBUTE);
                //获取 value 属性
                String attributeValue = attributeEle.getAttribute(VALUE_ATTRIBUTE);
                if (StringUtils.hasLength(attributeName) && StringUtils.hasLength(attributeValue)) {
                    //key - value 属性映射
                    BeanMetadataAttribute attribute =
                        new BeanMetadataAttribute(attributeName,attributeValue);
```

```
                    attribute.setSource(extractSource(attributeEle));
                    //添加 qualifier 属性值
                    qualifier.addMetadataAttribute(attribute);
                }
                else {
                    error("Qualifier 'attribute' tag must have a
'name' and 'value'",attributeEle);
                    return;
                }
            }
        }
        //添加 qualifier
        bd.addQualifier(qualifier);
    }
    finally {
        //移除阶段
        this.parseState.pop();
    }
}
```

这个解析过程就是对 qualifier 标签提取各个属性值再将其转换成 Java 对象,在 Spring 中 qualifier 标签对应的 Java 对象是 AutowireCandidateQualifier,下面来看经过解析后的数据,如图 5.22 所示。

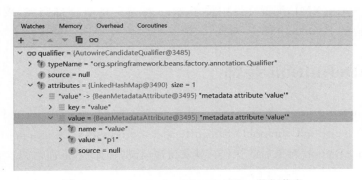

图 5.22　AutowireCandidateQualifier 数据信息

11. 设置 resource 和 source 属性

前文已经完成了 bean 标签的属性及下级标签的分析,最后还有两个属性需要设置,这两个属性分别是 resource 和 source,对于这两个属性的测试使用本章中提到的任何一个测试用例都可以进行测试。既然提到了 resource,那么什么是 resource(资源)呢？在 Java 工程中有一个 resources 文件夹,一般情况下认为该文件夹下的内容就是资源。在 Spring 中对于资源的定义可以简单地理解为 SpringXML 配置文件,不过还有其他的 resource,不只是配置文件,但是在此时需要设置的 resource 属性就是 SpringXML 配置文件,具体信息如图 5.23 所示。

在完成 resource 对象的设置后,Spring 对 source 进行了设置。在 Spring 中 source 的解析是交给 SourceExtractor 类进行处理的,下面了解一下 SourceExtractor 的类图,如图 5.24 所示。

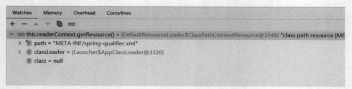

图 5.23 resource 对象信息

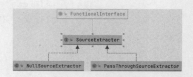

图 5.24 SourceExtractor 类图

通过类图可以发现 Spring 提供了以下两个关于 SourceExtractor 的实现。
（1）NullSourceExtractor 实现类直接将 null 作为解析结果进行返回。
（2）PassThroughSourceExtractor 实现类直接将需要解析的对象本身作为解析结果返回。

在 bean 标签解析的过程中，SourceExtractor 的具体实现类是 NullSourceExtractor，所以在设置 BeanDefinition 的 source 属性时设置的数据是 null，执行结果如图 5.25 所示。

图 5.25 执行后 source 的数据内容

至此，对于 SpringXML 配置文件中关于 bean 标签的解析全部完成，现在 Spring 得到了 bean 标签对应的 BeanDefinition 对象。

5.3　BeanDefinition 装饰

在得到 BeanDefinition 对象后，Spring 对这个对象进行了装饰操作，接下来将对该处理进行分析。在 DefaultBeanDefinitionDocumentReader#processBeanDefinition 方法中可以找到 Spring 获取 BeanDefinition 对象后的操作：对象装饰（数据补充）。下面请阅读处理代码。

```
protected void processBeanDefinition(Element ele,BeanDefinitionParserDelegate delegate) {
    //创建 BeanDefinition
    BeanDefinitionHolder bdHolder = delegate.parseBeanDefinitionElement(ele);
    if (bdHolder != null) {
        //BeanDefinition 装饰
        bdHolder = delegate.decorateBeanDefinitionIfRequired(ele,bdHolder);
        try {
            //注册 BeanDefinition
            BeanDefinitionReaderUtils.registerBeanDefinition(bdHolder,getReaderContext().getRegistry());
        }
        catch (BeanDefinitionStoreException ex) {
            getReaderContext().error("Failed to register bean definition with name '" +
                bdHolder.getBeanName() + "'",ele,ex);
        }
        //component 注册事件触发
```

```
        getReaderContext().fireComponentRegistered(new BeanComponentDefinition(bdHolder));
    }
}
```

在这段方法中真正的处理是交给 decorateBeanDefinitionIfRequired 方法进行的，进一步追踪源代码阅读 decorateBeanDefinitionIfRequired 方法内容：

```
public BeanDefinitionHolder decorateBeanDefinitionIfRequired(
        Element ele,BeanDefinitionHolder originalDef,@Nullable BeanDefinition containingBd) {

    BeanDefinitionHolder finalDefinition = originalDef;

    NamedNodeMap attributes = ele.getAttributes();
    for (int i = 0; i < attributes.getLength(); i++) {
        Node node = attributes.item(i);
        finalDefinition = decorateIfRequired(node,finalDefinition,containingBd);
    }

    NodeList children = ele.getChildNodes();
    for (int i = 0; i < children.getLength(); i++) {
        Node node = children.item(i);
        if (node.getNodeType() == Node.ELEMENT_NODE) {
            finalDefinition = decorateIfRequired(node,finalDefinition,containingBd);
        }
    }
    return finalDefinition;
}
```

在这段方法中可以整理出下面两种情况需要对 BeanDefinition 进行装饰。

(1) 当 bean 标签存在属性时。

(2) 当 bean 标签存在下级标签时。

在 decorateBeanDefinitionIfRequired 方法中确认了需要进行装饰的原因，最终进行装饰处理的方法是 decorateIfRequired，该方法就是需要分析的重点了，请先阅读源码。

```
public BeanDefinitionHolder decorateIfRequired(
        Node node,BeanDefinitionHolder originalDef,
@Nullable BeanDefinition containingBd) {

    //命名空间 url
    String namespaceUri = getNamespaceURI(node);
    if (namespaceUri != null && !isDefaultNamespace(namespaceUri)) {
        NamespaceHandler handler
 = this.readerContext.getNamespaceHandlerResolver().resolve(namespaceUri);
        if (handler != null) {
            //命名空间进行装饰
            BeanDefinitionHolder decorated =
                    handler.decorate(node,originalDef,new ParserContext(this.readerContext,
this,containingBd));
            if (decorated != null) {
                return decorated;
```

```
                }
            }
            else if (namespaceUri.startsWith("http://www.springframework.org/schema/")) {
                error("Unable to locate Spring NamespaceHandler for XML schema namespace [" +
namespaceUri + "]",node);
            }
            else {
                if (logger.isDebugEnabled()) {
                    logger.debug("No Spring NamespaceHandler found
for XML schema namespace [" + namespaceUri + "]");
                }
            }
        }
        return originalDef;
}
```

在这个方法的处理过程中可以发现比较熟悉的对象是 NamespaceHandler, 在第 4 章中对于 NamespaceHandler 接口做了关于自定义标签解析的相关内容,下面将延续第 4 章中的测试用例补充实现 decorate 方法,修改 UserXsdNamespaceHandler 类,具体修改后内容如下。

```
public class UserXsdNamespaceHandler extends NamespaceHandlerSupport {

    @Override
    public void init() {
        registerBeanDefinitionParser("user_xsd",new UserXsdParser());
    }

    @Override
    public BeanDefinitionHolder decorate ( Node node, BeanDefinitionHolder definition,
ParserContext parserContext) {
        BeanDefinition beanDefinition = definition.getBeanDefinition();
        beanDefinition.getPropertyValues().addPropertyValue("namespace","namespace");
        return definition;
    }
}
```

修改 SpringXML 配置文件,修改后内容如下。

```
<?xml version = "1.0" encoding = "UTF-8"?>
<beans xmlns = "http://www.springframework.org/schema/beans"
    xmlns:xsi = "http://www.w3.org/2001/XMLSchema-instance"
    xmlns:myname = "http://www.huifer.com/schema/user"
    xsi:schemaLocation = " http://www.springframework.org/schema/beans http://www.
springframework.org/schema/beans/spring-beans.xsd
    http://www.huifer.com/schema/user http://www.huifer.com/schema/user.xsd
">
    <bean id = "p1" class = "com.source.hot.ioc.book.pojo.PeopleBean">
        <myname:user_xsd id = "testUserBean" name = "huifer" idCard = "123"/>
    </bean>
</beans>
```

完成 SpringXML 配置文件编写后编写单元测试,具体单元测试代码如下。

```
@Test
void testXmlCustom() {
    ClassPathXmlApplicationContext context =
    new ClassPathXmlApplicationContext("META-INF/custom-xml.xml");
    UserXsd testUserBean = context.getBean("testUserBean",UserXsd.class);
    assert testUserBean.getName().equals("huifer");
    assert testUserBean.getIdCard().equals("123");
    context.close();
}
```

在第 4 章中已经讲述过从 namespaceUri 转换成 NamespaceHandler 的过程，相信读者对下面这段代码已经有了一定的了解。

```
BeanDefinitionHolder decorated = handler.decorate(node,originalDef,new
ParserContext(this.readerContext,this,containingBd));
```

这段代码的主要目的是调用在测试用例中所编写的 UserXsdNamespaceHandler#decorate 方法，下面来看经过装饰后的 BeanDefinition 对象信息，如图 5.26 所示。

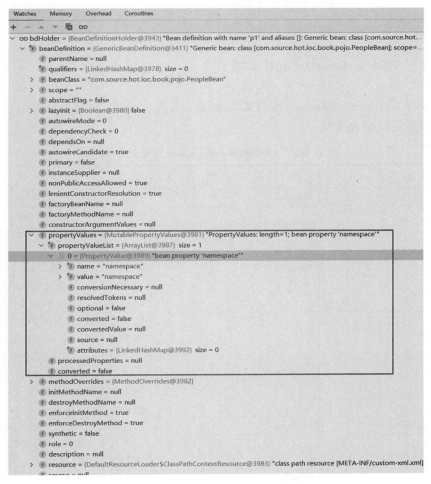

图 5.26　装饰后的 BeanDefinition

至此，对于 BeanDefinition 的装饰过程分析完成。

5.4 BeanDefinition 细节

本节将会对 BeanDefinition 的各个属性进行介绍，首先阅读 BeanDefinition 的类图，如图 5.27 所示。

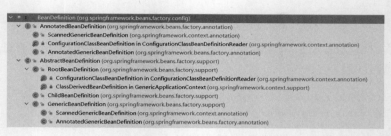

图 5.27　BeanDefinition 类图

从 BeanDefinition 的类图上可以知道所有 BeanDefinition 的根对象是 AbstractBeanDefinition，在这个对象中存有 BeanDefinition 的大量数据字段。

5.4.1　AbstractBeanDefinition 属性

下面来看 AbstractBeanDefinition 的属性，如表 5.2 所示，列举了 AbstractBeanDefinition 对象中属性对应的类型和含义。

表 5.2　AbstractBeanDefinition 对象属性表

属 性 名 称	属 性 类 型	属 性 含 义
beanClass	Object	存储 Bean 类型
scope	String	作用域，默认""，常见作用域有 singleton、prototype、request、session 和 globalsession
lazyInit	Boolean	是否懒加载
abstractFlag	boolean	是否是 abstract 修饰的
autowireMode	int	自动注入方式，常见的注入方式有 no、byName、byType 和 constructor
dependencyCheck	int	依赖检查级别，常见的依赖检查级别有 DEPENDENCY_CHECK_NONE、DEPENDENCY_CHECK_OBJECTS、DEPENDENCY_CHECK_SIMPLE 和 DEPENDENCY_CHECK_ALL
dependsOn	String[]	依赖的 BeanName 列表
autowireCandidate	boolean	是否自动注入，默认值：true
primary	boolean	是否是主要的，通常在同类型多个备案的情况下使用
instanceSupplier	Supplier	Bean 实例提供器

续表

属 性 名 称	属 性 类 型	属 性 含 义
nonPublicAccessAllowed	boolean	是否禁止公开访问
lenientConstructorResolution	String	
factoryBeanName	String	工厂 Bean 名称
factoryMethodName	String	工厂函数名称
constructorArgumentValues	ConstructorArgumentValues	构造函数对象,可以是 XML 中 constructor-arg 标签的解析结果,也可以是 Java 中构造函数的解析结果
propertyValues	MutablePropertyValues	属性列表
methodOverrides	MethodOverrides	重写的函数列表
initMethodName	String	Bean 初始化的函数名称
destroyMethodName	String	Bean 摧毁的函数名称
enforceInitMethod	boolean	是否强制执行 initMethodName 对应的 Java 方法
enforceDestroyMethod	boolean	是否强制执行 destroyMethodName 对应的 Java 方法
synthetic	boolean	合成标记
role	int	Spring 中的角色,一般有 ROLE_APPLICATION、ROLE_SUPPORT 和 ROLE_INFRASTRUCTURE
description	String	Bean 的描述信息
resource	Resource	资源对象

在 AbstractBeanDefinition 属性表中 beanClass 属性的类型是 Object,通常情况下会有 String 类型和 Class 类型,属性 scope 默认为单例,属性 propertyValues 含义为属性列表,它是一个 key-value 结构,用于存储开发者对 Bean 对象的属性定义,key 表示属性名称,value 表示属性值。

5.4.2　RootBeanDefinition 属性

RootBeanDefinition 属性详见表 5.3,列举了 RootBeanDefinition 对象中属性对应的类型和含义。

表 5.3　RootBeanDefinition 属性

属 性 名 称	属 性 类 型	属 性 含 义
constructorArgumentLock	Object	构造阶段的锁
postProcessingLock	Object	后置处理阶段的锁
stale	boolean	是否需要重新合并定义
allowCaching	boolean	是否缓存
isFactoryMethodUnique	boolean	工厂方法是否唯一
targetType	ResolvableType	目标类型
resolvedTargetType	Class	目标类型,Bean 的类型

续表

属性名称	属性类型	属性含义
isFactoryBean	Boolean	是否是工厂 Bean
factoryMethodReturnType	ResolvableType	工厂方法返回值
factoryMethodToIntrospect	Method	
resolvedConstructorOrFactoryMethod	Executable	执行器
constructorArgumentsResolved	boolean	构造函数的参数是否需要解析
resolvedConstructorArguments	Object[]	解析过的构造参数列表
preparedConstructorArguments	Object[]	未解析的构造参数列表
postProcessed	boolean	是否需要进行后置处理
beforeInstantiationResolved	Boolean	是否需要进行前置处理
decoratedDefinition	BeanDefinitionHolder	BeanDefinition 持有者
qualifiedElement	AnnotatedElement	qualified 注解信息
externallyManagedConfigMembers	Set<Member>	外部配置的成员
externallyManagedInitMethods	Set<String>	外部的初始化方法列表
externallyManagedDestroyMethods	Set<String>	外部的摧毁方法列表

在 RootBeanDefinition 成员变量中会有一些成对出现的内容，如 constructorArgumentsResolved、resolvedConstructorArguments 和 preparedConstructorArguments，它们用于对构造参数解析状态进行控制，externallyManagedConfigMembers、externallyManagedInitMethods 和 externallyManagedDestroyMethods，它们用于对外部成员依赖进行控制。

5.4.3 ChildBeanDefinition 属性

ChildBeanDefinition 属性详见表 5.4，列举了 ChildBeanDefinition 对象中属性对应的类型和含义。

表 5.4 ChildBeanDefinition 属性

属性名称	属性类型	属性含义
parentName	String	父 BeanDefinition 的名称

ChildBeanDefinition 对象是 AbstractBeanDefinition 的子类，具备 AbstractBeanDefinition 的所有属性，但是它比 AbstractBeanDefinition 增加了 parentName 属性用于指向父 BeanDefinition 对象。

5.4.4 GenericBeanDefinition 属性

GenericBeanDefinition 属性详见表 5.5，列举了 GenericBeanDefinition 对象中属性对应的类型和含义。

表 5.5 GenericBeanDefinition 属性

属性名称	属性类型	属性含义
parentName	String	父 BeanDefinition 的名称

GenericBeanDefinition 对象是 AbstractBeanDefinition 的子类，具备 AbstractBeanDefinition 的所有属性，但是它比 AbstractBeanDefinition 增加了 parentName 属性用于指向父 BeanDefinition 对象。

5.4.5　AnnotatedGenericBeanDefinition 属性

AnnotatedGenericBeanDefinition 属性详见表 5.6，该表列举了 AnnotatedGenericBeanDefinition 对象中属性对应的类型和含义。

表 5.6　AnnotatedGenericBeanDefinition 属性

属性名称	属性类型	属性含义
metadata	AnnotationMetadata	注解元信息
factoryMethodMetadata	MethodMetadata	工厂函数的元信息

AnnotatedGenericBeanDefinition 对象继承自 GenericBeanDefinition 对象，在 GenericBeanDefinition 的基础上增加了注解元信息和工厂函数的元信息变量，这两个变量为 Spring 注解模式开发中的注解 Bean 定义提供了充分支持。

小结

在本章中主要对 parseBeanDefinitionElement 方法进行分析（完整方法签名：org.springframework.beans.factory.xml.BeanDefinitionParserDelegate#parseBeanDefinitionElement(org.w3c.dom.Element，java.lang.String，org.springframework.beans.factory.config.BeanDefinition))，下面对整个处理过程进行总结，处理过程分为以下 12 步。

（1）处理 className 和 parent 属性。
（2）创建基本的 BeanDefinition 对象，具体类：AbstractBeanDefinition、GenericBeanDefinition。
（3）读取 bean 标签的属性，为 BeanDefinition 对象进行赋值。
（4）处理描述标签 description。
（5）处理 meta 标签。
（6）处理 lookup-override 标签。
（7）处理 replaced-method 标签。
（8）处理 constructor-arg 标签。
（9）处理 property 标签。
（10）处理 qualifier 标签。
（11）设置资源对象。
（12）设置 source 属性。

上述 12 个步骤是 Spring 中对于 bean 标签的解析细节，在完成 bean 标签的解析后 Spring 会对 BeanDefinition 对象进行装饰，具体装饰行为操作如下。

（1）读取标签所对应的 namespaceUri。
（2）根据 namespaceUri 在 NamespaceHandler 容器中寻找对应的 NamespaceHandler。

（3）调用 NamespaceHandler 所提供的 decorate 方法。

在完成 BeanDefinition 对象的创建及其属性赋值后需要对 BeanDefinition 对象进行注册，有关 BeanDefinition 对象的注册内容请查阅第 7 章。

最后将 bean 标签解析的入口方法的签名贴出，读者可以根据需求进行查阅。入口签名为 org.springframework.beans.factory.xml.DefaultBeanDefinitionDocumentReader#processBeanDefinition，入口方法如下。

```java
protected void processBeanDefinition(Element ele,BeanDefinitionParserDelegate delegate) {
    //创建 BeanDefinition
    BeanDefinitionHolder bdHolder = delegate.parseBeanDefinitionElement(ele);
    if (bdHolder != null) {
        //BeanDefinition 装饰
        bdHolder = delegate.decorateBeanDefinitionIfRequired(ele,bdHolder);
        try {
            //注册 BeanDefinition
            BeanDefinitionReaderUtils.registerBeanDefinition(bdHolder,getReaderContext().getRegistry());
        }
        catch (BeanDefinitionStoreException ex) {
            getReaderContext().error("Failed to register bean definition with name '" +
                    bdHolder.getBeanName() + "'",ele,ex);
        }
        //component 注册事件触发
        getReaderContext().fireComponentRegistered(new BeanComponentDefinition(bdHolder));
    }
}
```

第 6 章

Bean 的生命周期

在第 5 章中介绍了 bean 标签的使用和 bean 标签的解析流程,可以这么理解,在第 5 章中将 SpringXML 中的配置进行解析,将 bean 定义放入 Spring IoC 容器。本章将对 Bean 的生命周期进行分析。本章先对 Java 开发中 Java 对象的生命周期进行说明,从 Java 对象的生命周期出发映射到 Bean 的生命周期,此外还会对 Spring 中的生命周期相关接口进行说明。

6.1 Java 对象的生命周期

在开始讲述 Spring Bean 生命周期的内容之前先来了解一些关于 Java 对象的生命周期,在 Java 中对于一个对象的生命周期定义有下面 7 个阶段。

(1) 创建阶段(Created)。

(2) 应用阶段(In Use)。

(3) 不可见阶段(Invisible)。

(4) 不可达阶段(Unreachable)。

(5) 收集阶段(Collected)。

(6) 终结阶段(Finalized)。

(7) 对象空间重分配阶段(De-allocated)。

在上述 7 个生命周期状态中作为开发者可以操作的一般有第 1 个阶段和第 2 个阶段,其他 5 个阶段在 Java 程序中一般不受开发者控制。

接下来编写一个 Java 程序来描述生命周期的前两个阶段,测试类为 JavaBeanTest,详细代码如下。

```
public class JavaBeanTest {

    @Test
```

```
void javaBeanLifeCycle(){
    PeopleBean peopleBean = new PeopleBean();
    peopleBean.setName("zhangsan");

    System.out.println(peopleBean.getName());
}
```

在上述程序代码中，PeopleBean peopleBean＝new PeopleBean()；这一段代码就对应着第一个生命周期：创建阶段。在这行代码之后的 set 操作或者 get 操作都是在使用 peopleBean 对象，这些操作行为对应生命周期中的应用阶段。

6.2 浅看 Bean 生命周期

关于 Bean 的生命周期在 Spring 中可以通过 BeanFactory 接口上的注释文档进行了解，首先对 BeanFactory 初始化阶段的生命周期进行说明。

(1) BeanNameAware's setBeanName。

(2) BeanClassLoaderAware's setBeanClassLoader。

(3) BeanFactoryAware's setBeanFactory。

(4) EnvironmentAware's setEnvironment。

(5) EmbeddedValueResolverAware's setEmbeddedValueResolver。

(6) ResourceLoaderAware's setResourceLoader (only applicable when running in an application context)。

(7) ApplicationEventPublisherAware's setApplicationEventPublisher (only applicable when running in an application context)。

(8) MessageSourceAware's setMessageSource (only applicable when running in an application context)。

(9) ApplicationContextAware's setApplicationContext (only applicable when running in an application context)。

(10) ServletContextAware's setServletContext (only applicable when running in a web application context)。

(11) postProcessBeforeInitialization methods of BeanPostProcessors。

(12) InitializingBean's afterPropertiesSet。

(13) a custom init-method definition。

(14) postProcessAfterInitialization methods of BeanPostProcessors。

接下来介绍 BeanFactory 在关闭阶段的生命周期。

(1) postProcessBeforeDestruction methods of DestructionAwareBeanPostProcessors。

(2) DisposableBean's destroy。

(3) a custom destroy-method definition。

在开始分析 Bean 的生命周期相关内容之前，需要编写一个测试用例。首先编写一个 Java 对象，对象名为 LiveBean，详细代码如下。

```java
public class LiveBean implements BeanNameAware, BeanFactoryAware,
        ApplicationContextAware, InitializingBean,
DisposableBean, BeanClassLoaderAware,
        EnvironmentAware, EmbeddedValueResolverAware, ResourceLoaderAware,
        ApplicationEventPublisherAware, MessageSourceAware {

    private String address;

    public LiveBean() {
        System.out.println("init LiveBean");
    }

    @Override
    public void setBeanClassLoader(ClassLoader classLoader) {
        System.out.println("run setBeanClassLoader method.");
    }

    @Override
    public void setApplicationEventPublisher(ApplicationEventPublisher applicationEventPublisher) {
        System.out.println("run setApplicationEventPublisher method.");

    }

    @Override
    public void setEmbeddedValueResolver(StringValueResolver resolver) {
        System.out.println("run setEmbeddedValueResolver method.");

    }

    @Override
    public void setEnvironment(Environment environment) {
        System.out.println("run Environment method.");

    }

    @Override
    public void setMessageSource(MessageSource messageSource) {
        System.out.println("run setMessageSource method.");

    }

    @Override
    public void setResourceLoader(ResourceLoader resourceLoader) {
        System.out.println("run setResourceLoader method.");

    }

    @Override
    public void setBeanName(String name) {
        System.out.println("run setBeanName method.");
```

```java
    }

    @Override
    public void destroy() throws Exception {
        System.out.println("run destroy method.");
    }

    @Override
    public void afterPropertiesSet() throws Exception {
        System.out.println("run afterPropertiesSet method.");
    }

    @Override
    public void setApplicationContext(ApplicationContext applicationContext) throws BeansException {
    }

    @PostConstruct
    public void springPostConstruct() {
        System.out.println("@PostConstruct");
    }

    @PreDestroy
    public void springPreDestroy() {
        System.out.println("@PreDestroy");
    }


    public void myPostConstruct() {
        System.out.println("run myPostConstruct method.");
    }

    public void myPreDestroy() {
        System.out.println("run myPreDestroy method.");
    }


    public String getAddress() {
        return address;
    }

    public void setAddress(String address) {
       System.out.println("run setAddress method.");
       this.address = address;
    }


    @Override
    public void setBeanFactory(BeanFactory beanFactory) throws BeansException {
        System.out.println("run setBeanFactory method.");
    }
}
```

完成 JavaBean 的编写后需要编写 BeanPostProcessor 的实现类，实现类名称为

MyBeanPostProcessor,详细代码如下。

```java
package com.source.hot.ioc.book.live;

import org.springframework.beans.BeansException;
import org.springframework.beans.factory.config.BeanPostProcessor;

public class MyBeanPostProcessor implements BeanPostProcessor {

  public Object postProcessBeforeInitialization(Object bean, String beanName) throws BeansException {
      if (bean instanceof LiveBean) {
          System.out.println("run MyBeanPostProcessor postProcessBeforeInitialization method.");
      }
      return bean;
  }

  public Object postProcessAfterInitialization(Object bean, String beanName) throws BeansException {
      if (bean instanceof LiveBean) {
          System.out.println("run MyBeanPostProcessor postProcessAfterInitialization method.");
      }
      return bean;
  }
}
```

完成 BeanPostProcessor 的实现类编写后编写 SpringXML 配置文件,文件名为 live-bean.xml,文件内代码如下。

```xml
<?xml version="1.0" encoding="UTF-8"?>
<beans xmlns:xsi="http://www.w3.org/2001/XMLSchema-instance"
    xmlns="http://www.springframework.org/schema/beans"
    xsi:schemaLocation=" http://www.springframework.org/schema/beans http://www.springframework.org/schema/beans/spring-beans.xsd">

  <bean class="com.source.hot.ioc.book.live.MyBeanPostProcessor"/>
  <bean id="liveBean" class="com.source.hot.ioc.book.live.LiveBean"
      init-method="myPostConstruct" destroy-method="myPreDestroy">
      <property name="address" value="shangHai"/>
  </bean>
</beans>
```

完成 SpringXML 配置文件编写后编写测试类,测试类类名为 JavaBeanTest,详细代码如下。

```java
public class JavaBeanTest {
```

```
    @Test
    void testSpringBeanLive() {
        ClassPathXmlApplicationContext context =
new ClassPathXmlApplicationContext("META-INF/live-bean.xml");
        LiveBean liveBean = context.getBean("liveBean", LiveBean.class);
        context.close();
    }
}
```

执行测试方法会输出下面的结果。

```
init LiveBean
run setAddress method.
run setBeanName method.
run setBeanClassLoader method.
run setBeanFactory method.
run Environment method.
run setEmbeddedValueResolver method.
run setResourceLoader method.
run setApplicationEventPublisher method.
run setMessageSource method.
run MyBeanPostProcessor postProcessBeforeInitialization method.
run afterPropertiesSet method.
run myPostConstruct method.
run MyBeanPostProcessor postProcessAfterInitialization method.
run destroy method.
run myPreDestroy method.
```

6.3 初始化 Bean

在 Bean 的生命周期中首先需要关注的是初始化 Bean，即从 0 到 1 的过程，这一部分的处理可以简单理解为 Java 中的创建对象（关键字 new 的使用），但是在 Spring IoC 中对象的创建（Bean 的创建）并不是通过 new 关键字进行处理。下面是 SpringXML 中的一段配置。

```
<bean id="liveBean"
class="com.source.hot.ioc.book.live.LiveBean"
init-method="myPostConstruct" destroy-method="myPreDestroy">
    <property name="address" value="shangHai"/>
</bean>
```

在这段配置中并没有编写构造函数（构造函数标签 constructor-arg）相关的内容，如果存在构造函数标签相关内容，Bean 实例的创建会根据构造器进行，在第 6 章中介绍了关于构造函数标签有两种方式，第一种是 index 方式，第二种是 name 方式。当没有构造函数标签时会出现第三种方式——无参构造模式。下面将对这三种方式进行分析。

6.3.1 无构造标签

接下来介绍在 SpringXML 配置文件中不使用构造标签的情况，首先需要确认在这种情况下已经拥有的数据：类全路径。在 Java 中如何通过一个类全路径（字符串）将其转换成 Java 对象并进行使用？这个问题的解决方式就是无构造标签的解决方式。为了解决这个问题，必不可少的技术点是反射，通过反射可以比较方便地完成上述需求。

（1）定义一些变量，如 className 类名、contextClassLoader 类加载器，相关代码如下。

```
String className = "com.source.hot.ioc.book.live.LiveBean";
ClassLoader contextClassLoader = Thread.currentThread().getContextClassLoader();
```

（2）通过 ClassLoader 对象和 className 变量将其转换成 Class 对象，具体代码如下。

```
Class<?> aClass = contextClassLoader.loadClass(className);
```

（3）通过 Class 获得对象实例。获取对象实例有两种方式，第一种方式是通过 newInstance 方法直接获取，具体代码如下。

```
Object o2 = aClass.newInstance();
```

第二种方式是通过 getConstructor 方法获取构造器（Constructor），再通过构造器提供的 newInstance 方法进行对象实例的获取，具体代码如下。

```
Constructor<?> constructor = aClass.getConstructor();
Object o = constructor.newInstance();
```

下面是整个处理流程的完整代码。

```java
@Test
void testClass() throws Exception {
    String className = "com.source.hot.ioc.book.live.LiveBean";
    ClassLoader contextClassLoader = Thread.currentThread().getContextClassLoader();
    Class<?> aClass = contextClassLoader.loadClass(className);

    Object o2 = aClass.newInstance();

    Constructor<?> constructor = aClass.getConstructor();
    Object o = constructor.newInstance();
    if (o instanceof LiveBean) {
        LiveBean o1 = (LiveBean) o;
        o1.setAddress("shangHai");
        System.out.println(o1.getAddress());
    }
}
```

上述代码就是从类全路径（字符串）转换为 Java 对象的完整过程，在 Spring 中负责对象创建（对象初始化）的核心方法是 org.springframework.beans.BeanUtils#instantiateClass(java.lang.reflect.Constructor<T>,java.lang.Object…)，该方法的详细代码如下。

```java
//删除注释、异常处理和部分验证
public static <T> T instantiateClass(Constructor<T> ctor, Object... args)
throws BeanInstantiationException {

    ReflectionUtils.makeAccessible(ctor);
    if (KotlinDetector.isKotlinReflectPresent()
&& KotlinDetector.isKotlinType(ctor.getDeclaringClass())) {
```

```
            return KotlinDelegate.instantiateClass(ctor, args);
        }
        else {
            Class<?>[] parameterTypes = ctor.getParameterTypes();
            Assert.isTrue(args.length <= parameterTypes.length, "Can't specify more arguments than constructor parameters");
            Object[] argsWithDefaultValues = new Object[args.length];
            for (int i = 0; i < args.length; i++) {
                if (args[i] == null) {
                    Class<?> parameterType = parameterTypes[i];
                    argsWithDefaultValues[i] = (parameterType.isPrimitive() ? DEFAULT_TYPE_VALUES.get(parameterType) : null);
                }
                else {
                    argsWithDefaultValues[i] = args[i];
                }
            }
            return ctor.newInstance(argsWithDefaultValues);
        }
    }
```

在这段代码中需要忽略有关 Kotlin 的代码，主要关注 else 代码块中的内容。在这部分代码中可以看到前文对于已知类全路径如何获取 Java 实例讨论时遇到的类似代码 ctor.newInstance(argsWithDefaultValues)，这段代码和前文所编写的代码存在的差异是传递了构造参数列表（Object 数组）。

6.3.2 构造标签中的 index 模式和 name 模式

下面将介绍关于构造标签中的两种模式如何创建 Java 对象，在前文分析不使用构造标签时看到了一个关键信息 argsWithDefaultValues 变量，在这个对象中存储的数据是构造标签解析后的数据，下面以测试用例中 PeopleBean 对象举例，首先查看 PeopleBean 的代码。

```java
public class PeopleBean {
    private String name;
    public PeopleBean(String name) {
        this.name = name;
    }
}
```

其次，查看 SpringXML 中的配置内容。

```xml
<bean id="people" class="com.source.hot.ioc.book.pojo.PeopleBean">
    <constructor-arg index="0" type="java.lang.String" value="zhangsan">

    </constructor-arg>

</bean>
```

在第 5 章的分析中已经知道，构造标签（constructor-arg）解析后的数据是存储在 ConstructorArgumentValues 类中，在真正使用时需要将 ConstructorArgumentValues 中的数据转换成构造函数所需要的构造参数列表。对于这个问题的处理 Spring 将其交给 org.springframework.beans.factory.support.ConstructorResolver#autowireConstructor 方法进行，在 ConstructorResolver#autowireConstructor 方法中包含 Spring IoC 容器中的另一知识点构造函数的依赖注入，下面将对 ConstructorResolver#autowireConstructor 方法进行分析，首先关注以下三个变量的信息。

（1）Constructor<?> constructorToUse 表示需要被使用的构造函数。

（2）ArgumentsHolder argsHolderToUse 表示参数持有者，具体持有构造参数列表所需要的数据。

（3）Object[] argsToUse 表示需要被使用的构造函数的参数列表。

上述三个变量在创建对象过程中有着十分重要的地位。在 autowireConstructor 方法中对于变量 argsToUse 的确认提供了以下三种方式。

（1）直接将方法参数作为 argsToUse 的数据。

（2）通过 org.springframework.beans.factory.support.ConstructorResolver#resolvePreparedArguments 方法得到 argsToUse 对象。

（3）通过候选的构造函数对象（Constructor）进行推论。

在第三种方法中需要通过 Constructor 对象进行推论，具体的推论过程如下：在推论进行前，需要找到构造函数对象，获取方式是通过反射获取，具体的代码是 Class#getDeclaredConstructors 或者 Class#getConstructors。这两个方法都是获取所有的构造函数对象（返回值是构造函数列表），在得到构造函数列表后还需要确认具体的某一个构造函数，对于构造函数的确认需要通过三部分组成，第一部分是构造参数列表长度，第二部分是构造参数的参数类型，第三部分是参数名称，通过这三个条件可以从构造函数列表中确认唯一的一个构造函数，在确认完成构造函数对象后对于 argsToUse 对象也确认完成。不可否认的是，在对于 name 模式的处理过程中处理模式和前文相同，关于 index 模式和 name 模式的差异可以通过下面的代码确认。

```
/**
 * 构造函数信息
 * key: 索引
 * value: 标签数据
 */
private final Map<Integer, ValueHolder> indexedArgumentValues =
        new LinkedHashMap<>();

/**
 * name 模式的存储信息
 */
private final List<ValueHolder> genericArgumentValues = new ArrayList<>();
```

在这两个变量中，indexedArgumentValues 的存储结构是针对 index 模式，genericArgumentValues 的存储结构是针对 name 模式。

最终对于 argsToUse 数据的确定其实就是对于 ValueHolder 的数据提取（提取

ValueHolder 中的 convertedValue 对象数据),具体的提取方法可以看这个方法签名对应的方法:org.springframework.beans.factory.support.ConstructorResolver#createArgumentArray。

注意,在 ValueHolder 中有依赖注入相关的操作。

接下来修改 SpringXML 配置文件来模拟依赖注入,配置文件名称为 spring-constructor-arg.xml,修改后的完整代码如下。

```xml
<?xml version="1.0" encoding="UTF-8"?>
<beans xmlns:xsi="http://www.w3.org/2001/XMLSchema-instance"
       xmlns="http://www.springframework.org/schema/beans"
       xsi:schemaLocation=" http://www.springframework.org/schema/beans http://www.springframework.org/schema/beans/spring-beans.xsd">

    <bean id="people" class="com.source.hot.ioc.book.pojo.PeopleBean">
        <constructor-arg index="0" type="java.lang.String" value="zhangsan">

        </constructor-arg>

    </bean>
    <bean id="people2" class="com.source.hot.ioc.book.pojo.PeopleBean">
        <constructor-arg name="name" type="java.lang.String" value="zhangsan">

        </constructor-arg>
        <constructor-arg name="pb" ref="people"/>

    </bean>

</beans>
```

接下来通过调试工具观察此时的构造函数存储对象,构造函数信息如图 6.1 所示。

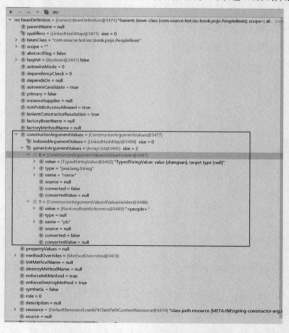

图 6.1 构造便签解析后数据

在这个测试用例中存在两种模式,第一种模式是字面量,第二种模式是引用量。当构造标签中使用 ref 属性时就是引用量,当使用 value 属性时就是字面量,但是在 value 中不能出现占位符。对于引用量的处理具体方法是 org.springframework.beans.factory.support.ConstructorResolver#resolveConstructorArguments。通过这个方法可以将 ref 变量转换成引用对象。至此,对于构造 Java 对象需要的所有内容都准备就绪,具体准备了构造函数对象,构造函数所需的参数列表,此时就可以通过 Constructor#newInstance(argsToUse) 方法进行创建 Java 对象。

6.3.3　Spring 中的实例化策略

接下来将介绍 Spring 中实例化策略相关内容。在 Java 中有关于代理的技术点(动态代理和静态代理),在 Java 中将对象分为两大类,第一类是非代理对象即普通对象,第二类是代理对象。对于这两种情况,Spring 通过策略模式实现了两个类分别对应不同的处理方式,这两个类的根接口是 InstantiationStrategy。Spring 中 InstantiationStrategy 类图信息如图 6.2 所示。

在 Spring 中有两种关于 Bean 的实例化方式,第一种是通过 BeanUtils.instantiateClass() 方法创建实例,第二种是通过 new CglibSubclassCreator(bd, owner).instantiate(ctor, args) 方法创建。这两种关于 Bean 的实例化方式分别在 SimpleInstantiationStrategy 和 CglibSubclassingInstantiationStrategy 中有所体现。这两种创建方式中第二种是关于 CGLIB 相关的创建,即动态代理对象创建,在某些特定的情况下会进行动态代理对象的创建。对于某些特定情况在 RootBeanDefinition 类中有一个方法作为主要判断依据,该方法是 hasMethodOverrides,具体代码如下。

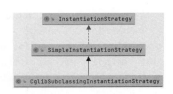

图 6.2　InstantiationStrategy 类图

```
public boolean hasMethodOverrides() {
    return !this.methodOverrides.isEmpty();
}
```

methodOverrides 变量存储的是需要重写的方法,在 SpringXML 环境下配置文件中关于方法重写有两种方式,第一种是使用 lookup-method 标签,第二种是使用 replaced-method 标签。如果有两个标签中的任意一个那么就会执行动态代理对象的创建。总结为一句话,当存在需要重写的方法(SpringXML 中 lookup-method 或者 replaced-method 标签的使用)时就会进行动态代理对象创建,反之则是普通对象创建。

6.4　Bean 属性设置

在完成 Bean 的实例化之后要对 Bean 的属性进行赋值,这一部分将对属性赋值做相关分析。首先介绍 Java 中是如何对属性赋值的,一般情况下如果是一个 JavaBean 对象,对于属性方法一般会有 getter 方法和 setter 方法(特殊情况下会有 is 开头的方法),开发者可以直接调用 setter 方法进行赋值,这是一种最常规的属性设置,再深入一些可以通过反射相关知识进行设置,具体代码如下。

```
void setField() throws Exception {
    LiveBean liveBean = new LiveBean();
    Field address = liveBean.getClass().getDeclaredField("address");
    address.setAccessible(true);
    address.set(liveBean, "shangHai");
    System.out.println(liveBean.getAddress());
}
```

在这段方法中使用了反射获取字段对象再进行数据设置,这种方式在 Spring 中以另一种类似的处理方式进行。下面先来编写一个测试用例,首先编写 SpringXML 配置,具体代码如下。

```
<?xml version = "1.0" encoding = "UTF-8"?>
<beans xmlns:xsi = "http://www.w3.org/2001/XMLSchema-instance"
    xmlns = "http://www.springframework.org/schema/beans"
    xsi:schemaLocation = " http://www.springframework.org/schema/beans http://www.springframework.org/schema/beans/spring-beans.xsd">

    <bean class = "com.source.hot.ioc.book.live.MyBeanPostProcessor"/>
    <bean id = "liveBean"
class = "com.source.hot.ioc.book.live.LiveBean"
init-method = "myPostConstruct" destroy-method = "myPreDestroy">
        <property name = "address" value = "shangHai"/>
    </bean>
</beans>
```

在本例中 address 属性是需要进行设置的属性字段,具体需要设置的数据是 shangHai,在 Spring 中对于属性设置的具体方法签名为 org.springframework.beans.factory.support.AbstractAutowireCapableBeanFactory#applyPropertyValues,首先介绍这个方法的参数。

(1) BeanName:Bean 名称。

(2) BeanDefinition:Bean 定义对象。

(3) BeanWrapper:Bean 实例的包装类。

(4) PropertyValues:属性值接口,本身包含 PropertyValue 对象。

在这四个变量中重点关注的是第四个变量 PropertyValues,该变量存储了 property 标签中的数据内容,这部分数据会被存储在 BeanDefinition 中,具体数据存储情况如图 6.3 所示。

在图 6.3 中可以看到,此时已经拥有了需要设置的属性名称和属性值,还缺少一个需要被进行属性设置的对象,这个对象是 Bean 实例,Bean 实例被存放在 BeanWrapper 接口中,BeanWrapper 类图如图 6.4 所示。

接下来将介绍 Spring 中对于属性设置的具体细节,首先必须要明确一点,设置属性必须通过属性名称进行设置,没有属性名称则没有办法设置属性。设置属性的过程是通过属性名称找到对应的设置属性方法,再通过属性设置方法进行属性设置。下面需要关注 PropertyValues 接口,它是迭代器(Iterable)的子类。由于它是迭代器子类,它可以迭代获

第6章　Bean的生命周期

图 6.3　BeanDefinition 中的 propertyValues

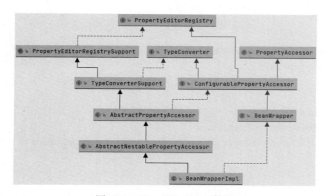

图 6.4　BeanWrapper 类图

取每一个数据元素,数据元素对象是 PropertyValue,在 PropertyValue 对象中的存储信息如表 6.1 所示。

表 6.1　PropertyValue 对象属性

字段名称	类　型	含　义
name	String	属性名称,对应 SpringXML 配置文件中 property 标签的 name 属性
value	Object	属性值,对应 SpringXML 配置文件中 property 标签的 value 属性

PropertyValue 对象表示属性表,name 和 value 在一个普通 Java 对象中可以这么理解,一个 User 对象中存在 username 字段(属性),通过 set 方法为 username 进行赋值,赋值张三,映射到 PropertyValue 对象 name 就是 username,value 就是张三。

6.4.1　BeanWrapper 创建

在使用 BeanWrapper 对象之前需要先对其进行创建。下面将介绍 BeanWrapper 的创建过程,负责进行创建 BeanWrapper 对象的方法是 org.springframework.beans.factory.support.AbstractAutowireCapableBeanFactory#createBeanInstance,具体代码如下。

```
protected BeanWrapper instantiateBean(final String beanName, final RootBeanDefinition mbd) {
    try {
```

```
            Object beanInstance;
            final BeanFactory parent = this;
            if (System.getSecurityManager() != null) {
                //获取实例化策略来进行实例化
                beanInstance = AccessController.doPrivileged(
                        (PrivilegedAction<Object>) () ->
                                getInstantiationStrategy().instantiate(mbd, beanName, parent),
                        getAccessControlContext()
                );
            }
            else {
                //获取实例化策略来进行实例化
                beanInstance = getInstantiationStrategy().instantiate(mbd, beanName, parent);
            }

            //beanWrapper 创建
            BeanWrapper bw = new BeanWrapperImpl(beanInstance);
            initBeanWrapper(bw);
            return bw;
        }
        catch (Throwable ex) {
            throw new BeanCreationException(
                    mbd.getResourceDescription(), beanName, "Instantiation of bean failed", ex);
        }
    }
```

在这段关于 BeanWrapper 对象的创建中包含两项操作,第一项是获取实例化策略接口(InstantiationStrategy),通过实例化接口创建 Bean 实例;第二项是将 Bean 实例封装为 BeanWrapperImpl 对象。通过这段方法就可以获得 BeanWrapper 接口对象。

6.4.2　BeanWrapper 属性设置

前文准备好了 BeanWrapper 对象,接下来就需要进行属性设置,负责属性设置的方法签名为 org.springframework.beans.AbstractNestablePropertyAccessor#setPropertyValue(org.springframework.beans.PropertyValue),详细代码如下。

```
@Override
public void setPropertyValue(PropertyValue pv) throws BeansException {
    //需要设置的字段名称
    PropertyTokenHolder tokens = (PropertyTokenHolder) pv.resolvedTokens;
    if (tokens == null) {
        //属性名称
        //获取属性名称
        String propertyName = pv.getName();
        AbstractNestablePropertyAccessor nestedPa;
        try {
            //属性访问器
            nestedPa = getPropertyAccessorForPropertyPath(propertyName);
        }
        catch (NotReadablePropertyException ex) {
```

```
            throw new NotWritablePropertyException(getRootClass(), this.nestedPath +
propertyName,
                    "Nested property in path '" + propertyName + "' does not exist", ex);
        }
        tokens = getPropertyNameTokens(getFinalPath(nestedPa, propertyName));
        if (nestedPa == this) {
            pv.getOriginalPropertyValue().resolvedTokens = tokens;
        }
        //设置属性
        nestedPa.setPropertyValue(tokens, pv);
    }
    else {
        setPropertyValue(tokens, pv);
    }
}
```

在这段关于属性设置的方法中需要着重关注三个变量。第一个是 nestedPa，第二个是 tokens，第三个是 pv。变量 nestedPa 的类型是 AbstractNestablePropertyAccessor，它主要用作于嵌套属性访问，注意它既可以访问属性也可以设置属性；变量 tokens 的类型是 PropertyTokenHolder，它主要存储属性名称等一些属性。

在进行最终的处理时会分别通过两个方法进行，一个是 processKeyedProperty 方法，另一个是 processLocalProperty 方法。在前文的测试用例中能够进入调试的方法是 processLocalProperty，下面是 processLocalProperty 的详细代码。

```
//删除日志,异常处理等代码
private void processLocalProperty(PropertyTokenHolder tokens, PropertyValue pv) {
    //ph 里面有对象
    PropertyHandler ph = getLocalPropertyHandler(tokens.actualName);
    Object oldValue = null;
    //获取 pv 的属性值
    Object originalValue = pv.getValue();
    Object valueToApply = originalValue;
    //是否需要
    if (!Boolean.FALSE.equals(pv.conversionNecessary)) {
        //是否能够转
        if (pv.isConverted()) {
            valueToApply = pv.getConvertedValue();
        }
        //不能转换
        else {
            if (isExtractOldValueForEditor() && ph.isReadable()) {
                oldValue = ph.getValue();
            }
            valueToApply = convertForProperty(tokens.canonicalName, oldValue, originalValue,
ph.toTypeDescriptor());
        }
        pv.getOriginalPropertyValue().conversionNecessary = (valueToApply !=
originalValue);
    }
    ph.setValue(valueToApply);
}
```

在这个方法中最关键的对象是 PropertyHandler，从上述代码中可以发现设置属性也和 PropertyHandler 对象有关，对于 getLocalPropertyHandler 方法的细节和 PropertyHandler 的类图都有值得研究的价值，PropertyHandler 类如图 6.5 所示。

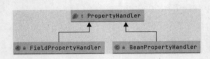

图 6.5　PropertyHandler 类图

当前正在对于 BeanWrapperImpl 对象进行分析，在这个类中 getLocalPropertyHandler 方法的返回值是 BeanPropertyHandler 类型，下面来看 BeanWrapperImpl # getLocalPropertyHandler 方法中的细节，具体代码如下。

```
@Override
@Nullable
protected BeanPropertyHandler getLocalPropertyHandler(String propertyName) {
 PropertyDescriptor pd =
getCachedIntrospectionResults().getPropertyDescriptor(propertyName);
 return (pd != null ? new BeanPropertyHandler(pd) : null);
}
```

在这个方法中需要了解 getCachedIntrospectionResults 方法的返回值，该返回值是成员变量 cachedIntrospectionResults 中的一些数据。

6.4.3　CachedIntrospectionResults 对象介绍

下面将介绍 CachedIntrospectionResults 对象的成员变量，成员变量信息见表 6.2。

表 6.2　CachedIntrospectionResults 对象的成员变量

变量名称	变量类型	变量含义
acceptedClassLoaders	Set < ClassLoader >	存储类加载器
strongClassCache	ConcurrentMap < Class <?>, CachedIntrospectionResults >	存储线程安全的 Bean 信息
softClassCache	ConcurrentMap < Class <?>, CachedIntrospectionResults >	存储线程不安全的 Bean 信息
shouldIntrospectorIgnoreBeaninfoClasses	Boolean	从配置文件中读取 spring.beaninfo.ignore 的信息作为变量的数据
beanInfoFactories	List < BeanInfoFactory >	Bean Factory 容器，存储多个 Bean Factory
beanInfo	BeanInfo	BeanInfo 接口是由 JDK 提供的一个关于 Bean 信息的接口，在这个接口中包含关于 Bean 的基本描述，如：PropertyDescriptor 属性描述，MethodDescriptor 方法描述等

续表

变量名称	变量类型	变量含义
propertyDescriptorCache	Map< String, PropertyDescriptor >	用于存储属性（字段）名称和属性（字段）描述的容器，key 为属性名称，value 为属性描述
typeDescriptorCache	ConcurrentMap< PropertyDescriptor, TypeDescriptor >	用于存储属性描述对象和属性类型的描述对象

了解 CachedIntrospectionResults 对象的成员变量数据后需要进一步对这个对象进行使用，在 CachedIntrospectionResults 类中提供了 forClass 方法，开发者可以直接调用该方法获取数据，该方法在 BeanWrapperImpl 中也有直接调用，具体调用方法如下。

```
private CachedIntrospectionResults getCachedIntrospectionResults() {
    if (this.cachedIntrospectionResults == null) {
        this.cachedIntrospectionResults = CachedIntrospectionResults.forClass(getWrappedClass());
    }
    return this.cachedIntrospectionResults;
}
```

图 6.6 是 LiveBean 通过计算后得到的 CachedIntrospectionResults 数据信息。

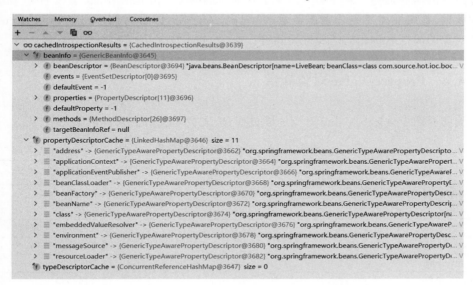

图 6.6　LiveBean 的 CachedIntrospectionResults 数据信息

还需要进一步搜索 propertyDescriptorCache 变量，重点关注 address，它是测试用例中需要设置的数据，address 键值对下的数据内容如图 6.7 所示。

通过图 6.7 的观察可以发现在属性描述对象中会存在 get 和 set 方法，这两个方法都是一个具体的 Method 对象，此时 address 属性的两个重要方法就已经准备就绪，这两个重要

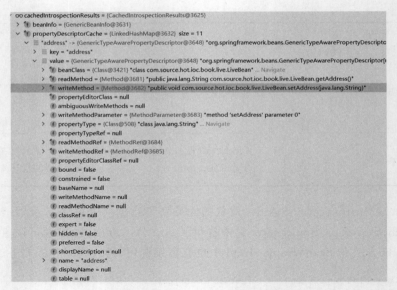

图 6.7　address 的键值对数据信息

方法分别是获取属性方法（getAddress）和设置属性方法（setAddress），通过这两个方法就可以进行数据设置、数据获取。

现在拥有了属性描述的缓存对象后需要和另一个对象（PropertyValue）进行处理，从而得到过滤后的属性描述。在 PropertyValue 中存储的是 SpringXML 配置文件中 property 标签的数据，在属性描述集合中需要通过 property 标签的 name 属性过滤得到 name 和属性描述对象的属性名称相同的属性描述符。

6.4.4　PropertyValue 对象介绍

下面将介绍 PropertyValue 对象的成员变量，成员变量信息见表 6.3。

表 6.3　PropertyValue 成员变量

变量名称	变量类型	变量含义
name	String	属性名称
value	Object	数据值
conversionNecessary	Boolean	是否需要进行转换
resolvedTokens	Object	需要解析的数据
optional	boolean	是否可选
converted	boolean	是否已转换
convertedValue	Object	转换后的结果

在 PropertyValue 对象中对于数据值的存储有两个变量：value 和 convertedValue。
现在拥有的数据量有一些多，下面是一些重要数据对象。

（1）PropertyHandler 具体是 BeanPropertyHandler，在 BeanPropertyHandler 中提供的 setValue 方法可以进行属性设置，在 PropertyHandler 中还有 set 方法。

（2）PropertyValue 中存储了真正的数据值。

6.4.5　最终的数据设置

下面将介绍 Bean 实例数据设置的最终步骤，处理方法是 org.springframework.beans.BeanWrapperImpl.BeanPropertyHandler#setValue，具体代码如下。

```java
@Override
public void setValue(final @Nullable Object value) throws Exception {
  final Method writeMethod = (this.pd instanceof GenericTypeAwarePropertyDescriptor ?
      ((GenericTypeAwarePropertyDescriptor) this.pd).getWriteMethodForActualAccess() :
      this.pd.getWriteMethod());
  if (System.getSecurityManager() != null) {
    AccessController.doPrivileged((PrivilegedAction<Object>) () -> {
      ReflectionUtils.makeAccessible(writeMethod);
      return null;
    });
    try {
      AccessController.doPrivileged((PrivilegedExceptionAction<Object>) () ->
          writeMethod.invoke(getWrappedInstance(), value), acc);
    }
    catch (PrivilegedActionException ex) {
      throw ex.getException();
    }
  }
  else {
    ReflectionUtils.makeAccessible(writeMethod);
    writeMethod.invoke(getWrappedInstance(), value);
  }
}
```

在这段代码中有以下三个关键信息。

（1）writeMethod：写函数，一般是 set 方法。

（2）value：设置的属性值。

（3）getWrappedInstance：该方法返回需要进行设置的对象。

writeMethod 数据对象是通过 PropertyDescriptor 对象进行获取。writeMethod 是一个 Method 对象，它可以直接进行调用将参数 value 设置给对应的属性。经过设置后的 BeanWrapper 对象如图 6.8 所示。

可以看到图 6.8 中 wrappedObject 对象的 address 属性被成功赋予数据。至此，关于 Bean 实例的属性赋值操作已经完成分析。

在属性设置中最关键的行为有两个，第一个是获取需要进行属性设置的对象，第二个是进行属性设置。这两个操作在常规 Java 开发中是很普通的操作，先 new 对象，再通过对象调用 set 方法进行属性设置。现在开发者将这部分操作都交给了 Spring 进行处理，这种处理模式引出了两个名词，第一个是控制反转，第二个是依赖注入。控制反转是将创建对象的过程交给 Spring 进行，包括数据设置。property 标签的使用就是一种依赖注入方式，明确

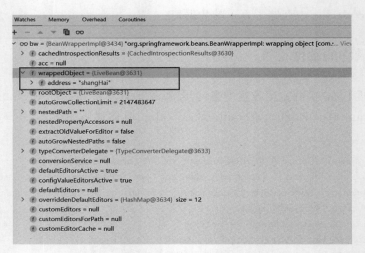

图 6.8　经过属性设置后的 BeanWrapper

依赖名称和依赖属性后具体设置交给 Spring 进行。简单理解，对 Bean 实例进行数据设置就是依赖注入。

6.5　Bean 生命周期值 Aware 接口

在完成 Bean 实例的属性设置之后 Spring 会做各类生命周期的拓展接口，有关这部分逻辑处理的方法签名是 org.springframework.beans.factory.support.AbstractAutowireCapableBeanFactory#initializeBean(java.lang.String,java.lang.Object,org.springframework.beans.factory.support.RootBeanDefinition)，

具体代码如下。

```
protected Object initializeBean(final String beanName, final Object bean, @ Nullable
RootBeanDefinition mbd) {
    //第一部分
    if (System.getSecurityManager() != null) {
        AccessController.doPrivileged((PrivilegedAction<Object>) () -> {
            invokeAwareMethods(beanName, bean);
            return null;
        }, getAccessControlContext());
    }
    else {
        //aware 接口执行
        invokeAwareMethods(beanName, bean);
    }

    //第二部分
    Object wrappedBean = bean;
    if (mbd == null || !mbd.isSynthetic()) {
        //BeanPostProcessor 前置方法执行
        wrappedBean =
```

```
        applyBeanPostProcessorsBeforeInitialization(wrappedBean, beanName);
    }

    //第三部分
    try {
        //执行实例化函数
        invokeInitMethods(beanName, wrappedBean, mbd);
    }
    catch (Throwable ex) {
        throw new BeanCreationException(
                (mbd != null ? mbd.getResourceDescription() : null),
                beanName, "Invocation of init method failed", ex
        );
    }

    //第四部分
    if (mbd == null || !mbd.isSynthetic()) {
        //BeanPostProcessor 后置方法执行
        wrappedBean =
 applyBeanPostProcessorsAfterInitialization(wrappedBean, beanName);
    }

    return wrappedBean;
}
```

在这段方法中处理了 Bean 生命周期相关的一些内容，主要是初始化、创建相关联的一些生命周期，在这段方法中主要关注下面四个方法。

（1）invokeAwareMethods：执行 Aware 相关的接口方法。

（2）applyBeanPostProcessorsBeforeInitialization：执行 BeanPostProcessor 中的 PostProcessBeforeInitialization 方法。

（3）invokeInitMethods：执行 InitializingBean 接口提供的方法和自定义配置的 init-method 方法（SpringXML bean 标签的属性）。

（4）applyBeanPostProcessorsAfterInitialization：执行 BeanPostProcessor 中的 postProcessAfterInitialization 方法。

下面围绕第一个方法 invokeAwareMethods 进行分析，该方法具体代码如下。

```
private void invokeAwareMethods(final String beanName, final Object bean) {
    if (bean instanceof Aware) {
        if (bean instanceof BeanNameAware) {
            ((BeanNameAware) bean).setBeanName(beanName);
        }
        if (bean instanceof BeanClassLoaderAware) {
            ClassLoader bcl = getBeanClassLoader();
            if (bcl != null) {
                ((BeanClassLoaderAware) bean).setBeanClassLoader(bcl);
            }
        }
        if (bean instanceof BeanFactoryAware) {
```

```
            ((BeanFactoryAware) bean).setBeanFactory(AbstractAutowireCapableBeanFactory.
this);
        }
    }
}
```

在这个方法中会执行三个和 Aware 相关的接口,第一个是 BeanNameAware,第二个是 BeanClassLoaderAware,第三个是 BeanFactoryAware,这些接口都可以是某一个 Bean 的实现接口,当某一个 Bean 拥有这三个接口时就会执行三个接口相关的内容,具体执行逻辑是在 Bean 的实现方法中,在这段方法中实施对三个 Aware 接口进行类型判断,然后再执行相关方法。

6.6　BeanPostProcessor♯postProcessBeforeInitialization

在完成 Aware 接口相关方法的处理后 Spring 会进行 applyBeanPostProcessorsBeforeInitialization 的处理。在这个方法中主要对 BeanPostProcessor 接口的实现类进行处理,且只会执行 postProcessBeforeInitialization 方法,一般称为 Bean 前置操作。下面是具体的处理代码。

```
@Override
public Object applyBeanPostProcessorsBeforeInitialization(Object existingBean, String beanName)
        throws BeansException {

    Object result = existingBean;
    for (BeanPostProcessor processor : getBeanPostProcessors()) {
        Object current = processor.postProcessBeforeInitialization(result, beanName);
        if (current == null) {
            return result;
        }
        result = current;
    }
    return result;
}
```

这段代码的主要行为是从容器中获取所有的 BeanPostProcessor 接口实现类,并调用实现类的 postProcessBeforeInitialization 方法,在这里可以发现所有的 Bean 实例都会被放入接口方法执行,在编写 BeanPostProcessor 接口实现类时开发者对于第一个参数 bean 一定要做好类型上的控制,否则会出现类型穿透的问题,导致作用域不正确。

6.7　InitializingBean 接口和自定义 init-method 方法

在完成 applyBeanPostProcessorsBeforeInitialization 之后就是执行 InitializingBean 和自定义 init-method 相关方法。在测试用例中 LiveBean 本身实现了 InitializingBean 接口,这里会直接执行 InitializingBean 中的 afterPropertiesSet 方法,这一点很好理解,下面还有一个关于自定义 init-method 的执行,首先需要知道 init-method 存储在哪里。在

AbstractBeanDefinition 中有一个属性 initMethodName，它就是存储 bean 标签中 init-method 属性的变量。存储信息如图 6.9 所示。

图 6.9　initMethod 在 BeanDefinition 中的存储信息

在图 6.9 中可以看到 SpringXML 配置文件中 init-method 属性的内容被存储在了 initMethodName 属性中，但是它现在还是一个字符串，字符串不具备方法调用功能，在使用时需要将字符串转换成 method 对象才可以调用相关方法。对于 initMethod 相关处理的代码如下。

```
protected void invokeInitMethods(String beanName, final Object bean, @Nullable
RootBeanDefinition mbd)
        throws Throwable {
    //第一部分
    //是不是 InitializingBean
    boolean isInitializingBean = (bean instanceof InitializingBean);
    //是否存在方法 afterPropertiesSet
    if (isInitializingBean && (mbd ==
null || !mbd.isExternallyManagedInitMethod("afterPropertiesSet"))) {
        if (logger.isTraceEnabled()) {
            logger.trace("Invoking afterPropertiesSet() on bean with name '" + beanName + "'");
        }
        if (System.getSecurityManager() != null) {
            try {
                //执行 afterPropertiesSet
                AccessController.doPrivileged((PrivilegedExceptionAction<Object>) () -> {
                    ((InitializingBean) bean).afterPropertiesSet();
                    return null;
                }, getAccessControlContext());
            }
            catch (PrivilegedActionException pae) {
                throw pae.getException();
            }
        }
        else {
            //执行 afterPropertiesSet
            ((InitializingBean) bean).afterPropertiesSet();
        }
    }

    //第二部分
    if (mbd != null && bean.getClass() != NullBean.class) {
        //获取 initMethod 字符串
        String initMethodName = mbd.getInitMethodName();
        if (StringUtils.hasLength(initMethodName) &&
                !(isInitializingBean && "afterPropertiesSet".equals(initMethodName)) &&
```

```
                !mbd.isExternallyManagedInitMethod(initMethodName)) {
            //自定义的 init method
            invokeCustomInitMethod(beanName, bean, mbd);
        }
    }
}
```

在 invokeInitMethods 方法中可以很明显地看到两个处理,这两个处理和这一节开始提到的一样,InitializingBean#afterPropertiesSet 执行和 invokeCustomInitMethod 自定义 init-method 执行。主要关注 invokeCustomInitMethod 方法的细节。InitializingBean#afterPropertiesSet 的处理逻辑是第一步进行类型推断,第二步类型推断成功后直接调用方法。

下面介绍 invokeInitMethods 中最值得关注的一个方法 invokeCustomInitMethod,具体代码如下。

```
protected void invokeCustomInitMethod(String beanName, final Object bean, RootBeanDefinition mbd) throws Throwable {
    //获取 initMethod 名称
    String initMethodName = mbd.getInitMethodName();
    Assert.state(initMethodName != null, "No init method set");
    //反射获取方法
    Method initMethod =
(mbd.isNonPublicAccessAllowed() ?
BeanUtils.findMethod(bean.getClass(),
initMethodName) : ClassUtils.getMethodIfAvailable(bean.getClass(), initMethodName));
    //方法是否存在判断
    if(initMethod == null) {
        if(mbd.isEnforceInitMethod()) {
            throw new BeanDefinitionValidationException("Could not find an init method named '" + initMethodName + "' on bean with name '" + beanName + "'");
        } else {
            return;
        }
    }
    //尝试获取接口方法
    Method methodToInvoke = ClassUtils.getInterfaceMethodIfPossible(initMethod);
    if(System.getSecurityManager() != null) {
        AccessController.doPrivileged((PrivilegedAction<Object>)()->{
            ReflectionUtils.makeAccessible(methodToInvoke);
            return null;
        });
        //反射调用
        AccessController.doPrivileged((PrivilegedExceptionAction<Object>)()->methodToInvoke.invoke(bean), getAccessControlContext());
    } else {
        //反射调用
        ReflectionUtils.makeAccessible(methodToInvoke);
        methodToInvoke.invoke(bean);
```

            }
        }

在这段方法中核心目标是找到 initMethodName 对应的方法，在这段方法中提供了以下三种找寻找对应方法的工具。

```
BeanUtils.findMethod(bean.getClass(), initMethodName)
ClassUtils.getMethodIfAvailable(bean.getClass(), initMethodName)
ClassUtils.getInterfaceMethodIfPossible(initMethod)
```

在上述三种方法找到 Method 对象之后，直接调用搜索到的方法即可完成 init-method 的生命周期处理。

6.8　BeanPostProcessor#postProcessAfterInitialization

在完成 InitializingBean 方法调用后会紧接着做 Bean 后置处理方法的调用，这段方法的调用和前置方法的调用类似，只是在执行 BeanPostProcessor 接口时对方法时做了更换，换成了 postProcessAfterInitialization 方法，具体处理代码如下。

```
@Override
public Object applyBeanPostProcessorsAfterInitialization(Object existingBean,
String beanName)
        throws BeansException {

    Object result = existingBean;
    for (BeanPostProcessor processor : getBeanPostProcessors()) {
        //执行 Spring 容器中的 BeanPostProcessor
        Object current = processor.postProcessAfterInitialization(result, beanName);
        if (current == null) {
            return result;
        }
        result = current;
    }
    return result;
}
```

6.9　Bean 的摧毁

关于 Bean 初始化（实例化）相关的一些生命周期内容都介绍完了，下面将介绍关于 Bean 摧毁阶段的一些生命周期内容。首先需要找到 Bean 进行摧毁的入口，入口方法是 org.springframework.context.support.AbstractApplicationContext#destroyBeans，具体代码如下。

```
protected void destroyBeans() {
    getBeanFactory().destroySingletons();
}
```

上述代码中摧毁操作只对单例 Bean 有效。

6.9.1 DefaultSingletonBeanRegistry 中的摧毁

接下来将介绍 DefaultSingletonBeanRegistry 中的摧毁方法，方法签名是 Default-SingletonBeanRegistry#destroySingletons，具体代码如下。

```java
public void destroySingletons() {
    if (logger.isTraceEnabled()) {
        logger.trace("Destroying singletons in " + this);
    }
    synchronized (this.singletonObjects) {
        this.singletonsCurrentlyInDestruction = true;
    }

    String[] disposableBeanNames;
    synchronized (this.disposableBeans) {
        disposableBeanNames = StringUtils.toStringArray(this.disposableBeans.keySet());
    }
    for (int i = disposableBeanNames.length - 1; i >= 0; i--) {
        destroySingleton(disposableBeanNames[i]);
    }

    this.containedBeanMap.clear();
    this.dependentBeanMap.clear();
    this.dependenciesForBeanMap.clear();

    clearSingletonCache();
}
```

在上述代码中可以看到代码对多个变量进行了清理和标记操作，了解每个变量的意义后对该方法的处理就理解了，变量分析见表 6.4。

表 6.4 DefaultSingletonBeanRegistry 摧毁时使用的变量说明

变量名称	变量类型	变量说明
singletonsCurrentlyInDestruction	boolean	用来标记是否正在进行清除（摧毁）状态
disposableBeanNames	String[]	记录需要进行摧毁的 BeanName
containedBeanMap	Map<String,Set<String>>	Bean 包含关系容器，key：BeanName，value：包含 key 的 BeanName 列表
dependentBeanMap	Map<String,Set<String>>	Bean 依赖关系容器，key：BeanName，value：依赖 key 的 BeanName 列表
dependenciesForBeanMap	Map<String,Set<String>>	Bean 依赖关系容器
singletonObjects	Map<String,Object>	单例对象容器，key：BeanName，value：Bean 实例
singletonFactories	Map<String,ObjectFactory<?>>	ObjectFactory 容器，key：BeanName，value：ObjectFactory

续表

变量名称	变量类型	变量说明
earlySingletonObjects	Map < String, Object >	早期暴露的 Bean 容器,key：BeanName,value：Bean 实例
registeredSingletons	Set < String >	已经注册的单例 Bean 的名称容器
disposableBeans	Map < String, Object >	会被摧毁的 Bean 容器

下面对方法处理细节进行分析。首先是方法参数，该方法的参数有两个，第一个是 beanName，类型是 String，表示需要摧毁的 BeanName；第二个是 Bean 实例，类型是 DisposableBean 接口，表示具有 DisposableBean 实现的 Bean 实例。

由于 DisposableBean 是一个接口，在参数传递的时候它具体是一个什么类型很重要，从 disposableBean =（DisposableBean）this. disposableBeans. remove（beanName）这段内容中可以发现，数据是从 disposableBeans 中获取的，但是 disposableBeans 的 value 类型是 Object，这里需要关注 registerDisposableBean 方法，在 registerDisposableBean 方法中核心目标是注册行为，注册代码如下。

```
public void registerDisposableBean(String beanName, DisposableBean bean) {
    synchronized (this.disposableBeans) {
        this.disposableBeans.put(beanName, bean);
    }
}
```

可以看到这里注册进入的就是一个 DisposableBean 实现类，因此 value 可以进行转换。接下来是最终的删除方法（DefaultSingletonBeanRegistry♯destroyBean），代码如下。

```
//删除异常处理和日志相关代码
protected void destroyBean(String beanName, @Nullable DisposableBean bean) {
  Set < String > dependencies;
  synchronized (this.dependentBeanMap) {
      //移除依赖 Bean
      dependencies = this.dependentBeanMap.remove(beanName);
  }
  if (dependencies != null) {
      //依赖列表中的也删除
      for (String dependentBeanName : dependencies) {
          destroySingleton(dependentBeanName);
      }
  }

  if (bean != null) {
      bean.destroy();
  }

  Set < String > containedBeans;
  synchronized (this.containedBeanMap) {
      //别名列表
      containedBeans = this.containedBeanMap.remove(beanName);
  }
```

```
    if (containedBeans != null) {
        //删除别名列表中的 BeanName
        for (String containedBeanName : containedBeans) {
            destroySingleton(containedBeanName);
        }
    }

    synchronized (this.dependentBeanMap) {
        //依赖 BeanMap 中删除 BeanName
        for (Iterator<Map.Entry<String, Set<String>>> it =
    this.dependentBeanMap.entrySet().iterator(); it.hasNext(); ) {
            Map.Entry<String, Set<String>> entry = it.next();
            Set<String> dependenciesToClean = entry.getValue();
            dependenciesToClean.remove(beanName);
            if (dependenciesToClean.isEmpty()) {
                it.remove();
            }
        }
    }

    this.dependenciesForBeanMap.remove(beanName);
}
```

在这个方法中,需要重点关注处理顺序。

(1) dependentBeanMap,Bean 依赖关系容器中先删除依赖项。

(2) 调用传入参数 Bean 的摧毁方法,具体实现类 DisposableBeanAdapter。

(3) containedBeanMap,Bean 包含关系容器中删除需要当前 Bean 依赖的对象。

(4) dependentBeanMap 重复进行处理。

(5) dependenciesForBeanMap 中删除当前 Bean 的依赖信息。

在上述五个操作过程中重点需要关注的是第二个,具体处理代码如下。

```
public void destroy() {
    if(!CollectionUtils.isEmpty(this.beanPostProcessors)) {
        //找到 DestructionAwareBeanPostProcessor
        for(DestructionAwareBeanPostProcessor processor: this.beanPostProcessors) {
            processor.postProcessBeforeDestruction(this.bean, this.beanName);
        }
    }
    if(this.invokeDisposableBean) {
        //如果 Bean 本身实现了 DisposableBean 接口会进行调用
        if(System.getSecurityManager() != null) {
            AccessController.doPrivileged((PrivilegedExceptionAction<Object>)() -> {
                ((DisposableBean) this.bean).destroy();
                return null;
            }, this.acc);
        } else {
            ((DisposableBean) this.bean).destroy();
        }
    }
    //自定义摧毁方法的调用
    if(this.destroyMethod != null) {
        invokeCustomDestroyMethod(this.destroyMethod);
```

```
        } else if(this.destroyMethodName != null) {
            Method methodToInvoke = determineDestroyMethod(this.destroyMethodName);
            if(methodToInvoke != null) {

invokeCustomDestroyMethod(ClassUtils.getInterfaceMethodIfPossible(methodToInvoke));
            }
        }
    }
```

在这段代码中有三个处理逻辑，分别如下。

（1）处理容器中 DestructionAwareBeanPostProcessor 的实现类。
（2）如果当前正在被摧毁的 Bean 是 DisposableBean 的实现类，则调用 destroy 方法。
（3）执行 bean 标签中 destroy-method 所对应的方法。

6.9.2　DefaultListableBeanFactory 中的摧毁

在 Spring 中还有另一个可以提供摧毁 Bean 能力的类，它是 DefaultListableBeanFactory，是 DefaultSingletonBeanRegistry 的子类，有关摧毁的代码如下。

```
@Override
public void destroySingleton(String beanName) {
    //摧毁方法的调用
    super.destroySingleton(beanName);
    //删除 manualSingletonNames 中的 BeanName
    removeManualSingletonName(beanName);
    clearByTypeCache();
}
```

在这个 Bean 摧毁方法中除了父类的操作以外还做了如下两个操作。
（1）在容器 manualSingletonNames 中删除当前的 BeanName。
（2）清除类型缓存。

在上述两个操作中需要关注清除类型缓存清除了哪些内容，主要清理两个容器，第一个是 allBeanNamesByType，它存储了所有 Bean；第二个是 singletonBeanNamesByType，它存储了单例的 Bean。

注意，当 Bean 的 scope 属性是 prototype 时是不会执行 DisposableBean # destroy 和 destroy-method 的。

小结

这一章中介绍了 SpringBean 的生命周期，大体上可以分为创建和摧毁。创建相关的入口是 org.springframework.beans.factory.support.AbstractAutowireCapableBeanFactory # doCreateBean，摧毁相关的入口是 org.springframework.context.support.AbstractApplicationContext # doClose。通过上述两个入口一步一步向下追踪源码，寻找在 BeanFactory 接口注释中提到的一些生命周期接口，了解每个生命周期接口在哪些地方使用。同时在分析时引出了 BeanInfo 对象，它是 java.beans 下面的内容，属于 JDK 原生，通过 java.beans 下面的一些类 Spring 得到了一个 Bean。

第7章

Bean 的获取

在第 5 章中讲述了 bean 标签的解析过程,在第 6 章中讲述了 Bean 的生命周期相关接口,在这两章中均未获取 Bean 的实例对象,在第 5 章中所得到的是 BeanDefinition 对象,在第 6 章中定义了 Bean 生命周期的接口,本章将对 Bean 的获取进行分析,主要对 doGetBean 方法和循环依赖进行分析。

7.1 Bean 获取方式配置

一般情况下,Spring 开发者会使用 BeanFactory 中提供的 getBean 方法来获取 Bean 实例,但是在配置 Bean 阶段会有多种方式,总共有以下四种形式。

(1) 直接给 bean 标签配置 id 和 class。
(2) 配置 bean 标签中的 factory-bean 和 factory-method。
(3) 通过配置 bean 标签的 factory-method。这种配置方式和第(2)种配置方式的区别是这里采用的 Class 并不是 Bean Class 而是提供静态方法的 Class。
(4) 通过实现 FactoryBean 接口和 SpringXML 来进行配置。

上述四种是 SpringXML 环境中 Bean 的创建方式,后续的分析也将围绕这四种形式进行。

7.2 Bean 获取的测试环境搭建

接下来准备 Bean 获取相关的测试用例,首先准备一个 JavaBean,类名是 BeanSourceFactory,具体代码如下。

```
public class BeanSourceFactory {
 public static BeanSource staticFactory() {
    BeanSource beanSource = new BeanSource();
```

```java
        beanSource.setType("StaticFactory");
        return beanSource;
    }

    public BeanSource noStaticFactory() {
        BeanSource beanSource = new BeanSource();
        beanSource.setType("noStaticFactory");
        return beanSource;
    }
}
```

完成基本 JavaBean 编写后再进一步创建 FactoryBean 的实现类,类名是 BeanSourceFactoryBean,代码如下。

```java
public class BeanSourceFactoryBean implements FactoryBean<BeanSource> {
    @Override
    public BeanSource getObject() throws Exception {
        BeanSource beanSource = new BeanSource();
        beanSource.setType("from factory bean.");
        return beanSource;
    }

    @Override
    public Class<?> getObjectType() {
        return BeanSource.class;
    }

    @Override
    public boolean isSingleton() {
        return true;
    }
}
```

接下来编写 SpringXML 配置文件,文件名为 get-bean.xml,具体代码如下。

```xml
<?xml version="1.0" encoding="UTF-8"?>
<beans xmlns:xsi="http://www.w3.org/2001/XMLSchema-instance"
    xmlns="http://www.springframework.org/schema/beans"
    xsi:schemaLocation="http://www.springframework.org/schema/beans http://www.springframework.org/schema/beans/spring-beans.xsd">

    <bean id="beanSource" class="com.source.hot.ioc.book.getbean.BeanSource">
        <property name="type" value="xml"/>
    </bean>

    <bean id="beanSourceFactory" class="com.source.hot.ioc.book.getbean.BeanSourceFactory"/>
    <bean id="beanSourceFromNoStatic" factory-bean="beanSourceFactory" factory-method="noStaticFactory"></bean>

    <bean id="beanSourceFromStatic"
```

```
        class = "com.source.hot.ioc.book.getbean.BeanSourceFactory"
              factory-method = "staticFactory"></bean>

    <bean id = "beanSourceFromFactoryBean"
        class = "com.source.hot.ioc.book.getbean.BeanSourceFactoryBean"></bean>
</beans>
```

完成 SpringXML 配置文件的编写后进行测试类的编写，类名是 GetBeanTest，具体代码如下。

```
class GetBeanTest {
  ClassPathXmlApplicationContext context = null;

  @BeforeEach
  void init() {
      context = new ClassPathXmlApplicationContext("META-INF/get-bean.xml");
  }

  @Test
  void fromBean() {
      BeanSource beanSource = context.getBean("beanSource", BeanSource.class);
      assert beanSource.getType().equals("xml");
  }

  @Test
  void fromStatic() {
      BeanSource beanSourceFromStatic =
context.getBean("beanSourceFromStatic", BeanSource.class);
      assert beanSourceFromStatic.getType().equals("StaticFactory");
  }

  @Test
  void fromNoStatic() {
      BeanSource beanSourceFromNoStatic =
context.getBean("beanSourceFromNoStatic", BeanSource.class);
      assert beanSourceFromNoStatic.getType().equals("noStaticFactory");

  }

  @Test
  void fromFactoryBean() {
      BeanSource beanSourceFromFactoryBean =
context.getBean("beanSourceFromFactoryBean", BeanSource.class);
      assert beanSourceFromFactoryBean.getType().equals("from factory bean.");
  }
}
```

7.3　doGetBean 分析

通常开发者在使用 BeanFactory 所提供的 getBean 方法时都会指向 doGetBean 方法，具体的方法签名是 org.springframework.beans.factory.support.AbstractBeanFactory#

doGetBean，具体代码如下。

```java
@SuppressWarnings("unchecked")
protected <T> T doGetBean(final String name, @Nullable final Class<T> requiredType,
        @Nullable final Object[] args, boolean typeCheckOnly)
        throws BeansException {
    //转换 BeanName
    final String beanName = transformedBeanName(name);
    Object bean;

    //Eagerly check singleton cache for manually registered singletons.
    //获取单例对象
    Object sharedInstance = getSingleton(beanName);
    //单例对象是否存在,参数是否为空
    if (sharedInstance != null && args == null) {
        if (logger.isTraceEnabled()) {
            if (isSingletonCurrentlyInCreation(beanName)) {
                logger.trace("Returning eagerly cached instance of singleton bean '" +
beanName + "' that is not fully initialized yet - a consequence of a circular reference");
            }
            else {
                logger.trace("Returning cached instance of singleton bean '" + beanName + "'");
            }
        }
        //获取 Bean 实例
        bean = getObjectForBeanInstance(sharedInstance, name, beanName, null);
    }

    else {
        //循环依赖的问题
        if (isPrototypeCurrentlyInCreation(beanName)) {
            throw new BeanCurrentlyInCreationException(beanName);
        }

        //获取父 Bean 工厂
        //从父 Bean 工厂中创建
        BeanFactory parentBeanFactory = getParentBeanFactory();
        //从父 Bean 工厂中查询
        if (parentBeanFactory != null && !containsBeanDefinition(beanName)) {
            //Not found -> check parent.
            //确定 BeanName
            String nameToLookup = originalBeanName(name);
            //父 Bean 工厂类型判断
            if (parentBeanFactory instanceof AbstractBeanFactory) {
                //再次获取
                return ((AbstractBeanFactory) parentBeanFactory).doGetBean(
                        nameToLookup, requiredType, args, typeCheckOnly);
            }
            else if (args != null) {
                return (T) parentBeanFactory.getBean(nameToLookup, args);
            }
```

```java
            else if (requiredType != null) {
                return parentBeanFactory.getBean(nameToLookup, requiredType);
            }
            else {
                return (T) parentBeanFactory.getBean(nameToLookup);
            }
        }

        //是否需要进行类型校验
        if (!typeCheckOnly) {
            //将 Bean 标记为已创建
            markBeanAsCreated(beanName);
        }

        try {
            //获取 Bean 定义
            final RootBeanDefinition mbd = getMergedLocalBeanDefinition(beanName);
            checkMergedBeanDefinition(mbd, beanName, args);

            //需要依赖的 Bean
            String[] dependsOn = mbd.getDependsOn();
            if (dependsOn != null) {
                for (String dep : dependsOn) {
                    //是否依赖
                    if (isDependent(beanName, dep)) {
                        throw new BeanCreationException(mbd.getResourceDescription(), beanName," Circular depends - on relationship between '" + beanName + "' and '" + dep + "'");
                    }
                    //注册依赖 Bean
                    registerDependentBean(dep, beanName);
                    try {
                        //获取 Bean
                        getBean(dep);
                    }
                    catch (NoSuchBeanDefinitionException ex) {
                        throw new BeanCreationException(mbd.getResourceDescription(), beanName, "'" + beanName + "' depends on missing bean '" + dep + "'", ex);
                    }
                }
            }

            //单例 Bean 创建
            //判断是否是单例
            if (mbd.isSingleton()) {
                //获取单例 Bean
                sharedInstance = getSingleton(beanName, () -> {
                    try {
                        //创建 Bean
                        return createBean(beanName, mbd, args);
```

```java
            }
            catch (BeansException ex) {
                //摧毁单例的 Bean
                destroySingleton(beanName);
                throw ex;
            }
        });
        //获取 Bean 实例
        bean = getObjectForBeanInstance(sharedInstance, name, beanName, mbd);
    }

    //原型模式创建
    //是否是原型模式
    else if (mbd.isPrototype()) {
        Object prototypeInstance = null;
        try {
            //创建之前的行为
            beforePrototypeCreation(beanName);
            //创建
            prototypeInstance = createBean(beanName, mbd, args);
        }
        finally {
            //创建后的行为
            afterPrototypeCreation(beanName);
        }
        //创建
        bean = getObjectForBeanInstance(prototypeInstance, name, beanName, mbd);
    }

    else {
        //获取作用域名称
        String scopeName = mbd.getScope();
        //从作用域容器中获取当前作用域名称对应的作用域接口 scope
        final Scope scope = this.scopes.get(scopeName);
        if (scope == null) {
            throw new IllegalStateException("No Scope registered for scope name '" + scopeName + "'");
        }
        try {
            //从 scope 接口获取
            Object scopedInstance = scope.get(beanName, () -> {
                //创建之前的行为
                beforePrototypeCreation(beanName);
                try {
                    //创建 Bean
                    return createBean(beanName, mbd, args);
                }
                finally {
                    //创建之后的行为
                    afterPrototypeCreation(beanName);
                }
```

```java
            });
            //获取 Bean 实例
            bean = getObjectForBeanInstance(scopedInstance, name, beanName, mbd);
        }
        catch (IllegalStateException ex) {
            throw new BeanCreationException(beanName, "Scope '" + scopeName + "' is not active for the current thread; consider " + "defining a scoped proxy for this bean if you intend to refer to it from a singleton",ex);
        }
    }
}
catch (BeansException ex) {
    //Bean 创建失败后的处理
    cleanupAfterBeanCreationFailure(beanName);
    throw ex;
}
    }

    //类型和需要的类型是否匹配
    if (requiredType != null && !requiredType.isInstance(bean)) {
        try {
            //获取类型转换器,通过类型转换器进行转换
            T convertedBean = getTypeConverter().convertIfNecessary(bean, requiredType);
            if (convertedBean == null) {
                throw new BeanNotOfRequiredTypeException(name, requiredType, bean.getClass());
            }
            return convertedBean;
        }
        catch (TypeMismatchException ex) {
            if (logger.isTraceEnabled()) {
                logger.trace("Failed to convert bean '" + name + "' to required type '" +
                        ClassUtils.getQualifiedName(requiredType) + "'", ex);
            }
            throw new BeanNotOfRequiredTypeException(name, requiredType, bean.getClass());
        }
    }
    return (T) bean;
}
```

在 doGetBean 方法中有下列 9 个操作。

(1) BeanName 转换得到真正的名字。

(2) 尝试从单例容器中获取。

(3) 从 FactoryBean 接口中获取实例。

(4) 尝试从父容器中获取。

(5) 进行 BeanName 标记。

(6) 非 FactoryBean 创建,单例模式。

(7) 非 FactoryBean 创建，原型模式。
(8) 非 FactoryBean 创建，既不是单例模式也不是原型模式。
(9) 类型转换器中获取 Bean。
接下来将对上述 9 个操作做详细分析。

7.3.1　BeanName 转换

在 Spring 中通过 BeanName 获取 Bean 实例是一个比较常用的方式。在获取 Bean 实例前都需要进行 BeanName 的转换，转换的原因是有别名标签或者注解的存在，具体的转换代码如下。

```
protected String transformedBeanName(String name) {
    //转换 BeanName
    //1. 通过 BeanFactoryUtils.transformedBeanName 求 BeanName
    //2. 如果是有别名的(方法参数是别名)，会从别名列表中获取对应的 BeanName
    return canonicalName(BeanFactoryUtils.transformedBeanName(name));
}
```

在转换 BeanName 时存在以下两个操作。

第一个操作是关于 FactoryBean 的操作，这里的操作是将 FactoryBean 的前置标记符号（&）删除，这部分提供者是 BeanFactoryUtils.transformedBeanName(name)，代码如下。

```
public static String transformedBeanName(String name) {
    Assert.notNull(name, "'name' must not be null");
    //名字不是以 & 开头，直接返回
    if (!name.startsWith(BeanFactory.FACTORY_BEAN_PREFIX)) {
        return name;
    }
    //截取字符串，再返回
    return transformedBeanNameCache.computeIfAbsent(name, beanName -> {
        do {
            beanName = beanName.substring(BeanFactory.FACTORY_BEAN_PREFIX.length());
        }
        while (beanName.startsWith(BeanFactory.FACTORY_BEAN_PREFIX));
        return beanName;
    });
}
```

第二个操作是关于别名的操作。Spring 会拿着第一次操作后的数据来别名容器中获取真正的 BeanName，方法提供者：SimpleAliasRegistry#canonicalName。

7.3.2　尝试从单例容器中获取

在得到 BeaNname 之后 Spring 会做获取操作，首先会尝试从单例容器中获取，Spring 中存储单例 Bean 的容器代码如下。

```
private final Map<String, Object> singletonObjects = new ConcurrentHashMap<>(256);
```

在这个容器中 key 是 BeanName，value 是 Bean 实例，从单例容器中获取就是从这个 Map 对象中获取。

7.3.3　从 FactoryBean 接口中获取实例

在 Spring 中当开发者在单例容器中根据 BeanName 获取 Bean 实例失败后并且参数列表为空，就会进行 getObjectForBeanInstance 方法调用。getObjectForBeanInstance 第一部分代码如下。

```
if (BeanFactoryUtils.isFactoryDereference(name)) {
   //类型判断
   if (beanInstance instanceof NullBean) {
      return beanInstance;
   }
   if (!(beanInstance instanceof FactoryBean)) {
      throw new BeanIsNotAFactoryException(beanName, beanInstance.getClass());
   }
   if (mbd != null) {
      mbd.isFactoryBean = true;
   }
   //返回实例
   return beanInstance;
}
```

在第一部分的代码中会做一个关于 FactoryBean 的处理。当正在处理的 Bean 是一个 FactoryBean 时会做出其他一些判断，当满足后会直接将 Bean 实例返回。注意这个 Bean 实例是从单例对象容器中获取的。

继续查看第二段代码。

```
//判断是不是 FactoryBean
if (!(beanInstance instanceof FactoryBean)) {
   return beanInstance;
}

Object object = null;
if (mbd != null) {
   mbd.isFactoryBean = true;
}
else {
   //缓存中获取
   object = getCachedObjectForFactoryBean(beanName);
}
if (object == null) {
   //如果还是 null,从 FactoryBean 中创建
   FactoryBean<?> factory = (FactoryBean<?>) beanInstance;
   if (mbd == null && containsBeanDefinition(beanName)) {
```

```
            mbd = getMergedLocalBeanDefinition(beanName);
        }
        boolean synthetic = (mbd != null && mbd.isSynthetic());
        //从 FactoryBean 中获取 Bean 实例
        object = getObjectFromFactoryBean(factory, beanName, !synthetic);
    }
    return object;
```

在第二部分代码中明面上可能看不出什么实际的操作信息，但是在 getCached-ObjectForFactoryBean 方法和 getObjectFromFactoryBean 方法中有一个共同的操作对象 factoryBeanObjectCache。有关 factoryBeanObjectCache 的数据结构如下。

```
private final Map<String, Object> factoryBeanObjectCache = new ConcurrentHashMap<>(16);
```

该容器主要存储 FactoryBean 的实例，getCachedObjectForFactoryBean 方法是直接从容器中获取，在 getObjectFromFactoryBean 方法中会有向容器中设置数据的具体操作。

在 getObjectFromFactoryBean 中对于 FactoryBean 的创建分为两种情况：单例和原型。

1. FactoryBean 的单例创建

下面将介绍从 FactoryBean 中进行单例对象的创建，当 FactoryBean#isSingleton()方法的返回值为 true 时会进入下面的代码。

```
synchronized (getSingletonMutex()) {
    //从工厂 Bean 的缓存中获取
    Object object = this.factoryBeanObjectCache.get(beanName);
    if (object == null) {

        //从 FactoryBean 接口中获取
        object = doGetObjectFromFactoryBean(factory, beanName);
        //从缓存 Map 中获取
        Object alreadyThere = this.factoryBeanObjectCache.get(beanName);
        if (alreadyThere != null) {
            //如果缓存中获取有值
            //object 覆盖
            object = alreadyThere;
        }
        else {
            //判断是否需要后置处理
            if (shouldPostProcess) {
                //是否处于创建中
                if (isSingletonCurrentlyInCreation(beanName)) {
                    return object;
                }
                //单例创建前的验证
                beforeSingletonCreation(beanName);
                try {
                    //从 FactoryBean 接口创建的,后置处理
                    object = postProcessObjectFromFactoryBean(object, beanName);
                }
```

```
                    catch (Throwable ex) {
                        throw new BeanCreationException(beanName, "Post-processing of
    FactoryBean's singleton object failed", ex);
                    }
                    finally {
                        //单例 Bean 创建之后
                        afterSingletonCreation(beanName);
                    }
                }
                //是否包含 BeanName
                if (containsSingleton(beanName)) {
                    //插入缓存
                    //后续使用时可以直接获取
                    this.factoryBeanObjectCache.put(beanName, object);
                }
            }
        }
        return object;
    }
```

在这段代码中会有四个细节操作,具体对应下面四个方法。

(1) doGetObjectFromFactoryBean。

(2) beforeSingletonCreation。

(3) postProcessObjectFromFactoryBean。

(4) afterSingletonCreation。

首先介绍 doGetObjectFromFactoryBean 方法,在这个方法中对于对象的创建是指调用 FactoryBean#getObject 方法直接获取,当获取失败时会返回 NullBean。其次是 beforeSingletonCreation 方法,在这个方法调用时会做一些数据验证,总共有两个验证,第一个验证是当前需要创建的 BeanName 是否处于排除的容器(inCreationCheckExclusions)中,第二个验证是当前需要创建的 BeanName 是否是正在创建的 Bean,通过 singletonsCurrentlyInCreation 容器判断,相关代码如下。

```
if (!this.inCreationCheckExclusions.contains(beanName)
 && !this.singletonsCurrentlyInCreation.add(beanName)) {
 throw new BeanCurrentlyInCreationException(beanName);
}
```

接下来是 postProcessObjectFromFactoryBean,这个方法的具体处理方法主要还是 BeanPostProcessor 的内容,即调用 BeanPostProcessor#postProcessAfterInitialization,具体实现有下面两个方法。

(1) org.springframework.beans.factory.support.AbstractAutowireCapableBeanFactory#postProcessObjectFromFactoryBean。

(2) org.springframework.beans.factory.support.FactoryBeanRegistrySupport#postProcessObjectFromFactoryBean。

最后是 afterSingletonCreation 方法,在这个方法中会先做数据验证再将对象返回,重点关注数据验证,验证规则有以下两条。

(1) 当前需要创建的 BeanName 是否处于排除的容器(inCreationCheckExclusions)中。
(2) 从 singletonsCurrentlyInCreation 中删除当前正在删除的 BeanName 是否成功。
具体处理代码如下。

```
if(!this.inCreationCheckExclusions.contains(beanName) && !this.
singletonsCurrentlyInCreation.remove(beanName)) {
  throw new IllegalStateException("Singleton '" + beanName + "' isn't currently in creation");
}
```

2. FactoryBean 的原型创建

接下来将介绍原型模式的创建,下面的代码是关于原型模式的创建(获取)。

```
//从 factoryBean 中创建
Object object = doGetObjectFromFactoryBean(factory, beanName);
//判断是否需要后置处理
if (shouldPostProcess) {
    try {
        //后置处理
        object = postProcessObjectFromFactoryBean(object, beanName);
    }
    catch (Throwable ex) {
        throw new BeanCreationException(beanName, "Post-processing of FactoryBean's object failed", ex);
    }
}
return object;
```

从这段代码中可以看到对于原型模式的 Bean 实例创建(获取)是通过 doGetObjectFromFactoryBean 方法获取的,在得到对象后进行数据验证,主体处理方法和单例模式相同。

7.3.4 尝试从父容器中获取

接下来将介绍从父容器中获取 Bean 实例,在 BeanFactory 中可以存在父子 BeanFactory,具体处理代码如下。

```
BeanFactory parentBeanFactory = getParentBeanFactory();
//从父 Bean 工厂中查询
if (parentBeanFactory != null && !containsBeanDefinition(beanName)) {
    //确定 BeanName
    String nameToLookup = originalBeanName(name);
    //父 Bean 工厂类型判断
    if (parentBeanFactory instanceof AbstractBeanFactory) {
        //再次获取
        return ((AbstractBeanFactory) parentBeanFactory).doGetBean(
            nameToLookup, requiredType, args, typeCheckOnly);
    }
    else if (args != null) {
```

```
                return (T) parentBeanFactory.getBean(nameToLookup, args);
            }
            else if (requiredType != null) {
                return parentBeanFactory.getBean(nameToLookup, requiredType);
            }
            else {
                return (T) parentBeanFactory.getBean(nameToLookup);
            }
        }
```

从父容器中获取本质上是同一段代码，只是能力提供者从自身变成了父类，处理逻辑完全相同。

7.3.5 BeanName 标记

接下来将介绍 BeanName 的标记。在 Spring 中进行非 FactoryBean 接口实现类创建（获取）Bean 会有 BeanName 标记这个动作，方法签名为 org.springframework.beans.factory.support.AbstractBeanFactory#markBeanAsCreated，在这个方法中标记了以下两个内容。

（1）将 BeanName 对应的 MergeBeanDefinition 中的 stale 设置为 true。
（2）往 alreadyCreated 容器放入当前的 BeanName。

在这两个标记行为中，第二个标记行为尤为重要，alreadyCreated 中存储的是正在创建的 BeanName，用来避免重复创建。具体处理标记的代码如下。

```
protected void markBeanAsCreated(String beanName) {
    //已创建的 BeanName 是否包含当前 BeanName
    if (!this.alreadyCreated.contains(beanName)) {
        synchronized (this.mergedBeanDefinitions) {
            if (!this.alreadyCreated.contains(beanName)) {
                //将属性 stale 设置为 true
                clearMergedBeanDefinition(beanName);
                //放入已创建集合中
                this.alreadyCreated.add(beanName);
            }
        }
    }
}
```

7.3.6 非 FactoryBean 的单例对象创建

下面将介绍非 FactoryBean 创建模式中关于单例 Bean 的创建，相关代码如下。

```
if (mbd.isSingleton()) {
    //获取单例 Bean
    sharedInstance = getSingleton(beanName, () -> {
        try {
```

```
            //创建 Bean
            return createBean(beanName, mbd, args);
        }
        catch (BeansException ex) {
            //摧毁单例的 Bean
            destroySingleton(beanName);
            throw ex;
        }
    });
    //获取 Bean 实例
    bean = getObjectForBeanInstance(sharedInstance, name, beanName, mbd);
}
```

这段代码是一个关于 Bean 单例模式的创建过程,在这个方法中比较熟悉的方法有 getObjectForBeanInstance。getObjectForBeanInstance 方法是前文提到的 FactoryBean 创建 Bean 实例。在这段代码中需要关注以下三个参数。

(1) mbd:合并之后的 BeanDefinition。

(2) beanName:BeanName。

(3) args:构造函数的参数列表。

除了上述三个参数以外还需要关注执行 createBean 方法失败后(创建 Bean 失败)进行摧毁 Bean 相关的一些信息。

在真正执行单例对象创建之前 Spring 还有部分准备工作。这部分准备工作单例和原型都需要进行。

1. 进行非 FactoryBean 创建之前的通用准备

接下来将介绍非 FactoryBean 创建之前的通用准备,下面是通用准备的代码。

```
//获取 Bean 定义
final RootBeanDefinition mbd = getMergedLocalBeanDefinition(beanName);
checkMergedBeanDefinition(mbd, beanName, args);

//需要依赖的 Bean
String[] dependsOn = mbd.getDependsOn();
if (dependsOn != null) {
    for (String dep : dependsOn) {
        //是否依赖
        if (isDependent(beanName, dep)) {
            throw new BeanCreationException(mbd.getResourceDescription(), beanName,
                    "Circular depends-on relationship between '" + beanName + "' and '" + dep + "'");
        }
        //注册依赖 Bean
        registerDependentBean(dep, beanName);
        try {
            //获取 Bean
            getBean(dep);
        }
        catch (NoSuchBeanDefinitionException ex) {
            throw new BeanCreationException(mbd.getResourceDescription(), beanName,
```

```
                    "'" + beanName + "' depends on missing bean '" + dep + "'", ex);
            }
        }
    }
```

进行非 FactoryBean 创建之前的通用准备分为下面三个步骤。
（1）通过当前的 BeanName 获取对应的 MergeBeanDefinition。
（2）对第一步获取的 MergeBeanDefinition 进行验证。
（3）MergeBeanDefinition 中对于依赖项的处理。

依赖项处理的细节有两个，第一个是注册依赖信息，第二个是创建依赖 Bean 的实例。

首先是 BeanDefinition 的合并操作，合并方法的签名是 org. springframework. beans. factory. support. AbstractBeanFactory # getMergedBeanDefinition（java. lang. String, org. springframework. beans. factory. config. BeanDefinition，org. springframework. beans. factory. config. BeanDefinition）。在这个方法中，重点关注合并的 BeanDefinition 是两个：当前 BeanFactory 中 BeanName 对应的 BeanDefinition 和当前 BeanFactory 的父 BeanFactory 中 BeanName 对应的 BeanDefinition。

另外一个关注点是具体的合并行为，在 Spring 中关于 BeanDefinition 的合并总共有两个方法：由 AbstractBeanDefinition 提供的 overrideFrom 方法和由 AbstractBeanFactory 提供的 copyRelevantMergedBeanDefinitionCaches 方法。

在这两个方法中属于大面积的字段合并处理，其他还有一些细节字段的处理在 org. springframework. beans. factory. support. AbstractBeanFactory # getMergedBeanDefinition（java. lang. String, org. springframework. beans. factory. config. BeanDefinition，org. springframework. beans. factory. config. BeanDefinition）方法中有补充。

接下来要看 BeanDefinition 合并后的验证，具体的验证方法是 org. springframework. beans. factory. support. AbstractBeanFactory # checkMergedBeanDefinition，具体代码如下。

```
protected void checkMergedBeanDefinition(RootBeanDefinition mbd, String beanName, @Nullable Object[] args)
        throws BeanDefinitionStoreException {

    if (mbd.isAbstract()) {
        throw new BeanIsAbstractException(beanName);
    }
}
```

在这段代码中如果 MergeBeanDefinition 对象的属性 abstract 属性为 true，就会抛出 BeanIsAbstractException 异常。

最后是 MergeBeanDefinition 相关的依赖处理，主要处理过程分为两步，第一步注册依赖信息，第二步创建所依赖的 Bean 实例。注册依赖信息就是往容器中放入数据，在这里使用到两个容器，第一个是 dependentBeanMap，第二个是 dependenciesForBeanMap。

2. doCreateBean 方法分析的第一部分——获取实例

接下来将进入 doCreateBean 的分析中，在 Spring 中 AbstractBeanFactory 类定义了抽

象方法 createBean，对应的实现只有一个 org.springframework.beans.factory.support.AbstractAutowireCapableBeanFactory#createBean(java.lang.String,.springframework.beans.factory.support.RootBeanDefinition,java.lang.Object[])。首先阅读 createBean 中的第一部分代码。

```
RootBeanDefinition mbdToUse = mbd;

//获取当前需要加载的类
Class<?> resolvedClass = resolveBeanClass(mbd, beanName);
//待处理的类不为空
//Bean 定义中含有 BeanClass
//ClassName 不为空
//满足上述三点的情况下会去创建 RootBeanDefinition
if (resolvedClass != null && !mbd.hasBeanClass() && mbd.getBeanClassName() != null) {
    //创建 BeanDefinition
    mbdToUse = new RootBeanDefinition(mbd);
    //设置 Bean Class
    mbdToUse.setBeanClass(resolvedClass);
}
```

在第一部分代码中主要是对 BeanDefinition 对象的准备，BeanDefinition 对象的定义存在以下两种情况。

(1) 直接使用参数作为后续操作对象（RootBeanDefinition mbdToUse＝mbd）。

(2) 满足下面三个条件会重新生成一个 BeanDefinition 对象进行后续使用。

① 当前 BeanName 对应的 BeanType 不为空，可以理解为 bean 标签中 class 属性存在。

② BeanDefinition 中 BeanClass 的类型是 Class。

③ BeanDefinition 中 BeanClass Name 存在。

在这个处理过程中还可以对 resolveBeanClass 方法进行关注，该方法是用来解析 BeanClass 的，主要推论方式有下面两个行为。

(1) 当 BeanDefinition 中存在 BeanClass 并且类型是 Class 时，立即使用该对象。

(2) 当 BeanDefinition 中 BeanClass 的类型是 String 时，Spring 会进行类加载，将字符串转换成具体的 Class 对象。

createBean 的第一部分分析结束，下面是第二部分的分析，首先是第二部分的代码。

```
//第二部分
try {
    //方法重写
    mbdToUse.prepareMethodOverrides();
}
catch (BeanDefinitionValidationException ex) {
    throw new BeanDefinitionStoreException(mbdToUse.getResourceDescription(),
            beanName, "Validation of method overrides failed", ex
    );
}
```

在这一部分中主要是对 BeanDefinition 对象中关于需要重写方法的处理，具体处理就是将 MethodOverride 的属性 overloaded 设置为 false。

第二部分中出现的一些核心方法和类如下。

（1）org.springframework.beans.factory.support.AbstractBeanDefinition#prepareMethodOverrides。

（2）org.springframework.beans.factory.support.AbstractBeanDefinition#prepareMethodOverride。

对象：org.springframework.beans.factory.support.MethodOverride。

在第二部分中实际处理操作不多，下面是第三部分代码。

```
try {
 //创建 Bean 之前的行为
 Object bean = resolveBeforeInstantiation(beanName, mbdToUse);
 if (bean != null) {
     return bean;
 }
}
catch (Throwable ex) {
 throw new BeanCreationException(mbdToUse.getResourceDescription(), beanName,
     "BeanPostProcessor before instantiation of bean failed", ex
 );
}
```

在第三部分中主要是关于 BeanPostProcessor 接口的相关操作，在 resolveBeforeInstantiation 方法中主要有以下两个关于 BeanPostProcessor 相关的操作。

（1）将容器中所有的 BeanPostProcessor 找出并且进行进一步类型确定，执行类型是 InstantiationAwareBeanPostProcessor 的前置方法 postProcessBeforeInstantiation。

（2）将容器中所有的 BeanPostProcessor 找出，然后循环每个 BeanPostProcessor 执行后置方法调用 postProcessAfterInitialization。

此时如果可以获取到 Bean 实例那么就会直接返回，如果 Bean 实例获取失败则会进入第四部分代码。

```
//第四部分
try {
  //创建 Bean
  Object beanInstance = doCreateBean(beanName, mbdToUse, args);
  if (logger.isTraceEnabled()) {
      logger.trace("Finished creating instance of bean '" + beanName + "'");
  }
  return beanInstance;
}
catch (BeanCreationException | ImplicitlyAppearedSingletonException ex) {
  throw ex;
}
catch (Throwable ex) {
  throw new BeanCreationException(
      mbdToUse.getResourceDescription(),beanName,"Unexpected exception during bean creation", ex);
}
```

在这段代码中核心就是 doCreateBean 了，经过了前文的三个操作终于到了这个方法，首先来看 doCreateBean 的第一部分代码。

```
//第一部分：获取实例
BeanWrapper instanceWrapper = null;
//是否单例
if (mbd.isSingleton()) {
    //BeanFactory 移除当前创建的 BeanName
    instanceWrapper = this.factoryBeanInstanceCache.remove(beanName);
}
//BeanWrapper 是否存在
if (instanceWrapper == null) {
    //创建 Bean 实例
    instanceWrapper = createBeanInstance(beanName, mbd, args);
}
```

这段代码的主要目的是获取 BeanWrapper 对象，获取方式有两种：第一种是 BeanDefinition 单例，从 factoryBeanInstanceCache 对象中移除当前正在处理的 BeanName 作为 BeanWrapper；第二种是在第一种操作的基础上没有获取到 BeanWrapper 则会进行 Bean 实例的创建，具体方法是 createBeanInstance。在 createBeanInstance 方法中主要执行了下面几个事项。

（1）进行 BeanClass 推论，确定 BeanClass。

（2）尝试从 BeanDefinition 中获取 Supplier 接口，并从 Supplier 接口中获取 Bean 实例。获取 BeanWrapper 的方式是调用 Supplier 接口的 get 方法再包装成 BeanWrapperImpl 对象。

（3）根据 FactoryBean＋FactoryMethod 创建 Bean，具体处理如下。

① 创建 BeanWrapperImpl 对象。

② 读取 BeanDefinition 中关于 FactorBeanName 的属性值，并将其转换成 Java 对象。

③ 通过第二步得到的 Java 对象执行 FactorMethod 获取。

（4）进行标记，主要标记数据有 autowireNecessary（含义是是否需要自动注入）。

（5）执行两个方法 autowireConstructor 和 instantiateBean，这两个方法都能够用来创建实例，前者是根据构造函数进行创建，后者是根据 Class 对象进行创建。

下面着重分析 autowireConstructor 创建 Bean 的过程。首先找到该方法的入口，方法签名为 org.springframework.beans.factory.support.ConstructorResolver#autowireConstructor，第一步认识参数，参数有以下四个。

（1）beanName：BeanName。

（2）mbd：合并后的 BeanDefinition。

（3）chosenCtors：可能的构造器列表。

（4）explicitArgs：构造参数列表。

在方法中关键调用是 instantiate 方法，它有两种调用行为：

```
instantiate(beanName, mbd, uniqueCandidate, EMPTY_ARGS)
instantiate(beanName, mbd, constructorToUse, argsToUse)
```

在这两个调用行为中都需要用到前面的四个参数，调用时对于各类参数的确认很重要，首先需要推论构造参数列表。构造参数的解析有两种：字面量，对于字面量的处理可以简单理解为从 BeanDefinition 对象中获取构造标签相关的数据即可；外部链接对象，对于外部链接对象可以在第一种处理模式的基础上添加一层通过 BeanId 或者 BeanName 获取 Bean 实例的过程。

完成构造函数的参数列表确认后需要确认构造函数，具体流程如下。

（1）在参数 chosenCtors 为 null 的情况下，Spring 会通过 BeanDefinition 中 BeanClass 来进行构造函数列表的获取，对应处理代码如下。

```
Constructor<?>[] candidates = chosenCtors;
//如果构造函数集合为空，直接获取 BeanClass 中的构造函数列表作为可选列表
if (candidates == null) {
    Class<?> beanClass = mbd.getBeanClass();
    candidates = (mbd.isNonPublicAccessAllowed() ?
            beanClass.getDeclaredConstructors() : beanClass.getConstructors());
}
```

（2）当候选的构造函数列表只有一个并且没有参数时则直接执行，对应处理代码如下。

```
if (candidates.length == 1
&& explicitArgs == null && !mbd.hasConstructorArgumentValues()) {
    Constructor<?> uniqueCandidate = candidates[0];
    if (uniqueCandidate.getParameterCount() == 0) {
        synchronized (mbd.constructorArgumentLock) {
            mbd.resolvedConstructorOrFactoryMethod = uniqueCandidate;
            mbd.constructorArgumentsResolved = true;
            mbd.resolvedConstructorArguments = EMPTY_ARGS;
        }
        bw.setBeanInstance(instantiate(beanName, mbd, uniqueCandidate, EMPTY_ARGS));
        return bw;
    }
}
```

（3）当构造函数列表中存在多个构造器的处理时，主要根据构造函数中参数列表的参数类型确认具体的某一个构造器，具体处理代码如下。

```
for (Constructor<?> candidate : candidates) {

    //构造函数的参数长度
    int parameterCount = candidate.getParameterCount();

    if (constructorToUse != null
&& argsToUse != null && argsToUse.length > parameterCount) {
        break;
    }
    //当前构造函数的参数长度小于最小构造函数的参数长度，不进行处理
    if (parameterCount < minNrOfArgs) {
        continue;
    }

    ArgumentsHolder argsHolder;
    //构造函数中关于类型和参数名称的处理
```

```java
        Class<?>[] paramTypes = candidate.getParameterTypes();
        if (resolvedValues != null) {
            try {
                String[] paramNames = ConstructorPropertiesChecker.evaluate(candidate, parameterCount);
                if (paramNames == null) {
                    ParameterNameDiscoverer pnd = this.beanFactory.getParameterNameDiscoverer();
                    if (pnd != null) {
                        paramNames = pnd.getParameterNames(candidate);
                    }
                }
                argsHolder = createArgumentArray(beanName, mbd, resolvedValues, bw, paramTypes, paramNames,

                        getUserDeclaredConstructor(candidate), autowiring, candidates.length == 1);
            }
            catch (UnsatisfiedDependencyException ex) {
                if (logger.isTraceEnabled()) {
                    logger.trace("Ignoring constructor [" + candidate + "] of bean '" + beanName + "': " + ex);
                }
                if (causes == null) {
                    causes = new LinkedList<>();
                }
                causes.add(ex);
                continue;
            }
        }
        else {
            if (parameterCount != explicitArgs.length) {
                continue;
            }
            argsHolder = new ArgumentsHolder(explicitArgs);
        }

        int typeDiffWeight = (mbd.isLenientConstructorResolution() ?
                argsHolder.getTypeDifferenceWeight(paramTypes) : argsHolder.getAssignabilityWeight(paramTypes));

        //确认使用的参数列表
        if (typeDiffWeight < minTypeDiffWeight) {
            constructorToUse = candidate;
            argsHolderToUse = argsHolder;
            argsToUse = argsHolder.arguments;
            minTypeDiffWeight = typeDiffWeight;
            ambiguousConstructors = null;
        }
```

```
            else if (constructorToUse != null && typeDiffWeight == minTypeDiffWeight) {
                if (ambiguousConstructors == null) {
                    ambiguousConstructors = new LinkedHashSet<>();
                    ambiguousConstructors.add(constructorToUse);
                }
                ambiguousConstructors.add(candidate);
            }
        }
```

最后当构造函数(构造器)、构造参数列表都准备完毕就可以开始创建 Bean。在创建 Bean 的时候有两种情况:非 CGLIB 对象创建,CGLIB 对象创建。这部分创建是交给 InstantiationStrategy 接口进行。

在 createBeanInstance 方法中应该关注的点是 Bean 实例创建的方法,如下所示。

(1) 通过 BeanDefinition 中 Supplier<?> instanceSupplier 属性来获取 Bean 实例。

(2) 通过 BeanDefinition 中 String factoryBeanName 和 String factoryMethodName 来获取 Bean 实例。

(3) 通过 autowireConstructor 方法,内部依赖 Constructor 构造 Bean 实例。

(4) 通过 instantiateBean 方法处理有以下两种形式。

① 内部依赖 BeanClass 找构造函数对象 Constructor 进行反射构造。

② CGLIB 代理,内部依赖还是 Constructor,不过此时的 Constructor 并不是原始的 BeanClass 所拥有的 Constructor 对象,而是通过 org.springframework.beans.factory.support.CglibSubclassingInstantiationStrategy.CglibSubclassCreator # createEnhancedSubclass 创建出来的 Class 拥有的 Constructor 对象。

3. doCreateBean 方法分析的第二部分——后置方法调用

接下来进入 doCreateBean 方法的第二部分——后置方法调用的分析,相关代码如下。

```
//第二部分:后置方法调用
//获取实例
final Object bean = instanceWrapper.getWrappedInstance();
//BeanWrapper 中存储的实例.class
Class<?> beanType = instanceWrapper.getWrappedClass();
if (beanType != NullBean.class) {
    mbd.resolvedTargetType = beanType;
}

synchronized (mbd.postProcessingLock) {
    if (!mbd.postProcessed) {
        try {
            //后置方法执行 BeanPostProcessor -> MergedBeanDefinitionPostProcessor
            applyMergedBeanDefinitionPostProcessors(mbd, beanType, beanName);
        }
        catch (Throwable ex) {
            throw new BeanCreationException(mbd.getResourceDescription(), beanName,
                "Post-processing of merged bean definition failed", ex
            );
        }
```

```
        mbd.postProcessed = true;
    }
}
```

在第二部分中主要操作的是容器中 BeanPostProcessor 的内容,注意这里主要处理 MergedBeanDefinitionPostProcessor 相关内容,在这个处理中只会处理 MergedBeanDefinitionPostProcessor 的方法,其他 BeanPostProcessor 不会进行处理。

4. doCreateBean 方法分析的第三部分——早期单例 Bean 的暴露

下面将进入 doCreateBean 方法的第三部分——早期单例 Bean 的暴露,相关代码如下。

```
//第三部分
//是否需要提前暴露
boolean earlySingletonExposure = (mbd.isSingleton() && this.allowCircularReferences &&
        isSingletonCurrentlyInCreation(beanName));

//单例对象暴露
if (earlySingletonExposure) {
    if (logger.isTraceEnabled()) {
        logger.trace("Eagerly caching bean '" + beanName +
                "' to allow for resolving potential circular references");
    }
    //添加单例工厂
    addSingletonFactory(beanName, () -> getEarlyBeanReference(beanName, mbd, bean));
}
```

这部分操作是为了循环依赖处理做准备,首先来看 earlySingletonExposure 变量的生成条件。

(1) BeanDefinition 中定义了这是一个单例的 Bean,当前 Bean 是一个单例 Bean。

(2) allowCircularReferences 是否为 true,这个数据表示了是否允许循环依赖,默认 true。

(3) isSingletonCurrentlyInCreation 方法判断当前 Bean 是否正处于创建状态。

继续向下分析两个方法,一个方法是 getEarlyBeanReference,另一个方法是 addSingletonFactory。

首先分析 getEarlyBeanReference 方法。在 getEarlyBeanReference 方法中主要操作目标是获取 Bean 对象,在这个方法中没有特别的处理,对于需要提前暴露的 Bean 实例可以分为下面两种情况。

(1) 在容器中没有 SmartInstantiationAwareBeanPostProcessor,实现直接使用参数中的 Bean 实例。

(2) 在容器中存在 SmartInstantiationAwareBeanPostProcessor,通过 SmartInstantiationAwareBeanPostProcessor 提供的方法进行获取。

getEarlyBeanReference 处理代码如下。

```
protected Object getEarlyBeanReference(String beanName, RootBeanDefinition mbd, Object bean) {
    Object exposedObject = bean;
```

```
        if (!mbd.isSynthetic() && hasInstantiationAwareBeanPostProcessors()) {
            for (BeanPostProcessor bp : getBeanPostProcessors()) {
                if (bp instanceof SmartInstantiationAwareBeanPostProcessor) {
                    SmartInstantiationAwareBeanPostProcessor ibp =
                    (SmartInstantiationAwareBeanPostProcessor) bp;
                    exposedObject = ibp.getEarlyBeanReference(exposedObject, beanName);
                }
            }
        }
        return exposedObject;
    }
```

下面分析 addSingletonFactory 方法,具体处理代码如下。

```
protected void addSingletonFactory(String beanName, ObjectFactory<?> singletonFactory) {
    Assert.notNull(singletonFactory, "Singleton factory must not be null");
    synchronized (this.singletonObjects) {
        //单例 Bean 容器中是否存在
        if (!this.singletonObjects.containsKey(beanName)) {
            //添加单例对象工厂
            this.singletonFactories.put(beanName, singletonFactory);
            //删除提前暴露的 BeanName
            this.earlySingletonObjects.remove(beanName);
            //注册单例 BeanName
            this.registeredSingletons.add(beanName);
        }
    }
}
```

在这段代码中只需要关注各类容器的存储结构即可,方法就是向容器中添加数据,容器信息见表 7.1。

表 7.1 容器信息

容器名称	存储结构	说明
singletonFactories	Map < String, ObjectFactory<?>>	key:BeanName,value:对象工厂,存储 BeanName 和对象工厂的绑定关系
singletonObjects	Map < String, Object >	key:BeanName,value:Bean 实例,存储 BeanName 和 Bean 实例的绑定关系
earlySingletonObjects	Map < String, Object >	key:BeanName,value:提前暴露的 Bean 实例
registeredSingletons	Set < String >	存储注册过的单例 Bean 对应的 Bean Name

在表 7.1 中需要关注它们存在递进关系,singletonFactories 是最开始的对象,对象还并没有从中提取,当数据提取后会先放入 singletonObjects,再放入 earlySingletonObjects。注意可以不用完整地依次放入 singletonObjects 和 earlySingletonObjects。

5. doCreateBean 方法分析的第四部分——属性设置、InitializingBean 和 init-method 调用

下面介绍 doCreateBean 方法的第四部分，具体处理代码如下。

```
//第四部分
//实例化 Bean
//Initialize the bean instance.
Object exposedObject = bean;
try {
    //设置属性
    populateBean(beanName, mbd, instanceWrapper);
    //实例化 Bean
    exposedObject = initializeBean(beanName, exposedObject, mbd);
}
catch (Throwable ex) {
    if (ex instanceof BeanCreationException && beanName.equals(((BeanCreationException) ex).getBeanName()))  {
        throw (BeanCreationException) ex;
    }
    else {
        throw new BeanCreationException(
                mbd.getResourceDescription(), beanName, "Initialization of bean failed", ex);
    }
}
```

在第四部分处理过程中主要是 Bean 生命周期的两个换阶属性设置和实例化 Bean。这部分内容可以参考第 8 章。

6. doCreateBean 方法分析的第五部分——依赖处理

接下来介绍 doCreateBean 方法的第五部分，具体代码如下。

```
if (earlySingletonExposure) {
    Object earlySingletonReference = getSingleton(beanName, false);
    if (earlySingletonReference != null) {
        if (exposedObject == bean) {
            exposedObject = earlySingletonReference;
        }
        else if (!this.allowRawInjectionDespiteWrapping && hasDependentBean(beanName)) {
            //当前 Bean 的依赖列表
            String[] dependentBeans = getDependentBeans(beanName);
            //当前 Bean 的依赖列表
            Set<String> actualDependentBeans =
new LinkedHashSet<>(dependentBeans.length);
            for (String dependentBean : dependentBeans) {
                if (!removeSingletonIfCreatedForTypeCheckOnly(dependentBean)) {
                    actualDependentBeans.add(dependentBean);
                }
            }
            if (!actualDependentBeans.isEmpty()) {
                throw new BeanCurrentlyInCreationException(
```

```
                        beanName,
                        "Bean with name '" + beanName + "' has been injected into other beans [" +
                        StringUtils.collectionToCommaDelimitedString(actualDependentBeans) +
                            "] in its raw version as part of a circular reference, but has
                            eventually been " + "wrapped. This means that said other beans do
                            not use the final version of the " + "bean. This is often the result of
                            over-eager type matching - consider using " + "'getBeanNamesOfType'
                            with the 'allowEagerInit' flag turned off, for example."
                    );
                }
            }
        }
    }
```

对于这部分依赖处理的关键方法是 removeSingletonIfCreatedForTypeCheckOnly，在 removeSingletonIfCreatedForTypeCheckOnly 方法中会移除单例容器中的部分数据，这部分数据是指当前 Bean 所依赖的 BeanName。Bean 创建时会创建对于 Bean 所依赖的 BeanName，在创建阶段使用完之后会被移除出容器。

7. doCreateBean 方法分析的第六部分——注册摧毁 Bean 信息

接下来介绍 doCreateBean 方法的第六部分，具体操作代码如下。

```
//第六部分
//Register bean as disposable.
try {
    //注册 Bean(一次性的 Bean)
    registerDisposableBeanIfNecessary(beanName, bean, mbd);
}
catch (BeanDefinitionValidationException ex) {
    throw new BeanCreationException(
            mbd.getResourceDescription(), beanName, "Invalid destruction signature", ex);
}
```

在 Spring 的管理中，Bean 的生命周期有些情况下并不是被 Spring 完整管理的，摧毁 Bean 这一生命周期只有单例对象拥有。

8. getSingleton 方法分析

下面将进行 getSingleton 方法分析，方法代码如下。

```
public Object getSingleton(String beanName, ObjectFactory<?> singletonFactory) {
    Assert.notNull(beanName, "Bean name must not be null");
    synchronized (this.singletonObjects) {
        //从单例对象缓存中获取
        Object singletonObject = this.singletonObjects.get(beanName);
        if (singletonObject == null) {
            if (this.singletonsCurrentlyInDestruction) {
                throw new BeanCreationNotAllowedException(beanName,
                        "Singleton bean creation not allowed while singletons of this factory are in
destruction " + "(Do not request a bean from a BeanFactory in a destroy method implementation!)");
            }
```

```java
            if (logger.isDebugEnabled()) {
                logger.debug("Creating shared instance of singleton bean '" + beanName + "'");
            }
            //单例创建前的验证
            beforeSingletonCreation(beanName);
            //是否是新的单例对象
            boolean newSingleton = false;
            //是否存在异常
            boolean recordSuppressedExceptions = (this.suppressedExceptions == null);
            if (recordSuppressedExceptions) {
                this.suppressedExceptions = new LinkedHashSet<>();
            }
            try {
                //从 ObjectFactory 中获取
                singletonObject = singletonFactory.getObject();
                newSingleton = true;
            }
            catch (IllegalStateException ex) {
                singletonObject = this.singletonObjects.get(beanName);
                if (singletonObject == null) {
                    throw ex;
                }
            }
            catch (BeanCreationException ex) {
                if (recordSuppressedExceptions) {
                    for (Exception suppressedException : this.suppressedExceptions) {
                        ex.addRelatedCause(suppressedException);
                    }
                }
                throw ex;
            }
            finally {
                if (recordSuppressedExceptions) {
                    this.suppressedExceptions = null;
                }
                //创建单例对象后的验证
                afterSingletonCreation(beanName);
            }
            if (newSingleton) {
                //添加到单例容器中
                addSingleton(beanName, singletonObject);
            }
        }
        return singletonObject;
    }
}
```

在这个方法中处理了两种情况：第一种是当单例对象容器中存在 BeanName 对应的 Bean 实例，在方法处理中会直接返回。第二种是当单例对象容器中不存在 BeanName 对应的 Bean 实例，则会做出更加详细的处理，处理分为四个阶段：第一阶段，单例 Bean 创建之前的验证，具体处理方法是 beforeSingletonCreation；第二阶段，从 ObjectFactory 中获取 Bean 实例，具体处理方法是 singletonFactory.getObject；第三阶段，单例 Bean 创建之后的验证，具体处理方法是 afterSingletonCreation；第四阶段，放入单例 Bean 存储容器中，具体处理方法是 addSingleton。

在 getSingleton 方法中还有一些方法需要进行分析，先对 beforeSingletonCreation 方法进行分析，该方法是在创建之前进行验证，具体验证代码如下。

```
protected void beforeSingletonCreation(String beanName) {
    if (!this.inCreationCheckExclusions.contains(beanName)
    && !this.singletonsCurrentlyInCreation.add(beanName)) {
        throw new BeanCurrentlyInCreationException(beanName);
    }
}
```

这段代码中做了两个验证，第一个是当前 BeanName 是否在排除的 BeanName 列表中出现，第二个是正在创建的 BeanName 列表中是否可以移除当前的 BeanName。

接下来将介绍 addSingleton 方法，该方法是将 BeanName 和 Bean 实例添加到单例容器中，处理代码如下。

```
protected void addSingleton(String beanName, Object singletonObject) {
    synchronized (this.singletonObjects) {
        //设置单例对象 Map
        this.singletonObjects.put(beanName, singletonObject);
        //删除单例的 BeanFactory
        this.singletonFactories.remove(beanName);
        //删除早期加载的 Bean
        this.earlySingletonObjects.remove(beanName);
        //放入已注册的 BeanName
        this.registeredSingletons.add(beanName);
    }
}
```

在这段代码中的主要操作是对于 Java 集合进行操作，在 addSingleton 使用的容器信息见表 7.2。

表 7.2 addSingleton 使用的容器

容 器 名 称	存 储 结 构	说 明
singletonFactories	Map < String , ObjectFactory <?>>	存储 BeanName 和 对象工厂的绑定关系
singletonObjects	Map < String , Object >	存储 BeanName 和 Bean 实例的绑定关系
earlySingletonObjects	Map < String , Object >	存储提前暴露的 Bean 实例
registeredSingletons	Set < String >	存储注册过的单例 Bean 对应的 BeanName

7.3.7 非 FactoryBean 的原型对象创建

接下来将对原型模式的创建做分析，首先阅读原型模式的处理代码。

```
else if (mbd.isPrototype()) {
    Object prototypeInstance = null;
    try {
        //创建之前的行为
        beforePrototypeCreation(beanName);
        //创建
        prototypeInstance = createBean(beanName, mbd, args);
    }
    finally {
        //创建之后的行为
        afterPrototypeCreation(beanName);
    }
    //创建
    bean = getObjectForBeanInstance(prototypeInstance, name, beanName, mbd);
}
```

在这段代码中可以看到很多熟悉的方法，这些方法在单例模式创建过程中也有使用，具体有三个方法：beforePrototypeCreation、createBean、afterPrototypeCreation，这三个方法在前文已有介绍和分析，这里不再赘述。

7.3.8　既不是单例模式也不是原型模式的非 FactoryBean 创建

接下来看最后一种情况，创建既不是单例模式也不是原型模式的非 FactoryBean，在这个模式下会采用 scope 进行处理，处理代码如下。

```
else {
    //获取作用域名称
    String scopeName = mbd.getScope();
    //从作用域容器中获取当前作用域名称对应的作用域接口 scope
    final Scope scope = this.scopes.get(scopeName);
    if (scope == null) {
        throw new IllegalStateException("No Scope registered for scope name '" + scopeName + "'");
    }
    try {
        //从 scope 接口获取
        Object scopedInstance = scope.get(beanName, () -> {
            //创建之前的行为
            beforePrototypeCreation(beanName);
            try {
                //创建 Bean
                return createBean(beanName, mbd, args);
            }
            finally {
                //创建之后的行为
                afterPrototypeCreation(beanName);
            }
        });
        //获取 Bean 实例
        bean = getObjectForBeanInstance(scopedInstance, name, beanName, mbd);
```

```
        }
        catch (IllegalStateException ex) {
            throw new BeanCreationException(beanName,
                    "Scope '" + scopeName + "' is not active for the current thread; consider " +
                            "defining a scoped proxy for this bean if you intend to refer to it from a singleton",ex);
        }
    }
}
```

在这段代码处理过程中,所使用到的方法在单例模式中也都有涉及,这里不再赘述。

7.3.9 类型转换器中获取 Bean

下面将介绍另一种获取 Bean 实例的方式:通过类型转换器获取 Bean 实例,具体处理代码如下。

```
if (requiredType != null && !requiredType.isInstance(bean)) {
    try {
        //获取类型转换器,通过类型转换器进行转换
        T convertedBean = getTypeConverter().convertIfNecessary(bean, requiredType);
        if (convertedBean == null) {
            throw new BeanNotOfRequiredTypeException(name, requiredType, bean.getClass());
        }
        return convertedBean;
    }
    catch (TypeMismatchException ex) {
        if (logger.isTraceEnabled()) {
            logger.trace("Failed to convert bean '" + name + "' to required type '" +
                    ClassUtils.getQualifiedName(requiredType) + "'", ex);
        }
        throw new BeanNotOfRequiredTypeException(name, requiredType, bean.getClass());
    }
}
```

在这段代码中引出了 TypeConverter 接口,在这个接口中有具体的处理方案。

对获取 Bean 做一个总结:首先需要确认入口,通常是以 getBean 方法作为入口,从这个入口继续挖掘可以找到最终的处理方法 org.springframework.beans.factory.support.AbstractBeanFactory#doGetBean,在这个方法中有下面几个操作。

(1) BeanName 转换。
(2) 从单例容器中获取,如果存在则返回 Bean 实例。
(3) 从 FactoryBean 接口中获取。
(4) 从父容器中获取(父 BeanFactory)。
(5) 非 FactoryBean 中获取,Spring 提供了三种处理:
 ① 单例 Bean 获取。
 ② 原型 Bean 获取。
 ③ 第一种和第二种以外的处理。

(6) 通过 TypeConverter 接口转换得到 Bean 实例。

上述六步就是 Spring 获取 Bean 实例的核心操作。

7.4 循环依赖

接下来将介绍 Spring 中的另一个重点：循环依赖。首先需要理解循环依赖，循环依赖是指需要互相使用对方所提供的能力（方法）。

7.4.1 Java 中的循环依赖

下面将介绍 Java 中的循环依赖。首先搭建一个测试环境，在这个测试环境搭建过程中需要创建两个对象，并让它们的成员变量互为彼此，这两个对象的定义代码如下。

```
class A {
public B innerB;
}

class B {
public A innerA;
}
```

此时 A 类和 B 类的依赖关系如图 7.1 所示。

在依赖关系确定后编写具体的测试方法，测试方法为 testCircularDependenceInJava。

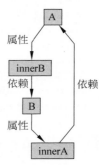

图 7.1 A 类和 B 类的依赖关系

```
@Test
void testCircularDependenceInJava() {

    A a = new A();
    B b = new B();
    a.innerB = b;
    b.innerA = a;
    System.out.println();
}
```

在这个测试方法中将两个对象的成员变量互相指向了彼此，通过这种方式来做一个简单的循环依赖模拟。这是一种最为基础的循环依赖处理，但是在实际操作中开发者应该不希望通过这样的方式每次都手动创建两个对象然后再进行数据设置，接下来对这个处理逻辑进行改进。

首先需要准备一个对象的存储容器，存储容器中 key 为类名，value 为对象实例，具体定义代码如下。

```
static Map<String, Object> nameMappingObjectInstance = new HashMap<>();
```

准备完成容器后需要编写一个获取实例的方法，获取实例的方法直接采用 Class 的 newInstance 进行获取，也可以使用寻找构造器的方式获取实例对象，这里选择的是通过寻找无参构造再进行实例化，具体代码如下。

```java
@NotNull
private static <T> T getInstance(Class<T> clazz) throws Exception {
    return clazz.getDeclaredConstructor().newInstance();
}
```

在获取完成对象后需要做属性设置，在设置属性时该属性可能是已经实例化的对象，即在容器中存在，此时可以从容器中直接获取。接下来修改 getBean 方法，修改后代码如下。

```java
public static <T> T getBean(Class<T> clazz) throws Exception {
    String className = clazz.getName();
    //容器中存在,直接获取
    if (nameMappingObjectInstance.containsKey(className)) {
        return (T) nameMappingObjectInstance.get(className);
    }
    //容器中不存在,手动创建对象
    //通过无参构造创建
    T objectInstance = getInstance(clazz);
    //存入容器
    nameMappingObjectInstance.put(className, objectInstance);
    //设置创建对象的数据
    setProperty(objectInstance);
    return objectInstance;
}
```

在这段代码中对于属性设置采取了一个新的方法 setProperty，该方法具体代码如下。

```java
private static <T> void setProperty(T objectInstance) throws Exception {
    Field[] fields = objectInstance.getClass().getDeclaredFields();
    for (Field field : fields) {
        field.setAccessible(true);
        //获取属性类型
        Class<?> fieldType = field.getType();
        String fieldClassName = fieldType.getName();
        //从容器中获取
        Object cache = nameMappingObjectInstance.get(fieldClassName);
        //容器中存在
        if (cache != null) {
            field.set(objectInstance, cache);
        }
        //容器中不存在
        else {
            Object bean = getBean(fieldType);
            field.set(objectInstance, bean);
        }

    }
}
```

这段代码的主要目的是进行属性设置，在这里会出现一个分支，即当前所需要的属性对象是否已经在容器中存在，这里就需要依赖 getBean 方法了。通过编写上述代码，在 Java 中处理循环依赖的简单版本已经完成，下面编写测试方法，方法为 testCircularDependenceInJavaAuto，具体代码如下。

```
@Test
void testCircularDependenceInJavaAuto() throws Exception {
    A aBean = getBean(A.class);
    B bBean = getBean(B.class);

    System.out.println();
}
```

在 Java 中处理循环依赖的流程如图 7.2 所示。

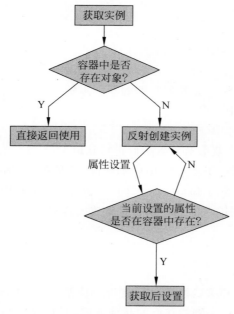

图 7.2　Java 中循环依赖的处理流程

7.4.2　Spring 中的循环依赖处理

前文介绍了 Java 中对于循环依赖的处理方式，下面将深入 Spring 了解 Spring 中是如何处理循环依赖的。在 Spring 中对于循环依赖的处理代码如下。

```
@Nullable
protected Object getSingleton(String beanName, boolean allowEarlyReference) {
    //尝试从单例缓存中获取
    Object singletonObject = this.singletonObjects.get(beanName);
    //单例对象是否 null
    //这个 BeanName 是否正在创建
    if (singletonObject == null && isSingletonCurrentlyInCreation(beanName)) {
        //锁
        synchronized (this.singletonObjects) {
```

```
        //从早期加载的 Map 中获取
        singletonObject = this.earlySingletonObjects.get(beanName);
        //对象是否空,是否允许提前应用
        if (singletonObject == null && allowEarlyReference) {
            //从对象工厂 Map 中获取对象工厂
            ObjectFactory<?> singletonFactory = this.singletonFactories.get(beanName);
            if (singletonFactory != null) {
                //对象获取后设置
                singletonObject = singletonFactory.getObject();
                this.earlySingletonObjects.put(beanName, singletonObject);
                this.singletonFactories.remove(beanName);
            }
        }
    }
    return singletonObject;
}
```

在这段代码中讲述了 Spring 获取实例的三个步骤。

（1）从单例对象容器中获取。

（2）从提前暴露的容器中获取。

（3）从对象工厂容器中获取对象工厂后进行获取。

在这段代码中需要关注 ObjectFactory 的来源,它是在 doCreateBean 中设置的,具体处理代码是 addSingletonFactory(beanName,() -> getEarlyBeanReference(beanName,mbd,bean)),通过这段方法的调用将提前暴露的对象放入容器中。

小结

至此,对于 Bean 获取相关的源码分析就告一段落了,在 Bean 获取的源码分析中主要探讨了 Spring 中对于 Bean 实例的获取的几种方式,并且对这些方式进行了分析。在了解获取方式之后进一步讨论了关于循环依赖的处理,从 Java 的一个基本循环依赖模型出发,简单实现了一个循环依赖处理的方法,完成 Java 中对循环依赖的处理后进一步分析了 Spring 中对于循环依赖的处理。

第8章 SpringXML模式下容器的生命周期

第 6 章对 Bean 的生命周期进行了相关分析,同样地,在 Spring 中 Spring 容器也具备生命周期。第 7 章对 Bean 的获取进行了相关分析,在 Spring 中 Bean 的存在依托于 Spring 容器。本章将分析 SpringXML 模式下容器的生命周期相关内容,主要介绍 SpringXML 的三个核心类 XmlBeanFactory、FileSystemXmlApplicationContext 和 ClassPathXmlApplicationContext。

8.1 SpringXML 模式下容器的生命周期测试环境搭建

首先需要对 SpringXML 模式下容器的生命周期进行测试环境搭建,该部分测试代码如下。

```
class FirstIoCDemoTest {
    @Test
    void testClassPathXmlApplicationContext() {
        ClassPathXmlApplicationContext context
                = new ClassPathXmlApplicationContext("META-INF/first-ioc.xml");

        PeopleBean people = context.getBean("people", PeopleBean.class);

        String name = people.getName();
        assumeTrue(name.equals("zhangsan"));
    }


    @Test
    void testFileSystemXmlApplicationContext() {
        FileSystemXmlApplicationContext context =
   new FileSystemXmlApplicationContext("D:\\desktop\\git_repo\\spring-ebk\\spring-framework-read\\spring-source-hot-ioc-book\\src\\test\\resources\\META-INF\\first-ioc.xml");
```

```java
        PeopleBean people = context.getBean("people", PeopleBean.class);

        String name = people.getName();
        assumeTrue(name.equals("zhangsan"));
        context.close();
    }

    @Test
    void testXmlBeanFactory() {
        XmlBeanFactory beanFactory =
                new XmlBeanFactory(new ClassPathResource("META-INF/first-ioc.xml"));

        PeopleBean people = beanFactory.getBean("people", PeopleBean.class);

        String name = people.getName();
        assumeTrue(name.equals("zhangsan"));
    }
}
```

对于三个核心类的分析将围绕它们的构造方法和摧毁方法进行。

8.2　XmlBeanFactory 分析

下面将对 XmlBeanFactory 对象的生命周期进行分析。生命周期是指从零到一再从一到零的过程，即创建对象-摧毁对象的过程。首先分析创建过程，即 XmlBeanFactory 对象的构造函数。XmlBeanFactory 构造函数的使用代码如下。

```java
XmlBeanFactory beanFactory = new XmlBeanFactory(new ClassPathResource("META-INF/first-ioc.xml"));
```

从这个方法进入真正的构造函数中，XmlBeanFactory 的构造函数如下。

```java
public XmlBeanFactory(Resource resource) throws BeansException {
    this(resource, null);
}
public XmlBeanFactory(Resource resource, BeanFactory parentBeanFactory) throws BeansException {
    super(parentBeanFactory);
    this.reader.loadBeanDefinitions(resource);
}
```

在这个构造函数中需要关注两个变量：第一个变量是 resource，表示一个资源对象；第二个变量是 parentBeanFactory，表示父容器（父 Bean 工厂）。在测试用例中使用 ClassPathResource 作为 resource 对象，该变量主要存储 SpringXML 的路径信息；parentBeanFactory 在测试用例中并没有具体的指向，传递的是 null。在第二个构造函数中有调用 super 方法，它的调用堆栈如下。

（1）XmlBeanFactory。

(2) DefaultListableBeanFactory。
(3) AbstractAutowireCapableBeanFactory。
(4) AbstractBeanFactory。

在上述四个调用堆栈中首先关注 AbstractAutowireCapableBeanFactory 的构造方法，具体代码如下。

```
public AbstractAutowireCapableBeanFactory(@Nullable BeanFactory parentBeanFactory) {
    this();
    setParentBeanFactory(parentBeanFactory);
}
```

在这个构造方法中主要进行的是父类构造方法调用和自身对父容器的设置。

继续向下追踪源码 this() 相关代码如下。

```
public AbstractAutowireCapableBeanFactory() {
    super();
    //添加忽略的依赖接口(即 add 方法调用)
    ignoreDependencyInterface(BeanNameAware.class);
    ignoreDependencyInterface(BeanFactoryAware.class);
    ignoreDependencyInterface(BeanClassLoaderAware.class);
}
```

在这个构造方法中 super 调用的是 AbstractBeanFactory 的构造方法，AbstractBeanFactory 没有特殊内容，就是一个无参数构造函数的调用。下面主要分析 ignoreDependencyInterface 方法，该方法的作用是将需要忽略的依赖放到集合中便于后续使用，放入的方法有三个，分别是 BeanNameAware、BeanFactoryAware 和 BeanClassLoaderAware。存储它们的容器是一个 Set 集合，具体数据结构如下。

```
private final Set<Class<?>> ignoredDependencyInterfaces = new HashSet<>();
```

下面分析设置父容器的方法 setParentBeanFactory，具体代码如下。

```
@Override
public void setParentBeanFactory(@Nullable BeanFactory parentBeanFactory) {
    if (this.parentBeanFactory != null && this.parentBeanFactory != parentBeanFactory) {
        throw new IllegalStateException("Already associated with parent BeanFactory: " + this.parentBeanFactory);
    }
    this.parentBeanFactory = parentBeanFactory;
}
```

在设置父容器的方法中对父容器做了验证，具体验证逻辑有两个，第一个是判断当前容器是否存在父容器，第二个是判断当前容器的父容器和参数父容器是否相同。

至此，对于 XmlBeanFactory 的构造函数分析完成，总结 XmlBeanFactory 的构造函数所执行的事项有两个，第一个是设置需要忽略的依赖接口，第二个是设置父容器。在 XmlBeanFactory 对象中并没有实现摧毁相关的内容，因此 XmlBeanFactory 对象的摧毁就是一个 Java 对象的摧毁，不会有其他事件进行。

8.3　FileSystemXmlApplicationContext 分析

下面将展开 FileSystemXmlApplicationContext 类的生命周期分析，值得注意的是，在 FileSystemXmlApplicationContext 中是存在关闭方法的。这个关闭方法是 ConfigurableApplicationContext，同时也存在构造方法。首先对构造方法进行分析，构造方法代码如下。

```
public FileSystemXmlApplicationContext(String configLocation) throws BeansException {
    this(new String[] {configLocation}, true, null);
}
```

上述代码是开发者在进行 Spring 开发时经常使用到的构造方式，它还需要进一步搜索找到 this 对应的构造函数，具体代码如下。

```
public FileSystemXmlApplicationContext(
        String[] configLocations, boolean refresh, @Nullable ApplicationContext parent)
        throws BeansException {

    super(parent);
    setConfigLocations(configLocations);
    if (refresh) {
        refresh();
    }
}
```

在这个函数中需要重点关注三个参数：第一个参数是 configLocations，它表示了 SpringXML 的配置文件路径；第二个参数是 refresh，它表示了是否需要刷新上下文；第三个参数是 parent，它表示了父上下文。构造方法围绕这个三个参数做了各自的处理。

8.3.1　父上下文处理

对于父上下文的处理可以在 AbstractApplicationContext 类中找到处理方案，具体处理代码如下。

```
public AbstractApplicationContext(@Nullable ApplicationContext parent) {
    this();
    setParent(parent);
}
```

在这段代码中可以看到，对于父上下文的处理其本质上就是一个成员变量的设置，这个方法中还需要关注 this 对应的代码，具体代码如下。

```
public AbstractApplicationContext() {
    this.resourcePatternResolver = getResourcePatternResolver();
}
```

```
protected ResourcePatternResolver getResourcePatternResolver() {
    return new PathMatchingResourcePatternResolver(this);
}
```

在这个构造函数中对于资源解析对象（resourcePatternResolver）做了初始化，具体的对象是 PathMatchingResourcePatternResolver 类型。

继续回到 setParent 方法中，在这个方法中处理了以下两个事项。

（1）成员变量的设置。

（2）当前上下文和父上下文的环境配置进行合并。

关于上述两个事项的处理代码如下。

```
@Override
public void setParent(@Nullable ApplicationContext parent) {
    this.parent = parent;
    if (parent != null) {
        //获取父上下文的环境信息
        Environment parentEnvironment = parent.getEnvironment();
        //当环境信息是 ConfigurableEnvironment,进行合并
        if (parentEnvironment instanceof ConfigurableEnvironment) {
            getEnvironment().merge((ConfigurableEnvironment) parentEnvironment);
        }
    }
}
```

在这个方法中进一步追寻源码找到 merge 的源代码，它是由 org.springframework.core.env.AbstractEnvironment#merge 提供的，具体处理代码如下。

```
@Override
public void merge(ConfigurableEnvironment parent) {
    for (PropertySource<?> ps : parent.getPropertySources()) {
        if (!this.propertySources.contains(ps.getName())) {
            this.propertySources.addLast(ps);
        }
    }
    String[] parentActiveProfiles = parent.getActiveProfiles();
    if (!ObjectUtils.isEmpty(parentActiveProfiles)) {
        synchronized (this.activeProfiles) {
            Collections.addAll(this.activeProfiles, parentActiveProfiles);
        }
    }
    String[] parentDefaultProfiles = parent.getDefaultProfiles();
    if (!ObjectUtils.isEmpty(parentDefaultProfiles)) {
        synchronized (this.defaultProfiles) {
            this.defaultProfiles.remove(RESERVED_DEFAULT_PROFILE_NAME);
            Collections.addAll(this.defaultProfiles, parentDefaultProfiles);
        }
    }
}
```

在这个合并方法中可以发现它围绕三个数据进行合并，分别如下。

（1）MutablePropertySources：存储了属性名称和属性值。对于这一项的处理逻辑是从父上下文中获取属性列表判断属性名称是否存在，如果不存在就追加到当前上下文的属性表中。

（2）activeProfiles：存放激活的 profile。对于这一项的处理逻辑是父上下文中的 activeProfiles 直接添加到当前上下文的 activeProfiles 中。

（3）defaultProfiles：存放默认的 profile。对于这一项的处理逻辑是父上下文中的 defaultProfiles 直接添加到当前上下文的 defaultProfiles 中。

8.3.2 配置文件路径解析

接下来将对配置地址的处理进行分析，处理代码如下。

```
public void setConfigLocations(@Nullable String... locations) {
    if (locations != null) {
        Assert.noNullElements(locations, "Config locations must not be null");
        this.configLocations = new String[locations.length];
        for (int i = 0; i < locations.length; i++) {
            this.configLocations[i] = resolvePath(locations[i]).trim();
        }
    }
    else {
        this.configLocations = null;
    }
}
```

在这段代码中可以做一个简单的理解：如果存在文件地址列表，那么循环处理每个文件列表，将解析结果放到成员变量 configLocations 中，如果没有就设置为 null。在这个方法中重点是解析方法 resolvePath。具体处理代码如下。

```
protected String resolvePath(String path) {
    return getEnvironment().resolveRequiredPlaceholders(path);
}
```

在这段方法调用中需要明确 Environment 的具体类型，在 Spring 中经常会用一些接口进行直接使用，而不是直接用具体的某一个实现类，因此需要确认 Environment 对象的真实类型。通过 getEnvironment 代码的阅读可以确认它是 StandardEnvironment 类型。具体论证代码可以参考下面的代码。

```
@Override
public ConfigurableEnvironment getEnvironment() {
    if (this.environment == null) {
        this.environment = createEnvironment();
    }
    return this.environment;
}

protected ConfigurableEnvironment createEnvironment() {
```

```
        return new StandardEnvironment();
}
```

在明确具体的类型后进一步进行搜索，找到 org.springframework.core.env.AbstractEnvironment 中提供的对应的处理方法，在这个方法中它明确了具体的解析能力提供者 PropertySourcesPropertyResolver。PropertySourcesPropertyResolver 对象就是提供配置文件地址解析的核心了，在 PropertySourcesPropertyResolver 的方法列表中有下面这段代码。

```
@Override
public String resolveRequiredPlaceholders(String text) throws IllegalArgumentException {
    if (this.strictHelper == null) {
        this.strictHelper = createPlaceholderHelper(false);
    }
    return doResolvePlaceholders(text, this.strictHelper);
}
```

在这个方法中负责配置文件地址解析的是 strictHelper 对象，下面简单介绍 strictHelper 对象的作用。对于 strictHelper 的理解需要先做出一个属性表，属性表内容如下。

a = 1
b = 2

拥有属性表后需要定义一个存在占位符的字符串，例如{a}{b}，这里占位符使用一对花括号表示，在定义完成这两个内容后经过 strictHelper 的解析会得到数据结果 12。strictHelper 的作用就是将占位符包裹的字符串在属性表中找到对应的数据值将其替换。

注意，strictHelper 只会对占位符进行解析，不会解析路径符号。

在测试用例中经过配置文件路径解析和设置后得到的数据信息如图 8.1 所示。

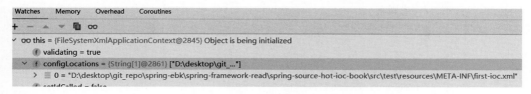

图 8.1 配置文件路径解析后结果

8.3.3 刷新操作

完成配置文件路径解析后会根据参数是否需要刷新来进行刷新操作的处理，处理代码如下。

```
@Override
public void refresh() throws BeansException, IllegalStateException {
    synchronized (this.startupShutdownMonitor) {
        //准备刷新此上下文
        prepareRefresh();
```

```java
//创建出 BeanFactory
ConfigurableListableBeanFactory beanFactory = obtainFreshBeanFactory();

//准备 BeanFactory，对 BeanFactory 进行数据设置等
prepareBeanFactory(beanFactory);

try {
    //BeanFactory 在子类中进行后置处理
    postProcessBeanFactory(beanFactory);

    //BeanFactoryPostProcessor 方法调用
    invokeBeanFactoryPostProcessors(beanFactory);

    //注册 beanPostProcessor
    registerBeanPostProcessors(beanFactory);

    //实例化 message source 相关信息
    initMessageSource();

    //实例化应用事件传播器
    initApplicationEventMulticaster();

    onRefresh();

    //注册监听器
    registerListeners();

    //完成 BeanFactory 的实例化
    finishBeanFactoryInitialization(beanFactory);

    //完成刷新
    finishRefresh();
}

catch (BeansException ex) {
    if (logger.isWarnEnabled()) {
        logger.warn("Exception encountered during context initialization - " +
            "cancelling refresh attempt: " + ex);
    }

    //摧毁 Bean
    destroyBeans();

    //取消刷新
    cancelRefresh(ex);

    throw ex;
}

finally {
    //重置通用缓存
```

```
            resetCommonCaches();
        }
    }
}
```

这个刷新方法是 Spring 容器启动中最重要的方法,后续各类上下文都会有类似的操作。这个刷新方法调用了很多方法,下面将对每个方法逐一分析。

1. prepareRefresh 方法分析

首先是 prepareRefresh 方法的分析,该方法是为刷新做前置准备工作,具体处理了下面六个事项。

(1) 设置开始时间。
(2) 设置关闭标记为 false。
(3) 设置激活标记为 true。
(4) 处理初始化属性、占位符、资源等数据。
(5) 进行数据必填项验证。
(6) 处理监听器列表,细分有以下两个小细节。
① 创建提前暴露的应用监听器列表。
② 创建应用监听器列表。
上述六步的处理代码如下。

```
protected void prepareRefresh() {
    //Switch to active.
    //设置开始时间
    this.startupDate = System.currentTimeMillis();
    //设置关闭标记为 false
    this.closed.set(false);
    //设置激活标记为 true
    this.active.set(true);

    //处理初始化属性、占位符、资源等数据
    //抽象方法,子类实现
    initPropertySources();

    //进行数据必填项验证
    getEnvironment().validateRequiredProperties();

    //处理监听器列表
    if (this.earlyApplicationListeners == null) {
        this.earlyApplicationListeners = new LinkedHashSet<>(this.applicationListeners);
    }
    else {
        this.applicationListeners.clear();
        this.applicationListeners.addAll(this.earlyApplicationListeners);
    }
    this.earlyApplicationEvents = new LinkedHashSet<>();
}
```

第四个操作对应的方法是 initPropertySources，它是一个抽象方法，它在 Web 应用上下文中会有相关实现，由于测试用例中并不是一个 Web 应用，因此该方法可以将其当作一个空方法进行理解。

继续向下有一个关于数据必填项的验证，验证方法的签名是 org.springframework.core.env.AbstractPropertyResolver#validateRequiredProperties，具体代码如下。

```
@Override
public void validateRequiredProperties() {
    //异常信息
    MissingRequiredPropertiesException ex = new MissingRequiredPropertiesException();
    for (String key : this.requiredProperties) {
        //判断 key 的属性是否存在,如果不存在则抛出异常
        if (this.getProperty(key) == null) {
            ex.addMissingRequiredProperty(key);
        }
    }
    if (!ex.getMissingRequiredProperties().isEmpty()) {
        throw ex;
    }
}
```

对于数据字段的必填验证在这个方法中操作比较简单：循环每个必填的数据 key，在属性表中获取，如果不存在对应的 value，那么会加入异常信息中，最后再统一抛出 MissingRequiredPropertiesException。

在 prepareRefresh 中除了上述两个方法以外，其他的都是一些成员变量的赋值操作，读者可以翻阅 8.5 节内容以对成员变量进行了解。

2. obtainFreshBeanFactory 方法分析

接下来对 obtainFreshBeanFactory 方法进行分析。该方法的作用是创建（获取）一个 BeanFactory 对象，具体处理代码如下。

```
protected ConfigurableListableBeanFactory obtainFreshBeanFactory() {
    //刷新 BeanFactory, 子类实现
    refreshBeanFactory();
    //获取 BeanFactory, 子类实现
    return getBeanFactory();
}
```

在这段方法中出现的两个方法 refreshBeanFactory 和 getBeanFactory 都是抽象方法，需要子类自行实现，本节以 FileSystemXmlApplicationContext 作为初始对象进行分析，上述两个方法会在 FileSystemXmlApplicationContext 的父类中出现。

首先对 refreshBeanFactory 方法进行分析，在 FileSystemXmlApplicationContext 的类图中可以找到一个具体的处理方法是 org.springframework.context.support.AbstractRefreshableApplicationContext#refreshBeanFactory，处理代码如下。

```
@Override
protected final void refreshBeanFactory() throws BeansException {
```

```
//是否存在 BeanFactory
if (hasBeanFactory()) {
    //如果存在 BeanFactory,则清空 Bean 相关信息
    //摧毁 Bean
    destroyBeans();
    //清空 BeanFactory
    closeBeanFactory();
}
try {
    //创建 BeanFactory
    DefaultListableBeanFactory beanFactory = createBeanFactory();
    //设置序列化 id
    beanFactory.setSerializationId(getId());
    //定制工厂的处理
    //设置两个属性值
    // 1. allowBeanDefinitionOverriding
    // 2. allowCircularReferences
    customizeBeanFactory(beanFactory);
    //加载 Bean 定义
    loadBeanDefinitions(beanFactory);
    //上锁设置 BeanFactory
    synchronized (this.beanFactoryMonitor) {
        this.beanFactory = beanFactory;
    }
}
catch (IOException ex) {
    throw new ApplicationContextException("I/O error parsing bean definition source for " + getDisplayName(), ex);
}
```

这段方法的处理逻辑如图 8.2 所示。

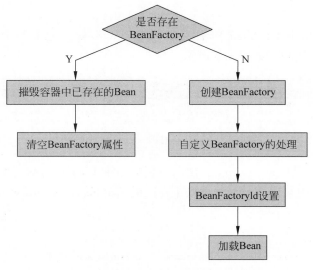

图 8.2　refreshBeanFactory 处理逻辑

了解处理逻辑后对处理逻辑中出现的几个方法进行说明。

（1）hasBeanFactory 方法的作用是判断是否存在 BeanFactory，判断方式是 this.beanFactory！=null。

（2）destroyBeans 方法的作用是摧毁 Bean。

（3）closeBeanFactory 方法的作用是清理 BeanFactory。清理内容：第一，将 BeanFactory 的序列化 ID 设置为空；第二，将 BeanFactory 设置为空。

（4）createBeanFactory 方法的作用是创建一个 BeanFactory。在这个方法中就是一个对象的创建（具体操作是 new 对象），这里的实质对象是 DefaultListableBeanFactory，在它的构造函数中有父 BeanFactory 作为参数。处理代码如下。

```
protected DefaultListableBeanFactory createBeanFactory() {
    return new DefaultListableBeanFactory(getInternalParentBeanFactory());
}
public DefaultListableBeanFactory(@Nullable BeanFactory parentBeanFactory) {
    super(parentBeanFactory);
}
```

（5）customizeBeanFactory 方法的作用是对 BeanFactory 的拓展，这里的拓展实质上是设置两个属性，第一个属性是 allowBeanDefinitionOverriding（是否允许 Bean 定义覆盖），第二个属性是 allowCircularReferences（是否允许循环引用）。具体处理代码如下。

```
protected void customizeBeanFactory(DefaultListableBeanFactory beanFactory) {
    //设置是否允许 Bean 定义覆盖
    if (this.allowBeanDefinitionOverriding != null) {
        beanFactory.setAllowBeanDefinitionOverriding(this.allowBeanDefinitionOverriding);
    }
    //设置是否允许循环引用
    if (this.allowCircularReferences != null) {
        beanFactory.setAllowCircularReferences(this.allowCircularReferences);
    }
}
```

（6）loadBeanDefinitions 方法的作用是加载 BeanDefinition，这是一个抽象方法，在 Spring 中有多种实现方式，实现类如图 8.3 所示。

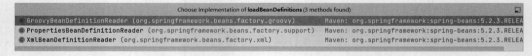

图 8.3　loadBeanDefinitions 实现类列表

至此，关于 refreshBeanFactory 方法的分析就结束了，下面将对 getBeanFactory 方法进行分析。在 Spring 中可以在 AbstractRefreshableApplicationContext 类中找到处理详情，具体处理代码如下。

```
@Override
public final ConfigurableListableBeanFactory getBeanFactory() {
    synchronized (this.beanFactoryMonitor) {
```

```
        if (this.beanFactory == null) {
            throw new IllegalStateException("BeanFactory not initialized or already closed - " +
                    "call 'refresh' before accessing beans via the ApplicationContext");
        }
        return this.beanFactory;
    }
}
```

在这段代码中做了成员变量的非空判断,通过判断会将变量返回,该方法的处理逻辑比较简单,可以直接理解为get方法。

3. prepareBeanFactory 方法分析

通过前文的分析现在已经得到了一个最基本的 BeanFactory 对象,在得到这个对象后 Spring 将它放入 prepareBeanFactory 方法进行数据补充,主要补充的是各种接口,具体处理代码如下。

```
protected void prepareBeanFactory(ConfigurableListableBeanFactory beanFactory) {
    //设置 classLaoder
    beanFactory.setBeanClassLoader(getClassLoader());
    //设置 EL 表达式解析器
    beanFactory.setBeanExpressionResolver(new StandardBeanExpressionResolver(beanFactory.getBeanClassLoader()));
    //添加属性编辑器注册工具
    beanFactory.addPropertyEditorRegistrar(new
ResourceEditorRegistrar(this, getEnvironment()));

    //添加 Bean 后置处理器
    beanFactory.addBeanPostProcessor(new ApplicationContextAwareProcessor(this));
    //添加忽略的接口
    beanFactory.ignoreDependencyInterface(EnvironmentAware.class);
    beanFactory.ignoreDependencyInterface(EmbeddedValueResolverAware.class);
    beanFactory.ignoreDependencyInterface(ResourceLoaderAware.class);
    beanFactory.ignoreDependencyInterface(ApplicationEventPublisherAware.class);
    beanFactory.ignoreDependencyInterface(MessageSourceAware.class);
    beanFactory.ignoreDependencyInterface(ApplicationContextAware.class);

    //注册依赖
    beanFactory.registerResolvableDependency(BeanFactory.class, beanFactory);
    beanFactory.registerResolvableDependency(ResourceLoader.class, this);
    beanFactory.registerResolvableDependency(ApplicationEventPublisher.class, this);
    beanFactory.registerResolvableDependency(ApplicationContext.class, this);

    //添加 Bean 后置处理器
    beanFactory.addBeanPostProcessor(new ApplicationListenerDetector(this));

    //判断是否存在 loadTimeWeaver Bean
    if (beanFactory.containsBean(LOAD_TIME_WEAVER_BEAN_NAME)) {
        //添加后置处理器
        beanFactory.addBeanPostProcessor(new LoadTimeWeaverAwareProcessor(beanFactory));
        //设置临时的 ClassLoader
```

```
        beanFactory.setTempClassLoader(new ContextTypeMatchClassLoader(beanFactory.
getBeanClassLoader()));
    }

    //environment Bean 注册
    if (!beanFactory.containsLocalBean(ENVIRONMENT_BEAN_NAME)) {
        beanFactory.registerSingleton(ENVIRONMENT_BEAN_NAME, getEnvironment());
    }
    //systemProperties Bean 注册
    if (!beanFactory.containsLocalBean(SYSTEM_PROPERTIES_BEAN_NAME)) {
         beanFactory.registerSingleton(SYSTEM_PROPERTIES_BEAN_NAME, getEnvironment().
getSystemProperties());
    }
    //systemEnvironment Bean 注册
    if (!beanFactory.containsLocalBean(SYSTEM_ENVIRONMENT_BEAN_NAME)) {
         beanFactory.registerSingleton(SYSTEM_ENVIRONMENT_BEAN_NAME, getEnvironment().
getSystemEnvironment());
    }
}
```

通过阅读上述代码可以发现,在这个方法中设置了下面六项内容。

(1) 设置类加载器。

(2) 设置 EL 表达式解析对象。

(3) 设置属性编辑器的注册工具。

(4) 设置 Bean 后置处理器。

(5) 设置需要忽略的依赖接口。

(6) 注册一些特定的依赖信息。

4. postProcessBeanFactory 方法分析

下面对 postProcessBeanFactory 方法进行分析。该方法是一个抽象方法,一般情况下这个方法会在 Web 应用中有相关处理,在测试用例中并不是一个 Web 应用,此时可以将这个方法当作一个空方法来理解。

5. invokeBeanFactoryPostProcessors 方法分析

接下来对 invokeBeanFactoryPostProcessors 方法进行分析。该方法主要是进行 BeanFactoryPostProcessor 接口的相关调度,下面是具体实现代码。

```
protected void invokeBeanFactoryPostProcessors(ConfigurableListableBeanFactory
beanFactory) {
    //后置处理器委托对象
    //调用 BeanFactoryPostProcessor 方法
    PostProcessorRegistrationDelegate.invokeBeanFactoryPostProcessors(beanFactory,
getBeanFactoryPostProcessors());

    //判断临时类加载器是否存在
    //是否包含 loadTimeWeaver Bean
    if (beanFactory.getTempClassLoader() == null &&
beanFactory.containsBean(LOAD_TIME_WEAVER_BEAN_NAME)) {
```

```
        //添加 Bean 后置处理器
        beanFactory.addBeanPostProcessor(new LoadTimeWeaverAwareProcessor(beanFactory));
        //添加临时类加载器
        beanFactory.setTempClassLoader(new ContextTypeMatchClassLoader(beanFactory.
getBeanClassLoader()));
    }
}
```

在这段代码中主要做了以下两件事。

（1）将容器中存在的 BeanPostProcessor 进行调度，即执行 BeanFactoryPostProcessor # postProcessBeanFactory 方法。

（2）处理 LoadTimeWeaver 相关的内容。

在处理第一件事时将具体的处理委托给了 PostProcessorRegistrationDelegate 对象进行处理。

6. registerBeanPostProcessors 方法分析

下面将对 registerBeanPostProcessors 方法进行分析。该方法的作用是进行 BeanPostProcessor 实例注册，具体处理代码如下。

```
protected void registerBeanPostProcessors(ConfigurableListableBeanFactory beanFactory) {
    //后置处理器委托类进行注册
    PostProcessorRegistrationDelegate.registerBeanPostProcessors(beanFactory, this);
}
```

在这个方法中，对于 BeanPostProcessor 的注册需要依赖 PostProcessorRegistrationDelegate 对象进行，具体分析在第 14 章中进行。

7. initMessageSource 方法分析

下面将对 initMessageSource 方法进行分析。该方法的主要目的是实例化 MessageSource 对象，具体处理方法如下。

```
protected void initMessageSource() {
    //获取 BeanFactory
    ConfigurableListableBeanFactory beanFactory = getBeanFactory();
    //判断容器中是否存在 messageSource 这个 BeanName
    //存在的情况
    if (beanFactory.containsLocalBean(MESSAGE_SOURCE_BEAN_NAME)) {
        //获取 messageSource 对象
        this.messageSource = beanFactory.getBean(MESSAGE_SOURCE_BEAN_NAME, MessageSource.
class);

        //设置父 MessageSource
        if (this.parent != null && this.messageSource instanceof HierarchicalMessageSource) {
            HierarchicalMessageSource hms = (HierarchicalMessageSource) this.messageSource;
            if (hms.getParentMessageSource() == null) {
                hms.setParentMessageSource(getInternalParentMessageSource());
            }
        }
    }
```

```
        //不存在的情况
        else {
            // MessageSource 实现类
            DelegatingMessageSource dms = new DelegatingMessageSource();
            //设置父 MessageSource
            dms.setParentMessageSource(getInternalParentMessageSource());
            this.messageSource = dms;
            //注册 MessageSource
            beanFactory.registerSingleton(MESSAGE_SOURCE_BEAN_NAME, this.messageSource);
        }
    }
```

在这段代码中有以下两种处理方式。

（1）当前容器中已经存在名称是 messageSource 的 Bean，经过判断确定是否需要设置父 messageSource。

（2）当前容器中不存在名称是 messageSource 的 Bean，此时会新建 DelegatingMessageSource 对象放入单例对象管理器中。

8. initApplicationEventMulticaster 方法分析

下面将分析 initApplicationEventMulticaster 方法。在这个方法中主要是将 ApplicationEventMulticaster 进行实例化，具体处理代码如下。

```
protected void initApplicationEventMulticaster() {
    ConfigurableListableBeanFactory beanFactory = getBeanFactory();
    //存在的情况
    if (beanFactory.containsLocalBean(APPLICATION_EVENT_MULTICASTER_BEAN_NAME)) {
        this.applicationEventMulticaster = beanFactory.getBean(APPLICATION_EVENT_MULTICASTER_BEAN_NAME, ApplicationEventMulticaster.class);
    }
    //不存在的情况
    else {
        this.applicationEventMulticaster = new SimpleApplicationEventMulticaster(beanFactory);
        //注册到容器
        beanFactory.registerSingleton(APPLICATION_EVENT_MULTICASTER_BEAN_NAME, this.applicationEventMulticaster);
    }
}
```

在这个方法中有以下两种处理方式。

（1）当前容器中已存在名称是 applicationEventMulticaster 的 Bean，从容器中获取 applicationEventMulticaster 赋值成员变量。

（2）当前容器中不存在名称是 applicationEventMulticaster 的 Bean，立即创建 SimpleApplicationEventMulticaster 对象并注册到 Bean 容器中。

9. onRefresh 方法分析

onRefresh 方法在非 Web 项目中可以当作一个空方法来看待。

10. registerListeners 方法分析

registerListeners 方法将各类 ApplicationListener 注册到容器中，具体处理代码如下。

```java
protected void registerListeners() {
    //获取应用监听器列表
    //Register statically specified listeners first.
    for (ApplicationListener<?> listener : getApplicationListeners()) {
        //获取事件广播器
        //添加应用监听器
        getApplicationEventMulticaster().addApplicationListener(listener);
    }

    //通过类型获取应用监听器名称列表
    String[] listenerBeanNames = getBeanNamesForType(ApplicationListener.class, true, false);
    //将应用监听器列表的名称注册到事件广播器中
    for (String listenerBeanName : listenerBeanNames) {
        getApplicationEventMulticaster().addApplicationListenerBean(listenerBeanName);
    }

    //早期应用事件发布
    Set<ApplicationEvent> earlyEventsToProcess = this.earlyApplicationEvents;
    this.earlyApplicationEvents = null;
    if (earlyEventsToProcess != null) {
        for (ApplicationEvent earlyEvent : earlyEventsToProcess) {
            //发布事件
            getApplicationEventMulticaster().multicastEvent(earlyEvent);
        }
    }
}
```

在这个方法中有三层处理逻辑：第一层是处理当前容器中存在的 ApplicationListener，将数据放入到 ApplicationEventMulticaster 中；第二层是在容器中通过类型将所有类型是 ApplicationListener 的 BeanName 获取后加入到 ApplicationEventMulticaster 中；第三层是处理 earlyApplicationEvents 相关事件，即发送事件进行相关处理。

11. finishBeanFactoryInitialization 方法分析

下面将介绍 finishBeanFactoryInitialization 方法，在这个方法中 BeanFactory 终于完成了最终的实例化和赋值操作，在这个方法之后 BeanFactory 的改动概率几乎为 0，具体的处理代码如下。

```java
protected void
finishBeanFactoryInitialization(ConfigurableListableBeanFactory beanFactory) {
    //判断是否存在转换服务
    //1. 转换服务的 BeanName 存在
    //2. 转换服务的 BeanName 和类型是否匹配
    if (beanFactory.containsBean(CONVERSION_SERVICE_BEAN_NAME) &&
            beanFactory.isTypeMatch(CONVERSION_SERVICE_BEAN_NAME, ConversionService.class)) {
        //注册转换服务
        beanFactory.setConversionService(
                beanFactory.getBean(CONVERSION_SERVICE_BEAN_NAME, ConversionService.
```

```
            class));
        }

        //添加嵌套值解析器,进行字符串解析
        if (!beanFactory.hasEmbeddedValueResolver()) {
            beanFactory.addEmbeddedValueResolver(strVal
                -> getEnvironment().resolvePlaceholders(strVal));
        }

        //将类型是 LoadTimeWeaverAware 的 Bean 全部初始化
        String[] weaverAwareNames =
beanFactory.getBeanNamesForType(LoadTimeWeaverAware.class, false, false);
        for (String weaverAwareName : weaverAwareNames) {
            getBean(weaverAwareName);
        }

        //删除临时类加载器
        beanFactory.setTempClassLoader(null);

        //冻结部分配置
        beanFactory.freezeConfiguration();

        //非懒加载的单例对象实例化
        beanFactory.preInstantiateSingletons();
}
```

在这段代码中处理了六个重点操作,分别如下。

(1) 设置转换服务接口。
(2) 设置字符串解析接口。
(3) 对类型是 LoadTimeWeaverAware 的 Bean 进行实例化。
(4) 删除临时类加载器。
(5) 冻结配置。
(6) 对非懒加载的 Bean 进行实例化。

下面对第(5)个和第(6)个操作进行详细分析。冻结配置其实是将配置信息做一个镜像,简单地说就是一个副本,具体处理代码如下。

```
@Override
public void freezeConfiguration() {
    this.configurationFrozen = true;
    this.frozenBeanDefinitionNames = StringUtils.toStringArray(this.beanDefinitionNames);
}
```

下面对非懒加载的 Bean 实例化相关代码进行阅读,具体代码如下。

```
@Override
public void preInstantiateSingletons() throws BeansException {
    if (logger.isTraceEnabled()) {
        logger.trace("Pre-instantiating singletons in " + this);
    }
```

```java
//BeanNames 内容是 BeanDefinition 的名称
List<String> beanNames = new ArrayList<>(this.beanDefinitionNames);

//对非懒加载的 Bean 进行实例化
for (String beanName : beanNames) {
    //获取 Bean 定义
    RootBeanDefinition bd = getMergedLocalBeanDefinition(beanName);
    //条件过滤
    //1. abstract 修饰
    //2. 是否单例
    //3. 是否懒加载
    if (!bd.isAbstract() && bd.isSingleton() && !bd.isLazyInit()) {
        //是否是工厂 Bean
        if (isFactoryBean(beanName)) {
            //获取 Bean
            Object bean = getBean(FACTORY_BEAN_PREFIX + beanName);
            //类型判断是不是 FactoryBean
            if (bean instanceof FactoryBean) {
                final FactoryBean<?> factory = (FactoryBean<?>) bean;
                //是否立即加载
                boolean isEagerInit;
                //计算 isEagerInit
                //1. 是不是 SmartFactoryBean
                //2. 执行 SmartFactoryBean 的 isEagerInit 方法
                if (System.getSecurityManager() != null && factory instanceof SmartFactoryBean) {
                    isEagerInit = AccessController.doPrivileged((PrivilegedAction<Boolean>)
                            ((SmartFactoryBean<?>) factory)::isEagerInit,
                            getAccessControlContext());
                }
                else {
                    isEagerInit = (factory instanceof SmartFactoryBean &&
                            ((SmartFactoryBean<?>) factory).isEagerInit());
                }
                if (isEagerInit) {
                    getBean(beanName);
                }
            }
        }
        else {
            getBean(beanName);
        }
    }
}

//Trigger post-initialization callback for all applicable beans...
//触发回调方法
for (String beanName : beanNames) {
    //获取单例对象的实例
    Object singletonInstance = getSingleton(beanName);
```

```
            //类型判断
            if (singletonInstance instanceof SmartInitializingSingleton) {
                //执行 SmartInitializingSingleton 方法 afterSingletonsInstantiated
                final SmartInitializingSingleton smartSingleton = (SmartInitializingSingleton) singletonInstance;
                if (System.getSecurityManager() != null) {
                    AccessController.doPrivileged((PrivilegedAction<Object>) () -> {
                        smartSingleton.afterSingletonsInstantiated();
                        return null;
                    }, getAccessControlContext());
                }
                else {
                    smartSingleton.afterSingletonsInstantiated();
                }
            }
        }
    }
}
```

这个方法的主要目的是对开发者自定义的非懒加载的单例 Bean 进行实例化，在这段代码中处理逻辑分为以下两种。

（1）获取非 Spring 内部的 BeanName，通过 getBean 的方式进行实例化（实例化是指对象创建和属性设置），在实例化上又会存在 FactoryBean 和非 FactoryBean 两种情况。

（2）通过步骤（1）得到了单例对象，在得到这些单例对象之后会对其做类型判断是否属于 SmartInitializingSingleton，如果是那么就执行 SmartInitializingSingleton # afterSingletonsInstantiated 方法。

下面来看两幅图，图 8.4 表示没有执行 finishBeanFactoryInitialization 时的 singletonObjects 对象情况，图 8.5 表示执行 finishBeanFactoryInitialization 后的 singletonObjects 对象情况，通过这两幅图的内容可以发现开发者自定义的一些非懒加载的单例 Bean 被实例化了。

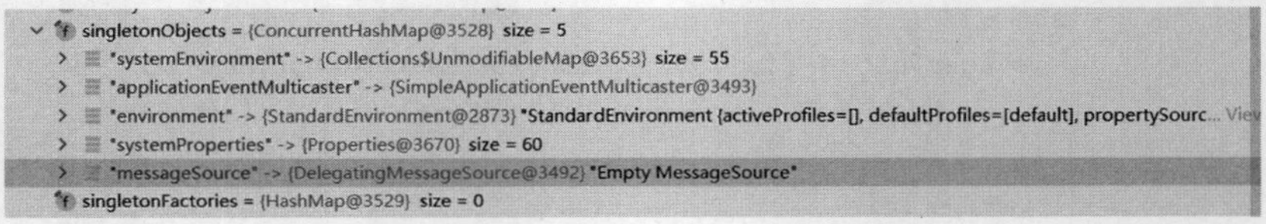

图 8.4　没有执行 finishBeanFactoryInitialization 时的 singletonObjects

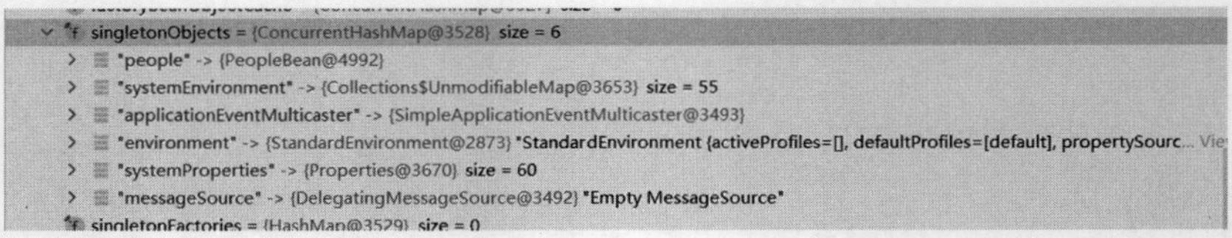

图 8.5　执行 finishBeanFactoryInitialization 时的 singletonObjects

12. finishRefresh 方法分析

接下来将对 finishRefresh 方法进行分析。该方法的作用是完成刷新方法的执行，具体处理代码如下。

```java
protected void finishRefresh() {
    //清空资源缓存
    clearResourceCaches();

    //实例化生命周期处理接口
    initLifecycleProcessor();

    //对生命周期处理接口进行刷新操作
    getLifecycleProcessor().onRefresh();

    //推送事件：上下文刷新事件
    publishEvent(new ContextRefreshedEvent(this));

    //注册应用上下文
    LiveBeansView.registerApplicationContext(this);
}
```

在这个方法中分别调用了其他五个方法，理解这五个方法后对这个方法就理解了。

（1）clearResourceCaches：该方法的作用是清理资源缓存。

（2）initLifecycleProcessor：该方法是用来进行 LifecycleProcessor 接口的实例化（生命周期处理接口），处理逻辑是如果容器中存在就直接从容器中获取，如果容器中不存在则立即创建一个默认的处理器，具体处理代码如下。

```java
protected void initLifecycleProcessor() {
    //获取 BeanFactory
    ConfigurableListableBeanFactory beanFactory = getBeanFactory();
    //判断 lifecycleProcessor beanName 是否有对应的 Bean 实例
    if (beanFactory.containsLocalBean(LIFECYCLE_PROCESSOR_BEAN_NAME)) {
        //设置 lifecycleProcessor
        this.lifecycleProcessor =
                beanFactory.getBean(LIFECYCLE_PROCESSOR_BEAN_NAME, LifecycleProcessor.class);
    }
    else {
        //创建默认的生命周期处理接口的实现对象
        DefaultLifecycleProcessor defaultProcessor = new DefaultLifecycleProcessor();
        //设置 BeanFactory
        defaultProcessor.setBeanFactory(beanFactory);
        //设置成员变量
        this.lifecycleProcessor = defaultProcessor;
        //注册
        beanFactory.registerSingleton(LIFECYCLE_PROCESSOR_BEAN_NAME, this.lifecycleProcessor);
    }
}
```

（3）getLifecycleProcessor().onRefresh()：该方法通过 LifecycleProcessor 接口来进行刷新的操作。

在这个调用链路中还需要第（2）个方法中生成的 LifecycleProcessor 对象实例来进行

onRefresh 操作,从第(2)个方法中可以确认具体实现类是 DefaultLifecycleProcessor,有关 DefaultLifecycleProcessor 的分析会在第 16 章进行详细分析。

(4) publishEvent(new ContextRefreshedEvent(this)):该方法的作用是进行推送事件。这里需要关注的是处理事件本体和事件处理者这两个对象,就目前而言,可以确定的是事件本体,ContextRefreshedEvent 事件处理者在 Spring 中是通过 ApplicationListener 来定义的,这些内容都是 Spring Event 中的知识点,相关分析会在第 15 章中进行。

(5) LiveBeansView.registerApplicationContext(this):在这个方法中尝试将当前的应用上下文放入 MBeanServer 中,对 MBeanServer 可以简单理解为也是一个容器。

13. destroyBeans 方法分析

接下来将分析 destroyBeans 方法。该方法进行 Bean 的摧毁。摧毁流程分为两步:第一步是判断正在摧毁的 Bean 是否实现了 DisposableBean 接口,如果实现了则执行 destroy 方法;第二步是执行自定义的摧毁方法。

14. cancelRefresh 方法分析

接下来将分析 cancelRefresh 方法。在这个方法中只做一件事,将 active 设置为 false,表示没有激活,具体处理代码如下。

```java
protected void cancelRefresh(BeansException ex) {
    this.active.set(false);
}
```

15. resetCommonCaches 方法分析

最后将对 resetCommonCaches 方法进行分析。在这个方法中会进行一些缓存的清理,具体操作代码如下。

```java
protected void resetCommonCaches() {
    //反射缓存
    ReflectionUtils.clearCache();
    //注解缓存
    AnnotationUtils.clearCache();
    //解析类型缓存
    ResolvableType.clearCache();
    CachedIntrospectionResults.clearClassLoader(getClassLoader());
}
```

在这段代码中总共清理三个缓存和一个类加载器,分别如下。
(1) 反射缓存。反射缓存需要使用到两个 Map,具体代码如下。

```java
private static final Map<Class<?>, Method[]> declaredMethodsCache =
    new ConcurrentReferenceHashMap<>(256);

private static final Map<Class<?>, Field[]> declaredFieldsCache =
    new ConcurrentReferenceHashMap<>(256);
```

(2) 注解缓存。注解缓存需要使用到三个 Map,具体代码如下。

```java
private static final Map<AnnotationFilter, Cache> standardRepeatablesCache =
```

```
new ConcurrentReferenceHashMap<>();

private static final Map<AnnotationFilter, Cache> noRepeatablesCache =
new ConcurrentReferenceHashMap<>();

private static final Map<AnnotatedElement, Annotation[]> declaredAnnotationCache =
new ConcurrentReferenceHashMap<>(256);

private static final Map<Class<?>, Method[]> baseTypeMethodsCache =
new ConcurrentReferenceHashMap<>(256);
```

(3)解析类型缓存。解析类型缓存需要使用两个容器,具体代码如下。

```
private static final ConcurrentReferenceHashMap<ResolvableType, ResolvableType> cache = new
ConcurrentReferenceHashMap<>(256);
static final ConcurrentReferenceHashMap<Type, Type> cache =
 new ConcurrentReferenceHashMap<>(256);
```

(4)类加载器。

8.3.4 关闭方法分析

前文介绍了关于 FileSystemXmlApplicationContext 对象的初始化相关内容,下面将介绍 FileSystemXmlApplicationContext 对象的关闭方法。在 Spring 中上下文关闭方法是由 org.springframework.context.ConfigurableApplicationContext 提供的,具体的处理代码如下。

```
@Override
public void close() {
    synchronized (this.startupShutdownMonitor) {
        //执行关闭
        doClose();
        if (this.shutdownHook != null) {
            try {
                //移除关闭线程
                Runtime.getRuntime().removeShutdownHook(this.shutdownHook);
            }
            catch (IllegalStateException ex) {
            }
        }
    }
}
```

在这段方法中可以确认最终的关闭方法是由 doClose 进行处理的,具体代码如下。

```
protected void doClose() {
    if (this.active.get() && this.closed.compareAndSet(false, true)) {
        if (logger.isDebugEnabled()) {
            logger.debug("Closing " + this);
        }
```

```java
            //在容器中移除当前上下文
            LiveBeansView.unregisterApplicationContext(this);

            try {
                //发布关闭上下文事件
                publishEvent(new ContextClosedEvent(this));
            }
            catch (Throwable ex) {
                logger.warn("Exception thrown from ApplicationListener
handling ContextClosedEvent", ex);
            }

            if (this.lifecycleProcessor != null) {
                try {
                    //生命周期处理器执行关闭函数
                    this.lifecycleProcessor.onClose();
                }
                catch (Throwable ex) {
                    logger.warn("Exception thrown from
LifecycleProcessor on context close", ex);
                }
            }

            //摧毁 Bean
            destroyBeans();

            //关闭 BeanFactory
            closeBeanFactory();

            //子类拓展关闭相关方法
            onClose();

            if (this.earlyApplicationListeners != null) {
                this.applicationListeners.clear();
                this.applicationListeners.addAll(this.earlyApplicationListeners);
            }

            //设置激活状态为 false
            this.active.set(false);
        }
    }
```

在这个方法中有八个处理流程,分别如下。

(1) MBeanServer 中移除当前上下文。
(2) 推送上下文关闭时间。
(3) 生命周期接口执行关闭相关操作。
(4) 摧毁 Beans。
(5) 清理 BeanFactory 相关数据。
(6) 子上下文的关闭操作,这个操作是拓展操作。

(7) 应用监听器相关处理。
(8) 设置激活状态为关闭。

在这八个处理流程中重点关注流程(5),相关代码如下。

```
@Override
protected final void closeBeanFactory() {
    synchronized (this.beanFactoryMonitor) {
        if (this.beanFactory != null) {
            this.beanFactory.setSerializationId(null);
            this.beanFactory = null;
        }
    }
}
```

在这段代码中将序列化 ID 设置为 null,同时将 BeanFactory 设置为 null,使其不再具备 BeanFactory 相关能力。

8.4 ClassPathXmlApplicationContext 分析

接下来将对 ClassPathXmlApplicationContext 的生命周期做分析。首先来看 ClassPathXmlApplicationContext 的构造方法。

```
public ClassPathXmlApplicationContext(
        String[] configLocations, boolean refresh, @Nullable ApplicationContext parent)
        throws BeansException {

    super(parent);
    setConfigLocations(configLocations);
    if (refresh) {
        refresh();
    }
}
```

通过阅读这段代码可以发现,它和 FileSystemXmlApplicationContext 的处理是一样的。本文将不再对 ClassPathXmlApplicationContext 对象的实例化和摧毁做相关分析,读者可以参考 FileSystemXmlApplicationContext 中的分析。

8.5 SpringXML 关键对象附表

下面将附上几个成员变量表,来对一些常用容器的成员变量进行说明。表 8.1 表示的是 AbstractAutowireCapableBeanFactory 对象中的成员变量。

表 8.1　AbstractAutowireCapableBeanFactory 成员变量

变量名称	变量类型	说明
ignoredDependencyTypes	Set < Class <?> >	需要忽略的依赖(特指类)
ignoredDependencyInterfaces	Set < Class <?> >	需要忽略的依赖(特指接口)

续表

变量名称	变量类型	说明
currentlyCreatedBean	NamedThreadLocal < String >	当前正在创建的 BeanName，Named-ThreadLocal 的本质是 ThreadLocal
factoryBeanInstanceCache	ConcurrentMap < String，BeanWrapper >	工厂 Bean 创建后的缓存，相关接口 FactoryBean key：BeanName value：Bean 实例（包装过的）
factoryMethodCandidateCache	ConcurrentMap < Class <?>，Method[ ]>	工厂方法的容器 key：工厂方法所在类 value：工厂方法列表
filteredPropertyDescriptorsCache	ConcurrentMap < Class <?>，PropertyDescriptor[ ]>	类的属性描述容器 key：类 value：属性描述对象集合
instantiationStrategy	InstantiationStrategy	Bean 的实例化策略
parameterNameDiscoverer	ParameterNameDiscoverer	用来提取参数名称的工具
allowCircularReferences	boolean	是否允许循环引用
allowRawInjectionDespiteWrapping	boolean	是否允许当前 Bean 的原始类型注入别的 Bean 中

在 AbstractAutowireCapableBeanFactory 的成员变量表中可以看到，currentlyCreatedBean 变量类型是 NamedThreadLocal，它是一个线程安全的类型。此外，在其他几个变量的类型中可以发现也有线程安全变量出现，特别需要关注的是 factoryBeanInstanceCache 变量，它是一个线程安全的容器，而且它存储了 BeanWapper 对象，从而可以说明 Bean 创建过程是线程安全的。

表 8.2 表示的是 AbstractApplicationContext 对象中的成员变量。

表 8.2　AbstractApplicationContext 成员变量

变量名称	变量类型	说明
MESSAGE_SOURCE_BEAN_NAME	String	接口 MessageSource 的 BeanName
LIFECYCLE_PROCESSOR_BEAN_NAME	String	接口 LifecycleProcessor 的 BeanName
APPLICATION_EVENT_MULTICASTER_BEAN_NAME	String	接口 ApplicationEventMulticaster 的 BeanName
beanFactoryPostProcessors	List < BeanFactoryPostProcessor >	存储 BeanFactoryPostProcessor 的容器
active	AtomicBoolean	标识是否处于激活状态
closed	AtomicBoolean	标识是否处于关闭状态
startupShutdownMonitor	Object	在执行 refresh 方法和 destroy 时的锁
applicationListeners	Set < ApplicationListener <?>>	应用监听器列表

续表

变量名称	变量类型	说明
id	String	上下文 ID
displayName	String	展示用名称
parent	ApplicationContext	父上下文
environment	ConfigurableEnvironment	环境配置
startupDate	long	上下文启动时间
shutdownHook	Thread	应用上下文关闭之后的线程
resourcePatternResolver	ResourcePatternResolver	资源解析器
lifecycleProcessor	LifecycleProcessor	容器生命周期的处理接口
messageSource	MessageSource	消息源接口，该接口主要用于国际化
applicationEventMulticaster	ApplicationEventMulticaster	事件广播接口
earlyApplicationListeners	Set<ApplicationListener<?>>	早期的应用事件监听器列表
earlyApplicationEvents	Set<ApplicationEvent>	早期的应用事件对象集合

在 AbstractApplicationContext 对象中定义了三个 BeanName，这三个 BeanName 将会对 MessageSource、LifecycleProcessor 和 ApplicationEventMulticaster 的实例化产生影响，具体影响是 BeanName 的约束，与这三个 BeanName 对应的变量是 messageSource、applicationEventMulticaster 和 lifecycleProcessor。

表 8.3 表示的是 DefaultListableBeanFactory 对象中的成员变量。

表 8.3　DefaultListableBeanFactory 成员变量

变量名称	变量类型	说明
serializableFactories	Map<String, Reference<DefaultListableBeanFactory>>	存储 BeanFactory 的容器
javaxInjectProviderClass	Class<?>	注入类型
resolvableDependencies	Map<Class<?>, Object>	解析后的依赖容器。 key：依赖类型 value：依赖对象
beanDefinitionMap	Map<String, BeanDefinition>	存储 Bean 定义的容器。 key：BeanName value：Bean 定义
allBeanNamesByType	Map<Class<?>, String[]>	类型的别名容器 key：类型 value：别名
singletonBeanNamesByType	Map<Class<?>, String[]>	单例 Bean 的别名容器 key：类型 value：别名
serializationId	String	序列化 id
allowBeanDefinitionOverriding	boolean	是否允许名字不同但是 Bean 定义相同
allowEagerClassLoading	boolean	延迟加载的 Bean 是否立即加载

续表

变量名称	变量类型	说明
dependencyComparator	Comparator < Object >	依赖比较器
autowireCandidateResolver	AutowireCandidateResolver	用来进行自动注入（自动装配）的解析类
BeanDefinitionNames	List < String >	BeanDefinition 名称列表
manualSingletonNames	Set < String >	按照注册顺序放入 单例的 BeanName，手动注册的单例对象
frozenBeanDefinitionNames	String[]	冻结的 BeanDefinition 的名称列表
configurationFrozen	boolean	是否需要给 Bean 元数据进行缓存

在 DefaultListableBeanFactory 成员变量中定义了大量的存储容器，这些容器可以理解为 Spring 中的数据库，在 Spring 开发或使用过程中大部分数据都从这些容器中来，比如别名容器 allBeanNamesByType 和 Bean 定义容器 beanDefinitionMap。

8.6 初识 LifecycleProcessor

Spring 中使用 LifecycleProcessor 对象是在完成刷新时，具体方法是 AbstractApplicationContext#finishRefresh，具体代码如下。

```
protected void finishRefresh() {
    //清空资源缓存
    clearResourceCaches();

    //实例化生命周期处理接口
    initLifecycleProcessor();

    //生命周期处理接口进行刷新操作
    getLifecycleProcessor().onRefresh();

    //推送事件：上下文刷新事件
    publishEvent(new ContextRefreshedEvent(this));

    //注册应用上下文
    LiveBeansView.registerApplicationContext(this);
}
```

在这段方法中可以看到 getLifecycleProcessor().onRefresh() 这个方法就是需要分析的目标方法，在 Spring 中对于 LifecycleProcessor 有以下三处处理。

(1) 在 finishRefresh 方法中执行 onRefresh 事件。
(2) 在 start 方法中执行 start 事件。
(3) 在 stop 方法中执行 stop 事件。

在 Spring 中 LifecycleProcessor 实现类目前只有一个 DefaultLifecycleProcessor，它是本章的分析对象。

8.7　LifecycleProcessor 测试环境搭建

接下来将对 LifecycleProcessor 的分析做出前置准备——搭建测试环境。首先需要编写 Lifecycle 接口的实现类，实现类类名为 HelloLifeCycle，具体代码如下。

```java
public class HelloLifeCycle implements Lifecycle {
    private volatile boolean running = false;

    @Override
    public void start() {
        System.out.println("lifycycle start");
        running = true;
    }

    @Override
    public void stop() {
        System.out.println("lifycycle stop");
        running = false;
    }

    @Override
    public boolean isRunning() {
        return running;
    }
}
```

完成实现类编写后需要编写 SpringXML 配置文件，文件名为 spring-lifecycle.xml，具体代码如下。

```xml
<?xml version="1.0" encoding="UTF-8"?>
<beans xmlns:xsi="http://www.w3.org/2001/XMLSchema-instance"
       xmlns="http://www.springframework.org/schema/beans"
       xsi:schemaLocation=" http://www.springframework.org/schema/beans http://www.springframework.org/schema/beans/spring-beans.xsd">

    <bean class="com.source.hot.ioc.book.lifecycle.HelloLifeCycle"/>
</beans>
```

完成 SpringXML 配置文件编写后需要进行测试用例编写，具体代码如下。

```java
public class LifeCycleTest {
    @Test
    void testLifeCycle() {
        ClassPathXmlApplicationContext context =
                new ClassPathXmlApplicationContext("META-INF/spring-lifecycle.xml");
        context.start();
        context.stop();
        context.close();
    }
}
```

执行测试用例会输出下面的结果。

```
lifycycle start
lifycycle stop
```

8.8　start 方法分析

本章所讨论的核心对象是 DefaultLifecycleProcessor，查看它的 start 方法，具体代码如下。

```java
public void start() {
    startBeans(false);
    this.running = true;
}
```

在这段代码中重点关注 startBeans 方法，具体代码如下。

```java
private void startBeans(boolean autoStartupOnly) {
    //获取生命周期 Bean 容器
    Map<String, Lifecycle> lifecycleBeans = getLifecycleBeans();
    //阶段容器
    Map<Integer, LifecycleGroup> phases = new HashMap<>();
    //循环生命周期 Bean 创建阶段对象
    lifecycleBeans.forEach((beanName, bean) -> {

        //1. autoStartupOnly 自动启动
        //2. 类型是否是 SmartLifecycle
        //3. SmartLifecycle 接口的方法 isAutoStartup 执行结果是否为 true
        if (!autoStartupOnly || (bean instanceof SmartLifecycle && ((SmartLifecycle) bean).isAutoStartup())) {
            //获取 Bean 的生命周期阶段
            int phase = getPhase(bean);
            //通过阶段值获取生命周期组
            LifecycleGroup group = phases.get(phase);
            if (group == null) {
                //创建生命周期组
                group = new LifecycleGroup(phase, this.timeoutPerShutdownPhase, lifecycleBeans, autoStartupOnly);
                phases.put(phase, group);
            }
            group.add(beanName, bean);
        }
    });
    if (!phases.isEmpty()) {
        //获取阶段数据
        List<Integer> keys = new ArrayList<>(phases.keySet());
        //阶段排序
        Collections.sort(keys);
        //顺序执行生命周期组的 start 方法
        for (Integer key : keys) {
```

```
            phases.get(key).start();
        }
    }
}
```

在这个方法中有以下三个处理事项。
（1）搜索容器中的 Lifecycle 类型对象，处理方式是通过 getBeanNamesForType 进行获取。
（2）创建阶段容器。
（3）根据阶段容器是否存在进一步执行 Lifecycle 接口的 start 方法。

阶段容器是 Phase 接口，在代码中可以看到（bean instanceof SmartLifecycle && ((SmartLifecycle) bean).isAutoStartup()）代码。在这段代码中 SmartLifecycle 接口也是继承 Phase 接口，但普通的 Lifecycle 接口不具备 Phase 接口的能力，因此在此时对于阶段数值的获取会有下面两种情况。
（1）默认取 0。
（2）从 SmartLifecycle 接口中获取数据。
查看测试用例中的阶段容器如图 8.6。

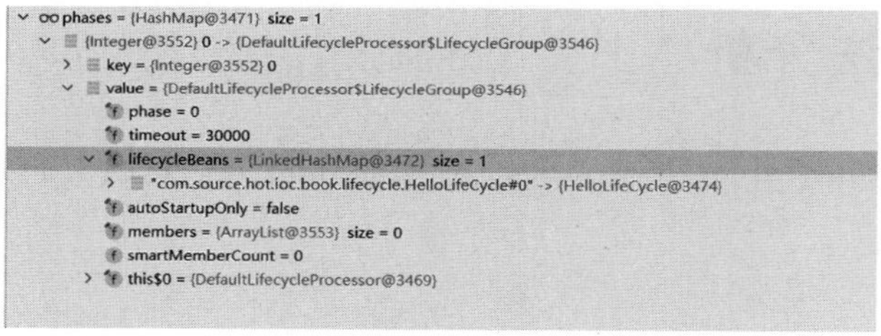

图 8.6　阶段容器

接下来对第(3)项处理进行分析，第(3)项的执行简单一些说就是调用方法，现在已经获取了 Lifecycle 实例，想要执行就很简单，直接执行 bean.start 即可。具体处理代码如下。

```
private void doStart(Map < String, ? extends Lifecycle > lifecycleBeans, String beanName,
boolean autoStartupOnly) {
    //从容器中删除当前处理的 BeanName
    //生命周期接口
    Lifecycle bean = lifecycleBeans.remove(beanName);

    if (bean != null && bean != this) {

        //BeanName 依赖的 BeanName 列表
        String[] dependenciesForBean =
getBeanFactory().getDependenciesForBean(beanName);

        //循环处理依赖的 Bean 生命周期
        for (String dependency : dependenciesForBean) {
            doStart(lifecycleBeans, dependency, autoStartupOnly);
        }
        if (!bean.isRunning() &&
                (!autoStartupOnly || !(bean instanceof SmartLifecycle) ||
```

```
                    ((SmartLifecycle) bean).isAutoStartup())) {
                try {
                    //执行生命周期的 start 方法
                    bean.start();
                }
                catch (Throwable ex) {
                    throw new ApplicationContextException("Failed to start bean '" + beanName +
"'", ex);
                }
            }
        }
    }
}
```

在这段代码中可以看到存在递归操作,这里是为了处理依赖项上的 Lifecycle#start 方法,注意这里会处理优先级,提供依赖的对象优先级永远高于本身,最后才是真正做执行的代码,就是 bean.start()。

8.9 stop 方法分析

接下来将对 DefaultLifecycleProcessor 对象的 stop 方法进行分析,具体处理代码如下。

```
public void stop() {
    stopBeans();
    this.running = false;
}

private void stopBeans() {
    Map<String, Lifecycle> lifecycleBeans = getLifecycleBeans();
    Map<Integer, LifecycleGroup> phases = new HashMap<>();
    lifecycleBeans.forEach((beanName, bean) -> {
        int shutdownPhase = getPhase(bean);
        LifecycleGroup group = phases.get(shutdownPhase);
        if (group == null) {
            group = new LifecycleGroup(shutdownPhase, this.timeoutPerShutdownPhase,
lifecycleBeans, false);
            phases.put(shutdownPhase, group);
        }
        group.add(beanName, bean);
    });
    if (!phases.isEmpty()) {
        List<Integer> keys = new ArrayList<>(phases.keySet());
        keys.sort(Collections.reverseOrder());
        for (Integer key : keys) {
            phases.get(key).stop();
        }
    }
}
```

这段代码的处理操作和 start 的操作基本相同,在这段代码中需要进一步找到 stop 方法,它才是具体的功能方法,具体处理代码如下。

```java
public void stop() {
    if (this.members.isEmpty()) {
        return;
    }
    if (logger.isDebugEnabled()) {
        logger.debug("Stopping beans in phase " + this.phase);
    }
    this.members.sort(Collections.reverseOrder());
    CountDownLatch latch = new CountDownLatch(this.smartMemberCount);
    Set<String> countDownBeanNames =
            Collections.synchronizedSet(new LinkedHashSet<>());
    Set<String> lifecycleBeanNames = new HashSet<>(this.lifecycleBeans.keySet());
    for (LifecycleGroupMember member : this.members) {
        if (lifecycleBeanNames.contains(member.name)) {
            doStop(this.lifecycleBeans, member.name, latch, countDownBeanNames);
        }
        else if (member.bean instanceof SmartLifecycle) {
            latch.countDown();
        }
    }
    try {
        latch.await(this.timeout, TimeUnit.MILLISECONDS);
        if (latch.getCount() > 0 &&
                !countDownBeanNames.isEmpty() && logger.isInfoEnabled()) {
            logger.info("Failed to shut down " + countDownBeanNames.size() + " bean" +
                    (countDownBeanNames.size() > 1 ? "s" : "") + " with phase value " +
                    this.phase + " within timeout of " + this.timeout + ": " +
                    countDownBeanNames);
        }
    }
    catch (InterruptedException ex) {
        Thread.currentThread().interrupt();
    }
}
```

在这段代码中会执行成员变量 members 保存的 Lifecycle 接口,并调用接口提供的 stop 方法。

8.10 LifecycleGroup 相关变量

接下来将对处理容器生命周期中的 LifecycleGroup 对象进行说明,关于 LifecycleGroup 对象重点在成员变量上,具体信息可查看表 8.4。

表 8.4 LifecycleGroup 成员变量

变量名称	变量类型	说 明
phase	int	从 Phase 接口中获取的数据，表示阶段值
timeout	long	超时时间
lifecycleBeans	Map < String,? extends Lifecycle >	存储 Lifecycle 接口实现类的容器。key：BeanName value：Lifecycle 实例
autoStartupOnly	boolean	是否自动启动
members	List < LifecycleGroupMember >	存储 Lifecycle 的容器
smartMemberCount	int	SmartLifecycle 实例的数量

在 LifecycleGroup 成员变量中还需要使用 LifecycleGroupMember 对象，关于 LifecycleGroupMember 的成员变量信息如下。

（1）name 属性表示 BeanName。

（2）bean 表示 Lifecycle 接口实例。

8.11 BeanPostProcessor 注册

接下来将分析 PostProcessorRegistrationDelegate 中关于 BeanPostProcessor 的注册方法，在 Spring 中关于 BeanPostProcessor 注册事件的唤醒在 AbstractApplicationContext#refresh 方法中有提及：

注册 BeanPostProcessor：org.springframework.context.support.AbstractApplicationContext#registerBeanPostProcessors。

调用 BeanFactoryPostProcessor：org.springframework.context.support.AbstractApplicationContext#invokeBeanFactoryPostProcessors。

最终负责注册方法的签名是：org.springframework.context.support.PostProcessorRegistrationDelegate#registerBeanPostProcessors（org.springframework.beans.factory.config.ConfigurableListableBeanFactory，org.springframework.context.support.AbstractApplicationContext）。

在注册方法中对 BeanPostProcessor 有下面四种类型。

（1）priorityOrderedPostProcessors：同时实现了 BeanPostProcessor 接口和 PriorityOrdered 接口的 BeanPostProcessor 实例。

（2）internalPostProcessors：同时实现了 BeanPostProcessor 接口、MergedBeanDefinitionPostProcessor 接口和 PriorityOrdered 接口的 BeanPostProcessor 实例。

（3）orderedPostProcessorNames：同时实现了 BeanPostProcessor 接口和 Ordered 接口的 BeanPostProcessor 实例的 BeanName。

（4）nonOrderedPostProcessorNames：只实现了 BeanPostProcessor 接口的 BeanPostProcessor 实例的 BeanName。

上述四种 BeanPostProcessor 的注册顺序如下。

（1）priorityOrderedPostProcessors。
（2）orderedPostProcessors。
（3）nonOrderedPostProcessors。
（4）internalPostProcessors。
（5）独立的 BeanPostProcessors 注册，类型是 ApplicationListenerDetector。

在注册之前除了 nonOrderedPostProcessors 以外其他 BeanPostProcessors 都会进行排序操作，这个排序操作是由 PriorityOrdered 和 Ordered 提供的，在这两个接口中有提供 getOrder 方法，排序就是依靠这个字段的数据内容进行排序。具体排序代码如下。

```java
private static void
sortPostProcessors(List<?> postProcessors, ConfigurableListableBeanFactory beanFactory) {
    Comparator<Object> comparatorToUse = null;
    if (beanFactory instanceof DefaultListableBeanFactory) {
        comparatorToUse =
((DefaultListableBeanFactory) beanFactory).getDependencyComparator();
    }
    if (comparatorToUse == null) {
        comparatorToUse = OrderComparator.INSTANCE;
    }
    postProcessors.sort(comparatorToUse);
}
```

在执行完成排序后就会进入注册阶段，注册方法是一个统一方法，具体代码如下。

```java
private static void registerBeanPostProcessors(
        ConfigurableListableBeanFactory beanFactory, List<BeanPostProcessor>
postProcessors) {

    for (BeanPostProcessor postProcessor : postProcessors) {
        beanFactory.addBeanPostProcessor(postProcessor);
    }
}
```

这个注册方法会将参数传递的 postProcessors 对象都注册到 AbstractBeanFactory♯beanPostProcessors 中。

8.12　BeanFactoryPostProcessor 方法调用

下面将对 BeanFactoryPostProcessor 接口方法调用进行分析。在执行 BeanFactoryPostProcessor 方法时会分别执行下面三个不同的 BeanDefinitionRegistryPostProcessor。
（1）同时实现 BeanDefinitionRegistryPostProcessor 和 PriorityOrdered。
（2）同时实现 BeanDefinitionRegistryPostProcessor 和 Ordered。
（3）只实现 BeanDefinitionRegistryPostProcessor。

上述三种不同的 BeanDefinitionRegistryPostProcessor 在执行时存在先后关系，具体的执行顺序是：最高 PriorityOrdered，其次 Ordered（需要根据接口数据值排序），最后没有实现的 Ordered。

这部分处理代码如下。

```java
public static void invokeBeanFactoryPostProcessors(
            ConfigurableListableBeanFactory
beanFactory, List<BeanFactoryPostProcessor> beanFactoryPostProcessors) {

    //需要处理的 BeanDefinitionRegistryPostProcessors 名称
    Set<String> processedBeans = new HashSet<>();

    //关于 BeanFactoryPostProcessor 的处理
    //BeanFactory 类型是 BeanDefinitionRegistry
    if (beanFactory instanceof BeanDefinitionRegistry) {
        BeanDefinitionRegistry registry = (BeanDefinitionRegistry) beanFactory;
        //BeanFactoryPostProcessor 接口列表
        List<BeanFactoryPostProcessor> regularPostProcessors = new ArrayList<>();
        //BeanDefinitionRegistryPostProcessor 接口列表
        List<BeanDefinitionRegistryPostProcessor> registryProcessors =
new ArrayList<>();

        //数据分类
        for (BeanFactoryPostProcessor postProcessor : beanFactoryPostProcessors) {
            if (postProcessor instanceof BeanDefinitionRegistryPostProcessor) {
                BeanDefinitionRegistryPostProcessor registryProcessor =
                        (BeanDefinitionRegistryPostProcessor) postProcessor;
                //将 beanDefinition 进行注册, Configuration 注解标注的对象
                registryProcessor.postProcessBeanDefinitionRegistry(registry);
                registryProcessors.add(registryProcessor);
            }
            else {
                regularPostProcessors.add(postProcessor);
            }
        }

        //Bean 定义后置处理器
        List<BeanDefinitionRegistryPostProcessor> currentRegistryProcessors = new
ArrayList<>();

        //处理 BeanDefinitionRegistryPostProcessor + PriorityOrdered
        //获取 BeanDefinitionRegistryPostProcessor 的 beanName
        String[] postProcessorNames =

beanFactory.getBeanNamesForType(BeanDefinitionRegistryPostProcessor.class,
true, false);
        //处理 BeanDefinitionRegistryPostProcessor beanName
        for (String ppName : postProcessorNames) {
            //类型是 PriorityOrdered
            if (beanFactory.isTypeMatch(ppName, PriorityOrdered.class)) {
                //添加到容器 currentRegistryProcessors
                currentRegistryProcessors.add(beanFactory.getBean(ppName,
BeanDefinitionRegistryPostProcessor.class));
                processedBeans.add(ppName);
```

```java
            }
        }

        //排序 BeanDefinitionRegistryPostProcessor 对象
        sortPostProcessors(currentRegistryProcessors, beanFactory);
        registryProcessors.addAll(currentRegistryProcessors);
        //执行 BeanDefinitionRegistryPostProcessor 的方法
        invokeBeanDefinitionRegistryPostProcessors(currentRegistryProcessors, registry);
        //清理数据
        currentRegistryProcessors.clear();

        //处理 BeanDefinitionRegistryPostProcessor + Ordered
        postProcessorNames =
beanFactory.getBeanNamesForType(BeanDefinitionRegistryPostProcessor.class, true, false);

        for (String ppName : postProcessorNames) {
            if (!processedBeans.contains(ppName)
&& beanFactory.isTypeMatch(ppName, Ordered.class)) {
                currentRegistryProcessors.add(beanFactory.getBean(ppName,
BeanDefinitionRegistryPostProcessor.class));
                processedBeans.add(ppName);
            }
        }
        sortPostProcessors(currentRegistryProcessors, beanFactory);
        registryProcessors.addAll(currentRegistryProcessors);
        invokeBeanDefinitionRegistryPostProcessors(currentRegistryProcessors, registry);
        currentRegistryProcessors.clear();

        //处理剩下的 BeanDefinitionRegistryPostProcessor
        boolean reiterate = true;
        while (reiterate) {
            reiterate = false;
            postProcessorNames =
beanFactory.getBeanNamesForType(BeanDefinitionRegistryPostProcessor.class, true, false);
            for (String ppName : postProcessorNames) {
                if (!processedBeans.contains(ppName)) {
                    currentRegistryProcessors.add(beanFactory.getBean(ppName,
BeanDefinitionRegistryPostProcessor.class));
                    processedBeans.add(ppName);
                    reiterate = true;
                }
            }
            sortPostProcessors(currentRegistryProcessors, beanFactory);
            registryProcessors.addAll(currentRegistryProcessors);
            invokeBeanDefinitionRegistryPostProcessors(currentRegistryProcessors, registry);
            currentRegistryProcessors.clear();
        }

        //BeanDefinitionRegistryPostProcessor 处理
        invokeBeanFactoryPostProcessors(registryProcessors, beanFactory);
        //BeanFactoryPostProcessor 处理
```

```java
                invokeBeanFactoryPostProcessors(regularPostProcessors, beanFactory);
            }

            //BeanFactory 其他类型的处理
            else {
                invokeBeanFactoryPostProcessors(beanFactoryPostProcessors, beanFactory);
            }

            //处理 BeanFactoryPostProcessor 类型的 Bean
            String[] postProcessorNames =
                    beanFactory.getBeanNamesForType(BeanFactoryPostProcessor.class, true, false);

            //分成两组进行处理
            //1. BeanFactoryPostProcessor + PriorityOrdered -> priorityOrderedPostProcessors
            //2. BeanFactoryPostProcessor + Ordered -> orderedPostProcessorNames
            //3. BeanFactoryPostProcessor -> nonOrderedPostProcessorNames
            List<BeanFactoryPostProcessor> priorityOrderedPostProcessors = new ArrayList<>();
            List<String> orderedPostProcessorNames = new ArrayList<>();
            List<String> nonOrderedPostProcessorNames = new ArrayList<>();
            //根据不同实现类进行分组
            for (String ppName : postProcessorNames) {
                if (processedBeans.contains(ppName)) {
                }
                else if (beanFactory.isTypeMatch(ppName, PriorityOrdered.class)) {
                    priorityOrderedPostProcessors.add(beanFactory.getBean(ppName, BeanFactoryPostProcessor.class));
                }
                else if (beanFactory.isTypeMatch(ppName, Ordered.class)) {
                    orderedPostProcessorNames.add(ppName);
                }
                else {
                    nonOrderedPostProcessorNames.add(ppName);
                }
            }

            //处理 BeanFactoryPostProcessors + PriorityOrdered 的类型
            sortPostProcessors(priorityOrderedPostProcessors, beanFactory);
            invokeBeanFactoryPostProcessors(priorityOrderedPostProcessors, beanFactory);

            //处理 BeanFactoryPostProcessors + Ordered 的类型
            List<BeanFactoryPostProcessor> orderedPostProcessors = new ArrayList<>(orderedPostProcessorNames.size());
            for (String postProcessorName : orderedPostProcessorNames) {
                orderedPostProcessors.add(beanFactory.getBean(postProcessorName, BeanFactoryPostProcessor.class));
            }
            sortPostProcessors(orderedPostProcessors, beanFactory);
            invokeBeanFactoryPostProcessors(orderedPostProcessors, beanFactory);
```

```
        //处理普通的 BeanFactoryPostProcessors 类型
        List < BeanFactoryPostProcessor > nonOrderedPostProcessors =
new ArrayList<>(nonOrderedPostProcessorNames.size());
        for (String postProcessorName : nonOrderedPostProcessorNames) {
            nonOrderedPostProcessors.add(beanFactory.getBean(postProcessorName,
BeanFactoryPostProcessor.class));
        }
        invokeBeanFactoryPostProcessors(nonOrderedPostProcessors, beanFactory);

        beanFactory.clearMetadataCache();
    }
```

小结

本章分析了关于 BeanFactoryPostProcessor 接口的注册和调用两个方法，负责处理这两个行为的对象是 PostProcessorRegistrationDelegate，采用了委派模式进行开发，在这两个方法使用中间接了解了 Spring 中关于 Bean 排序的操作，从整体上看，这里的调用和注册相对前几章提到的一些内容都简单了很多。

在本章中以 SpringXML 配置的形式展开，通过三种方式来进行容器创建，本章对创建的细节进行了追踪，了解其内部实现细节，在创建容器后进一步去查找了关于摧毁容器的相关实现。通过分析可以发现，XmlBeanFactory 在三个容器中是一个比较简单的 BeanFactory，相比较 FileSystemXmlApplicationContext 和 ClassPathXmlApplicationContext 继承关系更少一些，在后面分析了 FileSystemXmlApplicationContext 和 ClassPathXmlApplicationContext 的构造函数和 close 函数，经过分析发现这两者其实是一个同源的类在各自类设计上存在一定差异（成员变量）。此外，围绕 LifecycleProcessor 接口进行分析，通过 LifecycleProcessor 接口进一步了解 Spring 中关于容器生命周期的处理内容，处理了 start 和 stop 的生命周期。

第9章

Spring注解模式

本章开始进入 Spring 注解模式相关分析。在 Spring 中除了 SpringXML 启动模式以外还有另一种启动模式——Spring 注解模式（Spring 注解驱动），这项技术在后续的 SpringBoot 中大放异彩。本章主要介绍 Spring 注解模式的开发环境（源码测试环境）搭建和两种启动模式的分析。

9.1 注解模式测试环境搭建

本节将搭建一个 Spring 注解模式的测试环境，首先需要创建一个 JavaBean，类名为 AnnPeople，具体代码如下。

```
public class AnnPeople {
    private String name;

    public String getName() {
        return name;
    }

    public void setName(String name) {
        this.name = name;
    }
}
```

完成 JavaBean 编写后需要编写 Spring 注解模式中最重要的配置类，类名为 AnnBeans，具体代码如下。

```
@Component
public class AnnBeans {
    @Bean
    public AnnPeople annPeople() {
```

```
        AnnPeople annPeople = new AnnPeople();
        annPeople.setName("people");
        return annPeople;
    }
}
```

最后编写测试方法,具体代码如下。

```
public class AnnotationContextTest {

    @Test
    void testBasePackages(){
        AnnotationConfigApplicationContext context =
                new AnnotationConfigApplicationContext("com.source.hot.ioc.book.ann");
        AnnPeople bean = context.getBean(AnnPeople.class);
        assert bean.getName().equals("people");
    }

    @Test
    void testComponentClasses(){
        AnnotationConfigApplicationContext context =
                new AnnotationConfigApplicationContext(AnnBeans.class);
        AnnPeople bean = context.getBean(AnnPeople.class);
        assert bean.getName().equals("people");
    }
}
```

在测试用例中使用了两种模式进行应用上下文启动,第一种模式是 basePackages,第二种模式是 componentClasses。

9.2 basePackages 模式启动

接下来将对 basePackages 模式启动做相关分析,关于 basePackages 启动的构造函数如下。

```
public AnnotationConfigApplicationContext(String... basePackages) {
    this();
    scan(basePackages);
    refresh();
}
```

在这段代码中需要关注 this,它指向了 AnnotationConfigApplicationContext 对象的无参构造,在无参构造中有下面的代码。

```
public AnnotationConfigApplicationContext() {
    this.reader = new AnnotatedBeanDefinitionReader(this);
    this.scanner = new ClassPathBeanDefinitionScanner(this);
}
```

在这段代码中创建了两个核心对象,这两个对象可以读取资源,解析资源。

(1) AnnotatedBeanDefinitionReader 对象用来读取注解转换成 BeanDefinition 进行注册。

(2) ClassPathBeanDefinitionScanner 对象用来扫描某些指定包路径下的类转换成 BeanDefinition 进行注册。

9.2.1 scan 方法分析

通过无参构造函数创建两个对象后即可进入 scan 方法，scan 需要用到的对象是 ClassPathBeanDefinitionScanner，具体的扫描方法如下。

```
public int scan(String... basePackages) {
    //在执行扫描方法前 BeanDefinition 的数量
    int beanCountAtScanStart = this.registry.getBeanDefinitionCount();

    //真正的扫描方法
    doScan(basePackages);

    //是否需要注册注解的后置处理器
    if (this.includeAnnotationConfig) {
        //注册注解后置处理器
        AnnotationConfigUtils.registerAnnotationConfigProcessors(this.registry);
    }

    //当前 BeanDefinition 数量
    return (this.registry.getBeanDefinitionCount() - beanCountAtScanStart);
}
```

在这段扫描方法中处理了以下四个行为。
(1) 获取没有执行扫描前的 BeanDefinition 的数量。
(2) 进行扫描。
(3) 注册各类处理器(偏向注解配置的处理器)。
(4) 计算扫描后新增的 BeanDefinition 数量。

9.2.2 doScan 方法分析

在 ClassPathBeanDefinitionScanner#scan 方法中最关键的进行扫描的方法是 doScan 方法，具体处理代码如下。

```
protected Set<BeanDefinitionHolder> doScan(String... basePackages) {
    Assert.notEmpty(basePackages, "At least one base package must be specified");
    //Bean 定义持有器列表
    Set<BeanDefinitionHolder> beanDefinitions = new LinkedHashSet<>();
    //循环包路径进行扫描
    for (String basePackage : basePackages) {
        //搜索可能的组件,得到组件的 BeanDefinition
        Set<BeanDefinition> candidates = findCandidateComponents(basePackage);
```

```java
        //循环候选 Bean 定义
        for (BeanDefinition candidate : candidates) {
            //获取作用域元数据
            ScopeMetadata scopeMetadata =
this.scopeMetadataResolver.resolveScopeMetadata(candidate);
            //设置作用域
            candidate.setScope(scopeMetadata.getScopeName());
            //BeanName 生成
            String beanName = this.beanNameGenerator.generateBeanName(candidate,
this.registry);
            //类型判断 AbstractBeanDefinition
            if (candidate instanceof AbstractBeanDefinition) {
                //Bean 定义的后置处理
                postProcessBeanDefinition((AbstractBeanDefinition) candidate, beanName);
            }
            //类型判断 AnnotatedBeanDefinition
            if (candidate instanceof AnnotatedBeanDefinition) {
                //通用注解的处理
AnnotationConfigUtils.processCommonDefinitionAnnotations((AnnotatedBeanDefinition) candidate);
            }
            //候选检测
            if (checkCandidate(beanName, candidate)) {
                BeanDefinitionHolder definitionHolder =
new BeanDefinitionHolder(candidate, beanName);
                //作用域属性应用
                definitionHolder =
                        AnnotationConfigUtils.applyScopedProxyMode(scopeMetadata, definitionHolder,
this.registry);
                beanDefinitions.add(definitionHolder);
                //注册 Bean 定义
                registerBeanDefinition(definitionHolder, this.registry);
            }
        }
    }
    return beanDefinitions;
}
```

上述方法的处理流程如下。

（1）搜索某个包路径下的所有 BeanDefintion。

（2）对单个 BeanDefinition 进行处理，总共有以下 5 个处理。

① Scope 相关处理。

② BeanName 相关处理。

③ BeanDefinition 的补充处理，BeanDefinition 默认数据处理。Spring 中通用注解处理。

④ 代理相关的 Scope 设置。

⑤ 注册到容器。

接下来将对 findCandidateComponents 方法进行分析，具体处理代码如下。

```java
public Set<BeanDefinition> findCandidateComponents(String basePackage) {
    if (this.componentsIndex != null && indexSupportsIncludeFilters()) {
        return addCandidateComponentsFromIndex(this.componentsIndex, basePackage);
    }
    else {
        return scanCandidateComponents(basePackage);
    }
}
```

在这段代码中有两个处理方法。首先关注方法 scanCandidateComponents，它的具体代码信息如下。

```java
//删除日志和部分 else 代码
private Set<BeanDefinition> scanCandidateComponents(String basePackage) {
    //候选组件列表 BeanDefinition 列表
    Set<BeanDefinition> candidates = new LinkedHashSet<>();
    try {
        //classpath*: + replace(basePackage,'.','/') + / + **/*.class
        String packageSearchPath =
ResourcePatternResolver.CLASSPATH_ALL_URL_PREFIX +
                resolveBasePackage(basePackage) + '/' + this.resourcePattern;
        //转换成资源对象
        //这里会转换成 FileSystemResource
        Resource[] resources =
getResourcePatternResolver().getResources(packageSearchPath);

        //资源处理
        for (Resource resource : resources) {

            if (resource.isReadable()) {
                try {
                    //元数据读取器
                    MetadataReader metadataReader =
getMetadataReaderFactory().getMetadataReader(resource);
                    if (isCandidateComponent(metadataReader)) {
                        //Bean 定义扫描
                        ScannedGenericBeanDefinition sbd =
new ScannedGenericBeanDefinition(metadataReader);
                        //设置资源对象
                        sbd.setResource(resource);
                        //设置源对象
                        sbd.setSource(resource);
                        //判断是否是候选值
                        if (isCandidateComponent(sbd)) {

                            //加入容器
                            candidates.add(sbd);
                        }
                    }
                }
                catch (Throwable ex) {
                    throw new BeanDefinitionStoreException(
                            "Failed to read candidate component class: " + resource, ex);
                }
            }
        }
```

```
        }
        catch (IOException ex) {
            throw new BeanDefinitionStoreException("I/O failure during classpath scanning", ex);
        }
        return candidates;
    }
```

在这个方法中包含扫描的基本操作及思想,具体如下。

(1) 包路径转换。

(2) 将转换后的包路径解析成资源对象。

(3) 资源对象到 BeanDefinition 的转换。

包路径转换公式为"classpath*:" + basePackage.replace('.','/') + "/" + "**/*.class",经过步骤(1)计算后得到的数据如图9.1所示。

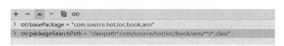

图 9.1　包路径转换结果

在得到一个以 classpath 开头的路径地址后 Spring 会进行数据收集,主要将各个可能的 class 都收集并将其组装成 FileSystemResource 对象(也可能有其他类型的 Resource 对象),负责处理该行为的核心类是 PathMatchingResourcePatternResolver。收集 Resource 的方法是 getResourcePatternResolver().getResources(packageSearchPath),收集后的数据结果如图9.2所示。

图 9.2　收集 Resource 结果

通过观察图 9.2 可以发现有两种数据形式，第一种是文件形式的资源对象，第二种是 Jar 形式的资源对象。这两种不同的数据形式都将通过一个方法进行转换，转换为 BeanDefinition 等待后续使用，转换方法是 getMetadataReaderFactory().getMetadataReader(resource)。

转换类是 metadataReader，它可以判断是否能够转换以及转换，具体判断是否能够转换的方法如下。

```
protected boolean isCandidateComponent(MetadataReader metadataReader)
throws IOException {
    for (TypeFilter tf : this.excludeFilters) {
        if (tf.match(metadataReader, getMetadataReaderFactory())) {
            return false;
        }
    }
    for (TypeFilter tf : this.includeFilters) {
        if (tf.match(metadataReader, getMetadataReaderFactory())) {
            return isConditionMatch(metadataReader);
        }
    }
    return false;
}
```

在这个方法中需要使用以下两个列表。
（1）includeFilters：需要的类型过滤器。
（2）excludeFilters：排除的类型过滤器。

在 includeFilters 对象中存放了关于注解 Component 的类型过滤器，具体信息如图 9.3 所示。

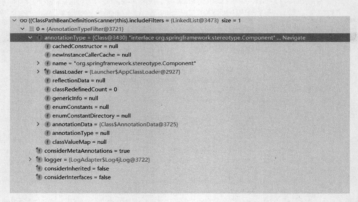

图 9.3　includeFilters 数据信息

在这段检测相关的代码中，其主要目的是判断是否存在 Component 注解，具体的判断方法是 tf.match(metadataReader, getMetadataReaderFactory())，在判断方法中通过 getMetadataReaderFactory 方法可以获取到一些元数据，具体元数据如图 9.4 所示。

经过数据检测后会进入注解 Conditional 的处理，测试用例中并未涉及该注解，不进入分析，如果没有标记 Conditional 注解会直接返回 true，完成上述操作后就可以进行 BeanDefinition 对象的创建，此时用到的 BeanDefinition 对象是 ScannedGenericBeanDefinition，经过处理的

BeanDefinition 对象数据如图 9.5 所示。

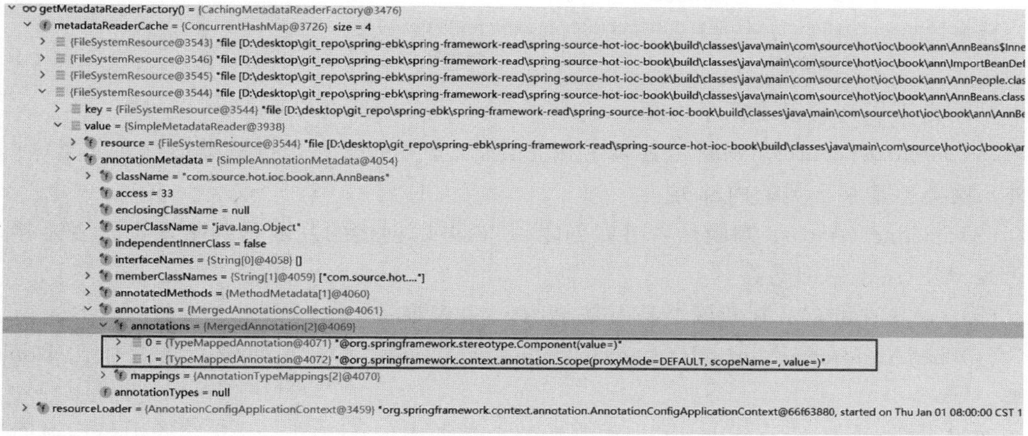

图 9.4 getMetadataReaderFactory 数据信息

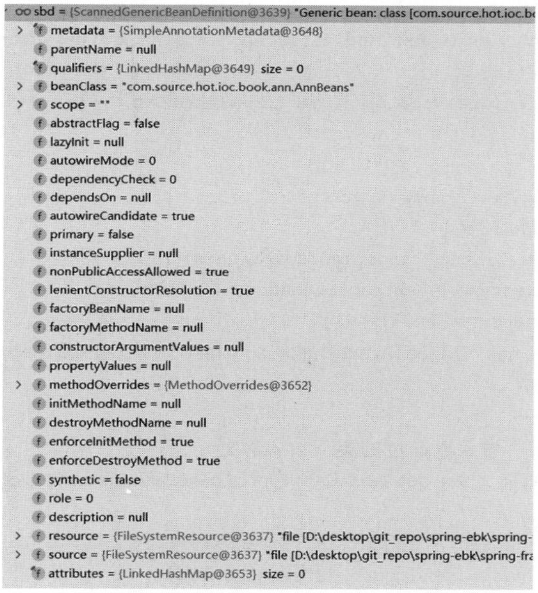

图 9.5 解析后的 BeanDefinition 对象

得到 BeanDefinition 对象后还需要进行候选判断，只有通过候选判断的 BeanDefinition 对象才可以放到结果集合中，具体的判断方法如下。

```
protected boolean isCandidateComponent(AnnotatedBeanDefinition beanDefinition) {
    //从注解的 Bean 定义中获取注解元信息
    AnnotationMetadata metadata = beanDefinition.getMetadata();
    //1. 是否独立
    //2. 是否可以创建
    //3. 是否 Abstract 修饰
    //4. 是否有 Lookup 注解
```

```
        return (metadata.isIndependent() && (metadata.isConcrete() ||
            (metadata.isAbstract()
    && metadata.hasAnnotatedMethods(Lookup.class.getName()))));
}
```

判断是否候选有以下四个方案。

(1) isIndependent()：判断是否属于独立类，独立类的概念是指该类不存在父类(Object 除外)，该类不是某个类的内部类。

(2) isConcreteu()：判断是否可以创建，是否可以创建的判断是判断该类是否是接口，是否是 Abstract 修饰的类。

(3) isAbstract()：判断该类是不是 Abstract 修饰。

(4) hasAnnotatedMethods(Lookup.class.getName())：判断该类是否存在 Lookup 注解。

通过 isCandidateComponent 中的判断就会被加入 BeanDefinition 容器中最后返回。

接下来对 addCandidateComponentsFromIndex 方法进行分析，具体处理代码如下。

```java
private Set<BeanDefinition>
addCandidateComponentsFromIndex(CandidateComponentsIndex index, String basePackage) {
    //候选 BeanDefinition
    Set<BeanDefinition> candidates = new LinkedHashSet<>();
    try {
        //类型列表
        Set<String> types = new HashSet<>();
        //导入的类型过滤器
        for (TypeFilter filter : this.includeFilters) {
            String stereotype = extractStereotype(filter);
            if (stereotype == null) {
                throw new IllegalArgumentException("Failed to extract stereotype from " +
filter);
            }
            //从组件索引中获取通过的类型放入容器
            types.addAll(index.getCandidateTypes(basePackage, stereotype));
        }
        boolean traceEnabled = logger.isTraceEnabled();
        boolean debugEnabled = logger.isDebugEnabled();
        //类型处理
        for (String type : types) {
            MetadataReader metadataReader =
getMetadataReaderFactory().getMetadataReader(type);
            //是否是候选组件
            if (isCandidateComponent(metadataReader)) {
                AnnotatedGenericBeanDefinition sbd =
new AnnotatedGenericBeanDefinition(
                    metadataReader.getAnnotationMetadata());

                //是否是候选组件
                if (isCandidateComponent(sbd)) {
                    if (debugEnabled) {
```

```
                    logger.debug("Using candidate component class from index: " + type);
                }
                candidates.add(sbd);
            }
            else {
                if (debugEnabled) {
                    logger.debug("Ignored because not a concrete top-level class: " + type);
                }
            }
        }
        else {
            if (traceEnabled) {
                logger.trace("Ignored because matching an exclude filter: " + type);
            }
        }
    }
}
catch (IOException ex) {
    throw new BeanDefinitionStoreException("I/O failure during classpath scanning", ex);
}
return candidates;
}
```

这个方法和 scanCandidateComponents 方法的差异只有一点，即 BeanDefinition 对象不同，其他操作都相同，在 scanCandidateComponents 方法中 BeanDefinition 的类型是 ScannedGenericBeanDefinition，在 addCandidateComponentsFromIndex 方法中 BeanDefinition 的类型是 AnnotatedGenericBeanDefinition。

9.2.3 处理单个 BeanDefinition

通过 findCandidateComponents 方法可以得到多个 BeanDefinition 信息，在得到这些 BeanDefinition 后需要对其进行处理，具体处理代码如下。

```
for (BeanDefinition candidate : candidates) {
    //获取作用域元数据
    ScopeMetadata scopeMetadata =
this.scopeMetadataResolver.resolveScopeMetadata(candidate);
    //设置作用域
    candidate.setScope(scopeMetadata.getScopeName());
    //BeanName 生成
    String beanName =
this.beanNameGenerator.generateBeanName(candidate, this.registry);
    //类型判断 AbstractBeanDefinition
    if (candidate instanceof AbstractBeanDefinition) {
        //Bean 定义的后置处理
        postProcessBeanDefinition((AbstractBeanDefinition) candidate, beanName);
    }
    //类型判断 AnnotatedBeanDefinition
    if (candidate instanceof AnnotatedBeanDefinition) {
```

```
        //通用注解的处理
AnnotationConfigUtils.processCommonDefinitionAnnotations((AnnotatedBeanDefinition) candidate);
    }
        //候选检测
    if (checkCandidate(beanName, candidate)) {
        BeanDefinitionHolder definitionHolder =
new BeanDefinitionHolder(candidate, beanName);
        //作用域属性应用
        definitionHolder =
            AnnotationConfigUtils.applyScopedProxyMode(scopeMetadata, definitionHolder,
this.registry);
        beanDefinitions.add(definitionHolder);
        //注册 Bean 定义
        registerBeanDefinition(definitionHolder, this.registry);
    }
}
```

(1) Scope 相关处理。具体处理会进行 Scope 注解解析,将解析结果放入 BeanDefinition 对象中,由于测试类中并没有包含 Scope 注解,因此解析结果为空字符串,具体解析结果如图 9.6 所示。

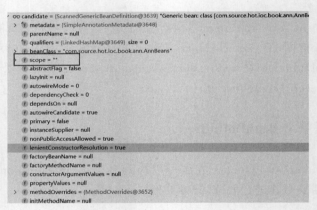

图 9.6 Scope 解析结果

(2) BeanName 相关处理。BeanName 的处理会通过 BeanNameGenerator 接口进行,具体的能力提供者是 AnnotationBeanNameGenerator 对象,具体处理方法如下。

```
@Override
public String generateBeanName(BeanDefinition definition, BeanDefinitionRegistry registry) {
    if (definition instanceof AnnotatedBeanDefinition) {
        //从注解中获取 BeanName
        //获取注解的 value 属性值
        String beanName = determineBeanNameFromAnnotation((AnnotatedBeanDefinition) definition);
        if (StringUtils.hasText(beanName)) {
            //如果存在直接返回
            return beanName;
        }
    }
```

```
    //默认 BeanName
    //类名,首字母小写
    return buildDefaultBeanName(definition, registry);
}
```

上述代码中的 BeanName 生成流程如下。

① 从 Spring 的注解中获取 value 数据作为 BeanName。

② 将类名首字母小写返回。

测试用例中的 BeanName 生成结果如图 9.7 所示。

图 9.7　BeanName 生成结果

（3）BeanDefinition 默认值处理。在处理完成 BeanName 内容后会对 BeanDefinition 的默认值进行填充，填充的数据有 lazyInit、autowireMode、dependencyCheck、initMethodName、enforceInitMethod、destroyMethodName 和 enforceDestroyMethod。除了这些填充以外，在方法 postProcessBeanDefinition 中还会对 autowireCandidate 属性进行补充。经过默认值填充的结果如图 9.8 所示。

图 9.8　BeanDefinition 默认值处理

（4）Spring 中通用注解的处理。在补充完成 BeanDefinition 的一些基础属性后会进入 Spring 注解模式下的注解值解析设置属性的阶段。主要处理方法是 AnnotationConfigUtils. processCommonDefinitionAnnotations((AnnotatedBeanDefinition) candidate)，处理代码如下。

```
static void
processCommonDefinitionAnnotations(AnnotatedBeanDefinition
abd, AnnotatedTypeMetadata metadata) {
    AnnotationAttributes lazy = attributesFor(metadata, Lazy.class);
    if (lazy != null) {
```

```
            abd.setLazyInit(lazy.getBoolean("value"));
        }
        else if (abd.getMetadata() != metadata) {
            lazy = attributesFor(abd.getMetadata(), Lazy.class);
            if (lazy != null) {
                abd.setLazyInit(lazy.getBoolean("value"));
            }
        }

        if (metadata.isAnnotated(Primary.class.getName())) {
            abd.setPrimary(true);
        }
        AnnotationAttributes dependsOn = attributesFor(metadata, DependsOn.class);
        if (dependsOn != null) {
            abd.setDependsOn(dependsOn.getStringArray("value"));
        }

        AnnotationAttributes role = attributesFor(metadata, Role.class);
        if (role != null) {
            abd.setRole(role.getNumber("value").intValue());
        }
        AnnotationAttributes description = attributesFor(metadata, Description.class);
        if (description != null) {
            abd.setDescription(description.getString("value"));
        }
    }
```

在这段代码中会进行通用注解的处理，方式是从注解元数据中提取对应的注解数据，再设置到对应的 BeanDefinition 属性上，具体处理的注解有 Lazy、DependsOn、Role 和 Description。

（5）代理相关的 Scope 设置。在处理完成 Spring 中的常用注解后，Spring 会创建 Scope Proxy BeanDefinition。在创建之前会有以下两层验证。

① checkCandidate：当前 BeanName 和 BeanDefiniton 对象是否是候选关系。

② scopedProxyMode.equals(ScopedProxyMode.NO)：ScopedProxyMode 属性是否是 NO，如果是 NO 就直接返回。

checkCandidate 方法代码如下。

```
protected boolean checkCandidate(String beanName, BeanDefinition beanDefinition) throws
        IllegalStateException {
    //当前注册器中是否包含 BeanName
    if (!this.registry.containsBeanDefinition(beanName)) {
        return true;
    }
    //注册器中的 BeanName 的 BeanInstance
    BeanDefinition existingDef = this.registry.getBeanDefinition(beanName);
    BeanDefinition originatingDef = existingDef.getOriginatingBeanDefinition();
    if (originatingDef != null) {
        existingDef = originatingDef;
    }
    //两个对象是否兼容
```

```java
        if (isCompatible(beanDefinition, existingDef)) {
            return false;
        }
        throw new ConflictingBeanDefinitionException("Annotation-specified bean name '" +
beanName + "' for bean class [" + beanDefinition.getBeanClassName() + "] conflicts with
existing, " + "non-compatible bean definition of same name and class [" + existingDef.
getBeanClassName() + "]");
    }
    protected boolean
isCompatible(BeanDefinition newDefinition, BeanDefinition existingDefinition) {
        //1. 是不是 ScannedGenericBeanDefinition 类型
        //2. source 是否相同
        //3. 参数是否相同
        return (!(existingDefinition instanceof ScannedGenericBeanDefinition) ||
                (newDefinition.getSource() !=
null && newDefinition.getSource().equals(existingDefinition.getSource())) ||
                newDefinition.equals(existingDefinition));
    }
```

该方法的作用是判断容器中 BeanName 对应的 BeanDefinition 和当前需要进行验证的 BeanDefinition 作用域是否相同，两个 BeanDefinition 的 equals 方法是否相同。

继续向下追踪 applyScopedProxyMode 方法，具体代码如下。

```java
static BeanDefinitionHolder applyScopedProxyMode(
        ScopeMetadata metadata, BeanDefinitionHolder definition, BeanDefinitionRegistry
registry) {

    ScopedProxyMode scopedProxyMode = metadata.getScopedProxyMode();
    if (scopedProxyMode.equals(ScopedProxyMode.NO)) {
        return definition;
    }
    boolean proxyTargetClass
= scopedProxyMode.equals(ScopedProxyMode.TARGET_CLASS);
    return ScopedProxyCreator.createScopedProxy(definition, registry, proxyTargetClass);
}
```

这里对于是否需要进行代理对象的创建（创建存在代理标记的 BeanDefinition Holder 对象）表示得很明确：作用域解析结果不是 NO 就进行代理对象创建。有关代理对象创建的代码如下。

```java
public static BeanDefinitionHolder createScopedProxy(BeanDefinitionHolder definition,
        BeanDefinitionRegistry registry, boolean proxyTargetClass) {

    //提取原始的 BeanName
    String originalBeanName = definition.getBeanName();
    //从 BeanDefinition Holder 中获取 BeanDefintion
    BeanDefinition targetDefinition = definition.getBeanDefinition();
    //计算 BeanName
    String targetBeanName = getTargetBeanName(originalBeanName);
```

```java
//创建代理的 BeanDefinition 对象,数据会从原始的 BeanDefinition 中迁移部分
RootBeanDefinition proxyDefinition =
    new RootBeanDefinition(ScopedProxyFactoryBean.class);
proxyDefinition.setDecoratedDefinition(new BeanDefinitionHolder(targetDefinition,
targetBeanName));
proxyDefinition.setOriginatingBeanDefinition(targetDefinition);
proxyDefinition.setSource(definition.getSource());
proxyDefinition.setRole(targetDefinition.getRole());

//添加属性 targetBeanName
proxyDefinition.getPropertyValues().add("targetBeanName", targetBeanName);
if (proxyTargetClass) {

    targetDefinition.setAttribute(AutoProxyUtils.PRESERVE_TARGET_CLASS_ATTRIBUTE,
Boolean.TRUE);
}
else {
    proxyDefinition.getPropertyValues().add("proxyTargetClass", Boolean.FALSE);
}

proxyDefinition.setAutowireCandidate(targetDefinition.isAutowireCandidate());
proxyDefinition.setPrimary(targetDefinition.isPrimary());
if (targetDefinition instanceof AbstractBeanDefinition) {
    proxyDefinition.copyQualifiersFrom((AbstractBeanDefinition) targetDefinition);
}

targetDefinition.setAutowireCandidate(false);
targetDefinition.setPrimary(false);

//将当前代理的对象注册到容器中
registry.registerBeanDefinition(targetBeanName, targetDefinition);

//创建返回对象
return new
BeanDefinitionHolder(proxyDefinition, originalBeanName, definition.getAliases());
}
```

在这个创建代理的 BeanDefinition Holder 过程中可以比较清晰地看到这里的操作是将原始 BeanDefiniiton 和目标 BeanDefinition 的属性复制的过程。

(6) BeanDefinition 注册到容器。具体负责进行注册的代码如下。

```java
public static void registerBeanDefinition(
        BeanDefinitionHolder definitionHolder, BeanDefinitionRegistry registry)
        throws BeanDefinitionStoreException {

    //获取 BeanName
    String beanName = definitionHolder.getBeanName();
    //注册 BeanDefinition
    registry.registerBeanDefinition(beanName, definitionHolder.getBeanDefinition());

    //别名列表
```

```
        String[] aliases = definitionHolder.getAliases();
        //注册别名列表
        if (aliases != null) {
            for (String alias : aliases) {
                registry.registerAlias(beanName, alias);
            }
        }
    }
```

对于 BeanDefinition 的注册操作这一点和 SpringXML 模式中是一模一样的。当完成 BeanDefinition 的注册后整个处理流程结束。

在完成 doScan 方法后会进行注册注解后置处理器操作,具体处理方法是 AnnotationConfigUtils. registerAnnotationConfigProcessors,在这个方法中会将这些 PostProcessor 进行注册: ConfigurationClassPostProcessor、AutowiredAnnotationBeanPostProcessor、CommonAnnotationBeanPostProcessor、 PersistenceAnnotationBeanPostProcessor、EventListenerMethodProcessor 和 DefaultEventListenerFactory。

完成 PostProcessor 注册后会计算本次扫描过程中注册了多少个 BeanDefinition 对象。

9.3　componentClasses 模式启动

接下来将对 componentClasses 模式进行分析,首先关注这个模式下的构造函数,具体代码如下。

```
public AnnotationConfigApplicationContext(Class<?>... componentClasses) {
    this();
    register(componentClasses);
    refresh();
}
```

在这段代码中 this 方法指向了无参构造,目的是创建 AnnotatedBeanDefinitionReader 和 ClassPathBeanDefinitionScanner 对象。

方法 register 是进行配置类注册的核心方法,具体代码如下。

```
@Override
public void register(Class<?>... componentClasses) {
    Assert.notEmpty(componentClasses, "At least one component class must be specified");
    this.reader.register(componentClasses);
}
```

继续向下追踪源代码,找到 org.springframework.context.annotation.AnnotatedBeanDefinitionReader#doRegisterBean 方法,具体代码如下。

```
private <T> void doRegisterBean(Class<T> beanClass, @Nullable String name,
        @Nullable Class<? extends Annotation>[] qualifiers, @Nullable
Supplier<T> supplier,
        @Nullable BeanDefinitionCustomizer[] customizers) {

    //带有注解的泛型 Bean 定义
    AnnotatedGenericBeanDefinition abd =
```

```java
        new AnnotatedGenericBeanDefinition(beanClass);
    //和条件注解相关的函数
    if (this.conditionEvaluator.shouldSkip(abd.getMetadata())) {
        return;
    }

    //设置实例提供者
    abd.setInstanceSupplier(supplier);
    //解析注解的 BeanDefinition 的作用域元数据
    ScopeMetadata scopeMetadata = this.scopeMetadataResolver.resolveScopeMetadata(abd);
    //设置作用域元数据
    abd.setScope(scopeMetadata.getScopeName());
    //BeanName 处理
    String beanName = (name != null ? name : this.beanNameGenerator.generateBeanName(abd, this.registry));

    //通用注解的处理
    AnnotationConfigUtils.processCommonDefinitionAnnotations(abd);
    if (qualifiers != null) {
        for (Class<? extends Annotation> qualifier : qualifiers) {
            if (Primary.class == qualifier) {
                abd.setPrimary(true);
            }
            else if (Lazy.class == qualifier) {
                abd.setLazyInit(true);
            }
            else {
                abd.addQualifier(new AutowireCandidateQualifier(qualifier));
            }
        }
    }
    //自定义的 BeanDefinition 处理
    if (customizers != null) {
        for (BeanDefinitionCustomizer customizer : customizers) {
            customizer.customize(abd);
        }
    }

    //创建 BeanDefinition Holder 后进行注册
    BeanDefinitionHolder definitionHolder = new BeanDefinitionHolder(abd, beanName);
    //应用作用域代理
    definitionHolder =
AnnotationConfigUtils.applyScopedProxyMode(scopeMetadata, definitionHolder, this.registry);
    BeanDefinitionReaderUtils.registerBeanDefinition(definitionHolder, this.registry);
}
```

该方法是处理单个 BeanClass 的方法,也就是处理单个配置类的方法,具体处理细节如下。

(1) 将 BeanClass 转换成 AnnotatedGenericBeanDefinition。

(2) 条件注解(@Conditional)的判断。

（3）Scope 元数据解析。
（4）注解模式下的 BeanName 处理。
（5）Spring 注解模式下通用注解处理。
（6）注解 @Qualifier 的处理。
（7）自定义的 BeanDefinition 处理。
（8）BeanDefinition 关于 Scope 代理处理。
（9）注册 BeanDefinition Holder。

关注第(1)项的处理代码，具体内容如下。

```
public AnnotatedGenericBeanDefinition(Class<?> beanClass) {
    setBeanClass(beanClass);
    this.metadata = AnnotationMetadata.introspect(beanClass);
}
```

在这段代码中可以看到 AnnotationMetadata 类，该类在注解模式下承担了极其重要的工作：整合注解数据。整合注解数据信息如图 9.9 所示。

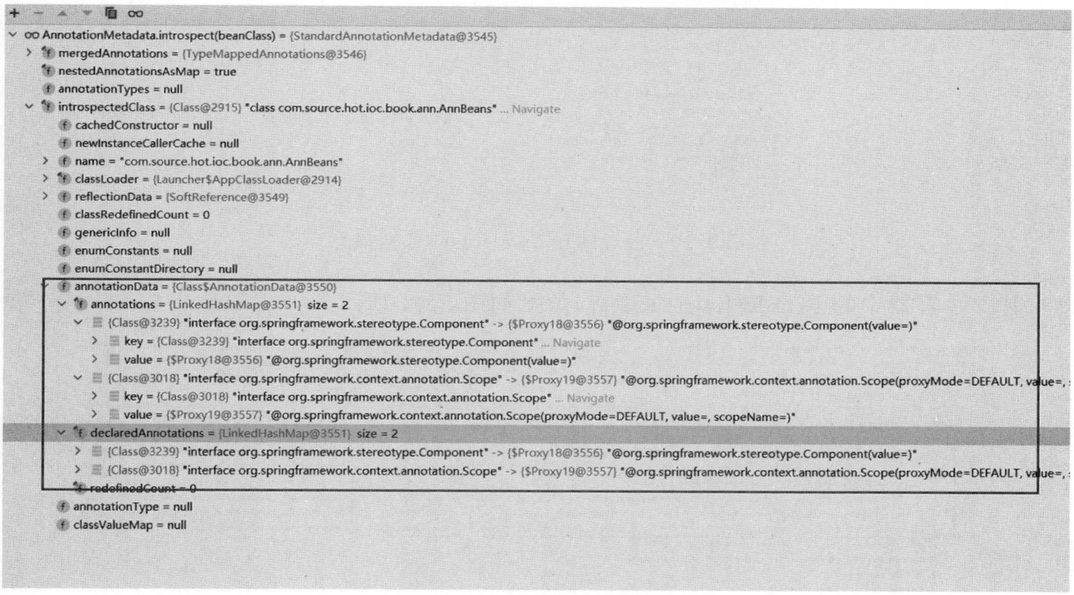

图 9.9　整合注解数据

通过该方法可以获取到 BeanDefinition 对象，具体类型是 AnnotatedGenericBeanDefinition，得到该对象后会进行下面八个操作。

（1）条件注解处理。
（2）设置实例提供者。
（3）解析 Scope 相关数据。
（4）设置 BeanName。
（5）通用 BeanDefinition 注解处理。
（6）创建 BeanDefinitionHolder。

(7) 将 BeanDefinitionHolder 应用代理。

(8) 注册 BeanDefinitionHolder 进行。

当完成上面八个操作后注册过程就结束了，下面关注几个处理过程中的对象，首先是 BeanDefinitionHolder，具体数据如图 9.10 所示。

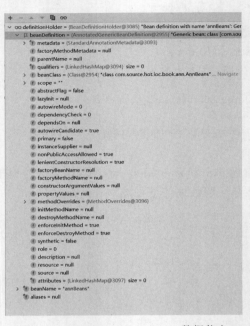

图 9.10　BeanDefinitionHolder 数据信息

其次是容器中的 BeanDefinition 集合，数据如图 9.11 所示。

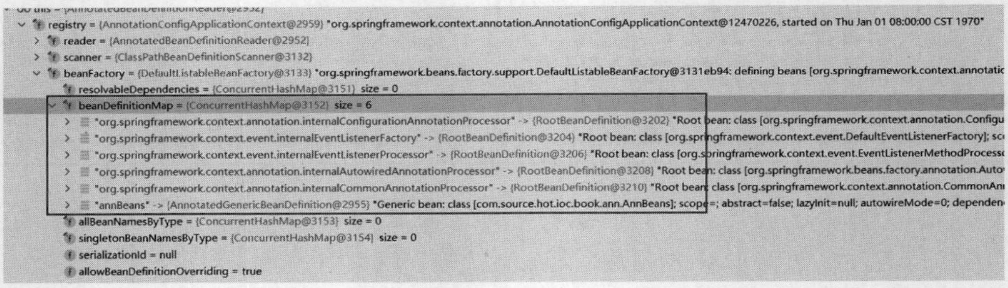

图 9.11　容器中的 BeanDefinition 集合

小结

Spring 在注解模式下提供了两种不同的启动方式（上下文对象创建的两种方式），第一种是通过包路径进行处理，第二种是通过传入指定的 Bean Class。Spring 为了处理这两种方式分别提供了两个不同的处理对象，处理第一种的对象是 ClassPathBeanDefinitionScanner，处理第二种的对象是 AnnotatedBeanDefinitionReader。针对这两种不同的处理方式还有两个不同的 BeanDefinition，具体是 ScannedGenericBeanDefinition 和 AnnotatedGenericBeanDefinition。

第 10 章

Spring 配置类解析

第 9 章介绍了 Spring 注解模式的开发方式,在 Spring 注解开发中最关键的对象莫过于 Spring 配置类,这样一个重要的类是如何进入 Spring 中并被 Spring 识别处理的,这个问题在本章中将会进行详细分析。

10.1 parse 方法分析

关于 Spring 配置类的解析在 ConfigurationClassPostProcessor#processConfigBeanDefinitions 方法中有一段关于配置类解析的代码,具体代码如下。

```
//..省略其他
ConfigurationClassParser parser = new ConfigurationClassParser(
        this.metadataReaderFactory, this.problemReporter, this.environment,
        this.resourceLoader, this.componentScanBeanNameGenerator, registry);

//候选的需要解析的 BeanDefinition 容器
Set<BeanDefinitionHolder> candidates = new LinkedHashSet<>(configCandidates);
//已经完成解析的 Spring Configuration Bean
Set<ConfigurationClass> alreadyParsed = new HashSet<>(configCandidates.size());
do {
    //解析候选 BeanDefinition Holder 集合
    parser.parse(candidates);
    //..省略其他
}
```

在这段代码中重点需要关注的是 parser.parse(candidates) 处理,具体处理代码如下。

```
public void parse(Set<BeanDefinitionHolder> configCandidates) {
    for (BeanDefinitionHolder holder : configCandidates) {
        BeanDefinition bd = holder.getBeanDefinition();
```

```
            try {
                if (bd instanceof AnnotatedBeanDefinition) {
                    parse(((AnnotatedBeanDefinition) bd).getMetadata(), holder.getBeanName());
                }
                else if (bd instanceof AbstractBeanDefinition &&
((AbstractBeanDefinition) bd).hasBeanClass()) {
                    parse(((AbstractBeanDefinition) bd).getBeanClass(), holder.getBeanName());
                }
                else {
                    parse(bd.getBeanClassName(), holder.getBeanName());
                }
            }
            catch (BeanDefinitionStoreException ex) {
                throw ex;
            }
            catch (Throwable ex) {
                throw new BeanDefinitionStoreException(
                        "Failed to parse configuration class [" + bd.getBeanClassName() + "]", ex);
            }
        }

        //处理 import 相关内容
        this.deferredImportSelectorHandler.process();
}
```

在 parse 方法中有以下两个至关重要的处理。

（1）对 Spring 配置类的解析（Spring Configuration Bean 解析）。

（2）关于 import 相关处理，主要是对 ImportSelector 接口的一些处理。

10.2 processConfigurationClass 方法分析

在前文提到了 parse 方法的两个处理，下面将对第一个处理进行详细分析，第一个处理的代码内容如下。

```
    protected final void parse(@Nullable String className, String beanName)
throws IOException {
        Assert.notNull(className,"No bean class name for configuration class bean definition");
        MetadataReader reader
= this.metadataReaderFactory.getMetadataReader(className);
        processConfigurationClass(new ConfigurationClass(reader, beanName));
    }
```

在这段方法中可以看到进一步调用了 processConfigurationClass 方法，该方法就是分析目标，具体处理代码如下。

```
    protected void processConfigurationClass(ConfigurationClass configClass) throws IOException {
        //条件注解解析，阶段为配置解析阶段
```

```java
    if (
this.conditionEvaluator.shouldSkip(configClass.getMetadata(), ConfigurationPhase.PARSE_
CONFIGURATION)) {
        return;
    }

    //尝试从配置类缓存中获取配置类
    ConfigurationClass existingClass = this.configurationClasses.get(configClass);
    //缓存中的配置类存在
    if (existingClass != null) {
        //判断配置类是不是导入的
        if (configClass.isImported()) {
            //判断缓存中的配置类是不是导入的
            if (existingClass.isImported()) {
                //合并配置信息
                existingClass.mergeImportedBy(configClass);
            }
            //Otherwise ignore new imported config class; existing non-imported class overrides it.
            return;
        }
        else {
            //从配置类缓存中移除当前配置类
            this.configurationClasses.remove(configClass);
            //从已知的配置类中移除当前的配置类
            this.knownSuperclasses.values().removeIf(configClass::equals);
        }
    }

    //将配置类转换成 SourceClass
    SourceClass sourceClass = asSourceClass(configClass);
    do {
        //解析配置类
        sourceClass = doProcessConfigurationClass(configClass, sourceClass);
    }
    while (sourceClass != null);

    //放入配置类缓存
    this.configurationClasses.put(configClass, configClass);
}
```

在这段代码中有比较多的处理操作,主要有如下几个操作。

(1) 对配置类进行条件注解的条件判断。

(2) 从配置类缓存中获取配置类对应的缓存配置类信息 existingClass。

(3) 如果 existingClass 存在,且当前需要处理的配置类是一个导入的类,如果缓存中的配置类 existingClass 是导入的,existingClass 对象将和当前正在处理的配置类进行合并操作。如果当前需要处理的配置类不是一个导入的类,会有以下两个细节操作。

① 从配置类缓存中移除当前配置类。

② 从已知的配置类中移除当前配置类。

(4) 将配置类转换成 SourceClass 对象。

(5) 解析配置类。

(6) 将配置类放入配置类缓存中。

在这个方法处理过程中出现了 3 个变量,关于该处理中所需要使用的变量信息可以查看表 10.1。

表 10.1　processConfigurationClass 处理时用到的变量

变 量 名 称	变 量 类 型	说　　　明
conditionEvaluator	ConditionEvaluator	条件注解的解析器
configurationClasses	Map < ConfigurationClass,ConfigurationClass >	存储配置类的缓存结构,key value 相同
knownSuperclasses	Map < String,ConfigurationClass >	已知的配置类 key:父类名称;value:配置类

接下来对上述三个变量中的后两个变量进行数据演示。首先编写一个父配置类,类名为 SupperConfiguration,具体代码如下。

```
@Configuration
public class SupperConfiguration {

}
```

其次编写一个子配置类,类名为 ExtendConfiguration,具体代码如下。

```
@Configuration
public class ExtendConfiguration extends SupperConfiguration {

}
```

配置类辨析完成后编写一个上下文类,类名为 ThreeVarBeans,具体代码如下。

```
@Configuration
@Import(ExtendConfiguration.class)
public class ThreeVarBeans {
}
```

最后编写单元测试方法,具体代码如下。

```
@Test
void testThreeVar() {
   AnnotationConfigApplicationContext context = new AnnotationConfigApplicationContext(ThreeVarBeans.class);
}
```

下面进入调试模式查看 configurationClasses 变量和 knownSuperclasses 变量的数据信息,数据如图 10.1 所示。

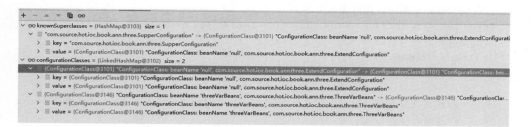

图 10.1　configurationClasses 和 knownSuperclasses 数据信息

10.3　doProcessConfigurationClass 方法分析

前文得到了一些结果对象,但是在得到结果对象前需要通过 doProcessConfigurationClass 方法进行处理才可以得到最终结果,该方法是负责对 Spring 配置类进行解析的方法,在 doProcessConfigurationClass 方法中可以分为以下七个核心处理操作,分别如下。

（1）处理 @Component 注解。
（2）处理 @PropertySource 和 @PropertySources 注解。
（3）处理 @ComponentScans 和 @ComponentScan 注解。
（4）处理 @Import 注解。
（5）处理 @ImportResource 注解。
（6）处理 @Bean 注解。
（7）处理父类配置。

接下来将上述七个处理流程单一拆解出来做详细的分析。

10.4　处理各类注解

本节将对 Spring 配置类中常用注解的处理过程进行分析。

10.4.1　处理 @Component 注解

首先对注解 Component 进行分析,关于它的处理具体代码如下。

```
//配置类是否存在 Component 注解
if (configClass.getMetadata().isAnnotated(Component.class.getName())) {
    processMemberClasses(configClass, sourceClass);
}
```

在这段代码中可以看到它会对配置类进行判断,如果存在 Component 注解才会进行后续的处理操作,后续的处理操作是 processMemberClasses 方法,processMemberClasses 方法详情如下。

```
private void processMemberClasses(ConfigurationClass configClass,
SourceClass sourceClass) throws IOException {
```

```
//找到当前配置类中存在的成员类
Collection<SourceClass> memberClasses = sourceClass.getMemberClasses();
//成员类列表不为空
if (!memberClasses.isEmpty()) {
    List<SourceClass> candidates = new ArrayList<>(memberClasses.size());
    for (SourceClass memberClass : memberClasses) {
        //成员类是否符合配置类候选标准,Component、ComponentScan、Import、ImportResource 注
        //解是否存在
        //成员类是否和配置类同名
        if (ConfigurationClassUtils.isConfigurationCandidate(memberClass.getMetadata()) &&
                !memberClass.getMetadata().getClassName().equals(configClass.getMetadata().getClassName())) {

            candidates.add(memberClass);
        }
    }
    //排序候选类
    OrderComparator.sort(candidates);
    for (SourceClass candidate : candidates) {
        //判断 importStack 中是否存在当前配置类
        if (this.importStack.contains(configClass)) {
            this.problemReporter.error(new
CircularImportProblem(configClass, this.importStack));
        }
        else {
            this.importStack.push(configClass);
            try {
                //解析成员类
                processConfigurationClass(candidate.asConfigClass(configClass));
            }
            finally {
                this.importStack.pop();
            }
        }
    }
}
```

这段代码的处理流程如下。

(1) 提取当前配置类中的内部类。

(2) 对内部类进行条件过滤,将符合条件的类进行收集,条件是内部类存在注解 Component、注解 ComponentScan、注解 Import、注解 ImportResource 和注解 Bean 中的任意一个。

(3) 对收集的内部类进行排序。

(4) 对符合条件的内部类进行配置解析,条件是判断 importStack 中是否存在当前配置类。

了解处理流程后需要编写一个可以论证这段处理流程的测试用例。首先定义一个配置类,类名为 BigConfiguration,在 BigConfiguration 类中定义两个内部类,其中一个被注解

Component 修饰，具体代码如下。

```
@Configuration
public class BigConfiguration {
    public class SmallConfigA {
    }
    @Component
    public class SmallConfigB {
    }
}
```

完成配置类编写后进一步编写单元测试，测试代码如下。

```
@Test
void testProcessMemberClasses(){
    AnnotationConfigApplicationContext context =
     new AnnotationConfigApplicationContext(BigConfiguration.class);

}
```

测试类编写完成后需要通过调试来进行一些关键变量的查看，首先要关注的变量是 memberClasses，该变量存储了当前配置类下拥有的内部类列表，memberClasses 数据信息如图 10.2 所示。

图 10.2　memberClasses 数据信息

从中可以发现它确实存储了当前配置类的两个内部类。继续向下关注的变量是 candidates，该对象表示通过条件判断得到的内部类，这些内部类会被当作 Spring Configuration Bean 处理，candidates 数据信息如图 10.3 所示。

图 10.3　candidates 数据信息

得到 candidates 数据信息后就会进行配置类解析，有关配置类的解析也就是本章所讨论的内容。

10.4.2　处理 @PropertySource 和 @PropertySources 注解

接下来将对 PropertySource 注解和 PropertySources 注解的处理进行分析，下面是处理这两个注解的代码。

```
for (AnnotationAttributes propertySource : AnnotationConfigUtils.attributesForRepeatable(
    sourceClass.getMetadata(), PropertySources.class,
```

```
                    org.springframework.context.annotation.PropertySource.class)) {
            if (this.environment instanceof ConfigurableEnvironment) {
                processPropertySource(propertySource);
            }
            else {
                logger.info("Ignoring @PropertySource annotation on
[" + sourceClass.getMetadata().getClassName() +
                    "]. Reason: Environment must implement ConfigurableEnvironment");
            }
        }
```

在这段代码中处理的是 PropertySource 注解,后者是多个 PropertySource 注解的集合,为了更好地对这个注解进行分析,编写一段测试用例来进行测试。首先编写一个配置文件,文件名为 data.properties,文件内容如下。

a = 123

编写完成配置文件后编写一个 Spring 配置类,类名为 PropertySourceTest,具体代码如下。

```
@PropertySource(value = "classpath:data.properties")
@Configuration
public class PropertySourceTest {
    @Test
    void testPropertySource() {
        AnnotationConfigApplicationContext context =
new AnnotationConfigApplicationContext(PropertySourceTest.class);
        assert context.getEnvironment().getProperty("a").equals("123");
    }
}
```

PropertySourceTest 类既作为配置类也作为一个测试类进行使用,下面进入源码分析,首先了解注解 PropertySource 的定义,具体代码如下。

```
@Target(ElementType.TYPE)
@Retention(RetentionPolicy.RUNTIME)
@Documented
@Repeatable(PropertySources.class)
public @interface PropertySource {

    String name() default "";

    String[] value();

    boolean ignoreResourceNotFound() default false;

    String encoding() default "";

    Class<? extends PropertySourceFactory> factory() default PropertySourceFactory.class;

}
```

其次了解 PropertySources 注解的定义,具体代码如下。

```
@Target(ElementType.TYPE)
@Retention(RetentionPolicy.RUNTIME)
@Documented
public @interface PropertySources {

    PropertySource[] value();

}
```

通过阅读注解 PropertySources，可以发现它包含一个 value 属性，该属性是注解 PropertySource 的集合，在确认这一点的情况下对于下面这段代码的理解会容易很多，具体代码如下。

```
AnnotationConfigUtils.attributesForRepeatable(
        sourceClass.getMetadata(), PropertySources.class,
        org.springframework.context.annotation.PropertySource.class)
```

这段代码将两个注解的数据都进行读取，得到一个 PropertySources 注解数据集合，为了进行这段代码的调试，可以修改配置类，修改后内容如下。

```
@PropertySource(value = "classpath:data.properties")
@PropertySources(value = @PropertySource("classpath:data.properties"))
@Configuration
public class PropertySourceTest {
    @Test
    void testPropertySource() {
        AnnotationConfigApplicationContext context =
new AnnotationConfigApplicationContext(PropertySourceTest.class);
        assert context.getEnvironment().getProperty("a").equals("123");
    }
}
```

接下来通过调试来看经过解析得到的 PropertySource 集合，具体信息如图 10.4 所示。

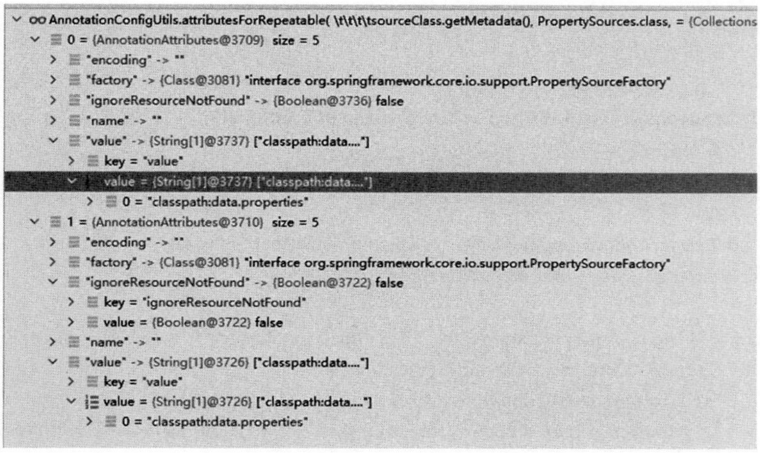

图 10.4 PropertySource 解析集合

从图 10.4 中可以看到,注解信息被解析得到了。接下来就会循环处理每一个 AnnotationAttributes 数据,即处理每一个 PropertySource 注解的元数据,具体的处理方法是 processPropertySource,处理代码如下。

```java
private void processPropertySource(AnnotationAttributes propertySource) throws IOException {
    //提取注解属性中的 name 属性
    String name = propertySource.getString("name");
    if (!StringUtils.hasLength(name)) {
        name = null;
    }
    //提取注解属性中的 encoding 属性
    String encoding = propertySource.getString("encoding");
    if (!StringUtils.hasLength(encoding)) {
        encoding = null;
    }
    //提取注解属性中的 value 属性
    String[] locations = propertySource.getStringArray("value");
    Assert.isTrue(locations.length > 0, "At least one @PropertySource(value) location is required");

    //提取注解属性中的 ignoreResourceNotFound 属性
    boolean ignoreResourceNotFound
            = propertySource.getBoolean("ignoreResourceNotFound");

    //提取注解属性中的 factory 属性
    Class<? extends PropertySourceFactory> factoryClass
            = propertySource.getClass("factory");
    PropertySourceFactory factory = (factoryClass == PropertySourceFactory.class ?
            DEFAULT_PROPERTY_SOURCE_FACTORY : BeanUtils.instantiateClass(factoryClass));

    //循环处理每个配置文件
    for (String location : locations) {
        try {
            //解析配置文件地址
            String resolvedLocation
                    = this.environment.resolveRequiredPlaceholders(location);
            //资源加载
            Resource resource = this.resourceLoader.getResource(resolvedLocation);
            //加入配置
            addPropertySource(factory.createPropertySource(name,
                    new EncodedResource(resource, encoding)));
        }
        catch (IllegalArgumentException | FileNotFoundException | UnknownHostException ex) {
            if (ignoreResourceNotFound) {
                if (logger.isInfoEnabled()) {
                    logger.info("Properties location [" + location + "] not resolvable: " + ex.getMessage());
                }
            }
```

```
            else {
                throw ex;
            }
        }
    }
}
```

这段代码中的处理过程有两个，第一个是提取注解元数据中与 PropertySource 注解相关的数据信息，第二个是对提取得到的 locations 数据进行处理，对于这一项处理又分为以下三点。

（1）将 locations 中的单个元素进行路径解析。
（2）将 locations 中的单个元素的解析结果转换成 Resource 对象。
（3）添加属性，加入到配置列表中。

有关 locations 的单一元素解析可以查看第 12 章的内容，有关 locations 中单一元素解析成为 Resource 对象的内容可以查看第 18 章。

最后是添加属性方法的分析，在分析添加属性方法之前需要理解下面这段代码。

```
factory.createPropertySource(name, new EncodedResource(resource, encoding))
```

这段代码是通过 factory 进行处理，在源代码中可以看到 factory 变量是通过反射创建出来的，具体处理代码如下。

```
PropertySourceFactory factory = (factoryClass == PropertySourceFactory.class ?
        DEFAULT_PROPERTY_SOURCE_FACTORY : BeanUtils.instantiateClass(factoryClass));
```

在这段代码中可以看到它是根据 factoryClass 属性进行不同的操作。

（1）如果 factory 属性不存在，则采用默认的 PropertySourceFactory，具体类是 org.springframework.core.io.support.DefaultPropertySourceFactory。
（2）如果 factory 属性存在，通过 BeanUtils 反射创建该对象。

前文准备的测试用例符合第一种情况，下面是 DefaultPropertySourceFactory 的实现代码。

```
public class DefaultPropertySourceFactory implements PropertySourceFactory {

    @Override
    public PropertySource<?> createPropertySource(@Nullable String name, EncodedResource resource) throws IOException {
        return (name != null ? new ResourcePropertySource(name, resource) :
 new ResourcePropertySource(resource));
    }

}
```

在这段代码中直接创建了一个对象 ResourcePropertySource，继续追踪源代码，查看 ResourcePropertySource 的构造函数：

```
public ResourcePropertySource(String name, EncodedResource resource)
        throws IOException {
```

```
    //设置 name + map 对象
    //map 对象是资源信息
    super(name, PropertiesLoaderUtils.loadProperties(resource));
    //获取 resource name
    this.resourceName = getNameForResource(resource.getResource());
}
```

在这个构造函数中可以发现进行了加载操作,这部分加载操作就是为了加载配置文件的内容。这段操作同样出现在 ResourcePropertySource 的构造函数中,具体代码如下。

```
public ResourcePropertySource(EncodedResource resource) throws IOException {
    //设置 key: name, resource 的 name
    //设置 value: resource 资源信息
    super(getNameForResource(resource.getResource()), PropertiesLoaderUtils.loadProperties
(resource));
    this.resourceName = null;
}
```

这两段代码的共同点就是 PropertiesLoaderUtils.loadProperties 方法,它就是配置文件(外部配置文件)加载的核心。它将资源文件解析成 Properties 对象。

在得到需要进行添加属性操作的对象后进行添加操作,具体处理方法是 addPropertySource,相关代码如下。

```
private void addPropertySource(PropertySource<?> propertySource) {
    //获取配置名称
    String name = propertySource.getName();
    //提取环境对象中的 MutablePropertySources 对象
    MutablePropertySources propertySources =
((ConfigurableEnvironment) this.environment).getPropertySources();

    //当前处理的配置名称是否在 propertySourceNames 中存在
    if (this.propertySourceNames.contains(name)) {
        //从 propertySources 中获取当前配置名称对应的 PropertySource
        PropertySource<?> existing = propertySources.get(name);
        if (existing != null) {
            PropertySource<?> newSource =
(propertySource instanceof ResourcePropertySource ?
                    ((ResourcePropertySource)propertySource).withResourceName():propertySource);
            if (existing instanceof CompositePropertySource) {
                ((CompositePropertySource) existing).addFirstPropertySource(newSource);
            }
            else {
                if (existing instanceof ResourcePropertySource) {
                    existing = ((ResourcePropertySource) existing).withResourceName();
                }
                CompositePropertySource composite =
new CompositePropertySource(name);
                composite.addPropertySource(newSource);
                composite.addPropertySource(existing);
                propertySources.replace(name, composite);
```

```
            }
            return;
        }
    }

    //如果 propertySourceNames 为空
    if (this.propertySourceNames.isEmpty()) {
        propertySources.addLast(propertySource);
    }
    else {
        String firstProcessed =
this.propertySourceNames.get(this.propertySourceNames.size() - 1);
        propertySources.addBefore(firstProcessed, propertySource);
    }
    this.propertySourceNames.add(name);
}
```

在这段代码中操作的实际对象有两个,第一个是 environment 中的 Mutable-PropertySources 属性,第二个是 propertySourceNames。对于这两个对象需要了解它们的存储结构。propertySourceNames 存储结构相对简单,存储形式为 List＜String＞,存储内容是配置名称；MutablePropertySources 中存储了多个 PropertySource 对象,具体结构如下。

```
public class MutablePropertySources implements PropertySources {

    private final List< PropertySource <?>> propertySourceList = new CopyOnWriteArrayList<>();
}
```

addPropertySource 的处理逻辑主要是围绕配置名称是否存在于 propertySourceNames 中和配置对象的类型进行处理,下面是整个方法的处理逻辑。

（1）提取配置名称。

（2）从环境配置中提取 MutablePropertySources 对象。

（3）配置名称集合中存在当前需要处理的配置名称。

（4）从 MutablePropertySources 中提取当前配置名称对应的 PropertySource,命名为 existing。如果 existing 存在。会分为下面两种处理情况。

① 对当前需要处理的 PropertySource 进行类型判断并转换,判断是不是 ResourcePropertySource 类型。

② 对 existing 进行类型判断。类型如果是 CompositePropertySource,采用头插法插入数据,其他类型则进行数据替换。

（5）如果 propertySourceNames 为空,使用尾插法插入数据。

（6）如果 propertySourceNames 不为空,进行指定标记后往前插入。

（7）将配置名称加入 propertySourceNames 中。

经过上述几个步骤的操作后在 propertySourceNames 对象中的存储信息如图 10.5 所示。

在 environment 中的数据信息如图 10.6 所示。

图 10.5 propertySourceNames 数据信息

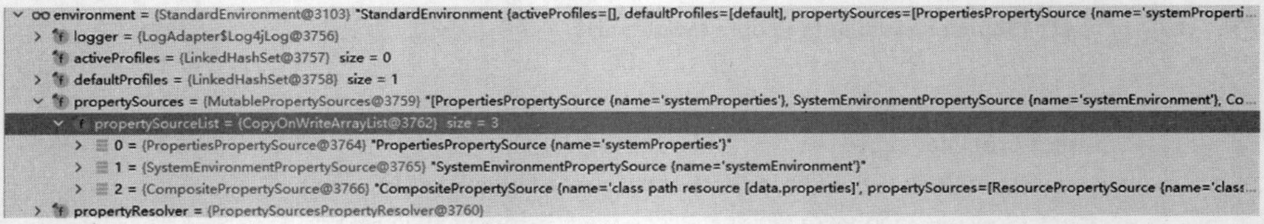

图 10.6 environment 数据信息

完成数据对象的设置后关于 PropertySources 注解和 PropertySource 注解的处理就完成了。

10.4.3 处理 @ComponentScans 和 @ComponentScan 注解

接下来将介绍 ComponentScans 注解和 ComponentScan 注解的处理，它们的处理代码如下。

```
//处理 ComponentScans 和 ComponentScan 注解
Set < AnnotationAttributes > componentScans =
AnnotationConfigUtils.attributesForRepeatable(
        sourceClass.getMetadata(), ComponentScans.class, ComponentScan.class);
if (!componentScans.isEmpty() &&
        !this.conditionEvaluator.shouldSkip(sourceClass.getMetadata(), ConfigurationPhase.
REGISTER_BEAN)) {
    for (AnnotationAttributes componentScan : componentScans) {
        Set < BeanDefinitionHolder > scannedBeanDefinitions =
                this.componentScanParser.parse(componentScan, sourceClass.getMetadata().
getClassName());
        for (BeanDefinitionHolder holder : scannedBeanDefinitions) {
            BeanDefinition bdCand =
holder.getBeanDefinition().getOriginatingBeanDefinition();
            if (bdCand == null) {
                bdCand = holder.getBeanDefinition();
            }
            if
(ConfigurationClassUtils.checkConfigurationClassCandidate(bdCand,this.
metadataReaderFactory)) {
                parse(bdCand.getBeanClassName(), holder.getBeanName());
            }
        }
    }
}
```

为了更好地理解调试代码,需要编写一个测试环境,首先需要编写一个配置类,类名为 ScanBeanA,具体代码如下。

```java
@Configuration
public class ScanBeanA {
    @Bean
    public AnnPeople annPeople() {
        AnnPeople annPeople = new AnnPeople();
        annPeople.setName("scanBeanA.people");
        return annPeople;
    }
}
```

其次编写一个测试用例。在这个测试用例中希望 ScanBeanA 配置类会被扫描到,具体代码如下。

```java
@ComponentScan(basePackageClasses = ScanBeanA.class)
@Configuration
public class ComponentScanTest {
    @Test
    void scan(){
        AnnotationConfigApplicationContext context =
new AnnotationConfigApplicationContext(ComponentScanTest.class);
        AnnPeople bean = context.getBean(AnnPeople.class);
        assert bean.getName().equals("scanBeanA.people");
    }
}
```

完成基本测试环境搭建后来看 ComponentScans 注解和 ComponentScan 注解的关系。从 ComponentScans 注解的代码中可以发现,它的 value 属性存储了多个 ComponentScan 注解,它们的代码如下。

```java
@Retention(RetentionPolicy.RUNTIME)
@Target(ElementType.TYPE)
@Documented
public @interface ComponentScans {

    ComponentScan[] value();

}
@Retention(RetentionPolicy.RUNTIME)
@Target(ElementType.TYPE)
@Documented
@Repeatable(ComponentScans.class)
public @interface ComponentScan {

    @AliasFor("basePackages")
    String[] value() default {};

    @AliasFor("value")
    String[] basePackages() default {};
```

```java
    Class<?>[] basePackageClasses() default {};

    Class<? extends BeanNameGenerator> nameGenerator()
default BeanNameGenerator.class;

    Class<? extends ScopeMetadataResolver> scopeResolver()
default AnnotationScopeMetadataResolver.class;

    ScopedProxyMode scopedProxy() default ScopedProxyMode.DEFAULT;

    String resourcePattern()
default
ClassPathScanningCandidateComponentProvider.DEFAULT_RESOURCE_PATTERN;

    boolean useDefaultFilters() default true;

    Filter[] includeFilters() default {};

    Filter[] excludeFilters() default {};

    boolean lazyInit() default false;

    @Retention(RetentionPolicy.RUNTIME)
    @Target({})
    @interface Filter {

        FilterType type() default FilterType.ANNOTATION;

        @AliasFor("classes")
        Class<?>[] value() default {};

        @AliasFor("value")
        Class<?>[] classes() default {};

        String[] pattern() default {};

    }

}
```

了解两个注解之间的关系后对于这段代码的理解会比较容易,具体代码如下。

```
Set<AnnotationAttributes> componentScans =
AnnotationConfigUtils.attributesForRepeatable(
    sourceClass.getMetadata(), ComponentScans.class, ComponentScan.class);
```

这段代码会将 ComponentScan 注解和 ComponentScans 注解都进行解析得到 AnnotationAttributes 对象。AnnotationAttributes 中存储了单个 componentScan 注解的属性,进入调试可以看到测试类中的数据信息,如图 10.7 所示。

下面将介绍单个 componentScan 注解数据的处理过程,总共分为以下两步。

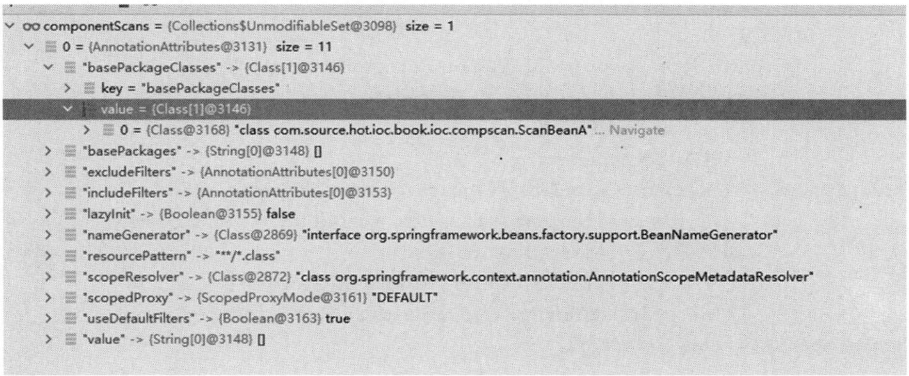

图 10.7　componentScans 数据信息

（1）通过 componentScanParser 进行解析，解析后得到 BeanDefinitionHolder。

（2）将解析得到的 BeanDefinitionHolder 根据条件筛选后进行注册：

第一步的处理代码是 org.springframework.context.annotation.ComponentScanAnnotationParser#parse。在这个方法中做了两件事，这里仅做一个简单概要。

（1）创建 ClassPathBeanDefinitionScanner 对象并从注解属性中提取数据赋值到对应字段中。

（2）通过 ClassPathBeanDefinitionScanner 进行解析。

10.4.4　处理 @Import 注解

下面将对 Import 注解的处理进行分析，负责处理的代码如下。

```
private void processImports(ConfigurationClass configClass,
SourceClass currentSourceClass,
      Collection<SourceClass> importCandidates, boolean checkForCircularImports) {

   //判断是否存在需要处理的 ImportSelector 集合，如果不需要则不处理直接返回
   if (importCandidates.isEmpty()) {
      return;
   }

   //判断是否需要进行 import 循环检查
   //判断当前配置类是否在 importStack 中
   if (checkForCircularImports && isChainedImportOnStack(configClass)) {
      this.problemReporter.error(new
CircularImportProblem(configClass, this.importStack));
   }
   else {
      //向 importStack 中加入当前正在处理的配置类
      this.importStack.push(configClass);
      try {
         //循环处理参数传递过来的 importSelector
         for (SourceClass candidate : importCandidates) {
```

```java
                //判断
                if (candidate.isAssignable(ImportSelector.class)) {
                    Class<?> candidateClass = candidate.loadClass();
                    //将 class 转换成对象
                    ImportSelector selector
 = ParserStrategyUtils.instantiateClass(candidateClass, ImportSelector.class,
                            this.environment, this.resourceLoader, this.registry);
                    //如果类型是 DeferredImportSelector
                    if (selector instanceof DeferredImportSelector) {
                        this.deferredImportSelectorHandler.handle(configClass,
 (DeferredImportSelector) selector);
                    }
                    else {
                        //获取需要导入的类
                        String[] importClassNames
 = selector.selectImports(currentSourceClass.getMetadata());
                        //将需要导入的类从字符串转换成 SourceClass
                        Collection<SourceClass> importSourceClasses
 = asSourceClasses(importClassNames);
                        //递归处理需要导入的类
                        processImports(configClass, currentSourceClass, importSourceClasses, false);
                    }
                }

                //处理类型是 ImportBeanDefinitionRegistrar 的情况
                else if (candidate.isAssignable(ImportBeanDefinitionRegistrar.class)) {
                    Class<?> candidateClass = candidate.loadClass();
                    ImportBeanDefinitionRegistrar registrar =
                            ParserStrategyUtils.instantiateClass(candidateClass,
 ImportBeanDefinitionRegistrar.class,this.environment, this.resourceLoader, this.registry);
                    configClass.addImportBeanDefinitionRegistrar(registrar, currentSourceClass.
 getMetadata());
                }
                //其他情况
                else {
                    this.importStack.registerImport(
                            currentSourceClass.getMetadata(), candidate.getMetadata().getClassName());
                    processConfigurationClass(candidate.asConfigClass(configClass));
                }
            }
        }
        catch (BeanDefinitionStoreException ex) {
            throw ex;
        }
        catch (Throwable ex) {
            throw new BeanDefinitionStoreException(
                    "Failed to process import candidates for configuration class [" +
                    configClass.getMetadata().getClassName() + "]", ex);
        }
        finally {
```

```
            //从 importStack 删除配置类
            this.importStack.pop();
        }
    }
}
```

在这段代码中的处理细节如下。

(1) 判断参数 importCandidates 是否存在数据,如果不存在数据就不进行后续处理。

(2) 判断参数 checkForCircularImports 是否为 true,同时验证 importStack 中是否存在当前参数 configClass。如果 checkForCircularImports 为 true 同时验证通过就会抛出异常。

(3) 处理 importCandidates 集合中的每个 ImportSelector 对象,处理方式有以下三点。

① 类型是 ImportSelector,将 SourceClass 转换成 ImportSelector 对象,得到 ImportSelector 后分为两种情况处理:第一种,类型是 DeferredImportSelector,重复 handler 方法进行处理;第二种,类型不是 DeferredImportSelector,调用 ImportSelector#selectImports 方法得到需要导入的类,将类名转换成 SourceClass,最后解析每个 SourceClass 处理方法是本方法(processImports),递归调用。

② 类型是 ImportBeanDefinitionRegistrar,将 SourceClass 转换成 ImportBeanDefinitionRegistrar 实例,将这个实例放入 importBeanDefinitionRegistrars 容器。

③ 类型既不是 ImportSelector 也不是 ImportBeanDefinitionRegistrar,将这个对象当作配置类进行处理。

10.4.5 处理 @ImportResource 注解

接下来对 ImportResource 注解的处理进行分析,对应处理代码如下。

```
AnnotationAttributes importResource =
    AnnotationConfigUtils.attributesFor(sourceClass.getMetadata(),ImportResource.class);
if (importResource != null) {
    String[] resources = importResource.getStringArray("locations");
    Class<? extends BeanDefinitionReader> readerClass =
importResource.getClass("reader");
    for (String resource : resources) {
        String resolvedResource = this.environment.resolveRequiredPlaceholders(resource);
        configClass.addImportedResource(resolvedResource, readerClass);
    }
}
```

这段代码的处理是将注解元数据中关于 ImportResource 的属性进行提取,得到属性后将其转换成 importedResources 中的数据,在这个过程中需要关注 importedResources 的存储结构,它是一个 map 对象,key 表示资源文件路径地址,value 表示 BeanDefinition Reader 对象,具体存储容器代码如下。

```
private final Map<String, Class<? extends BeanDefinitionReader>> importedResources =
    new LinkedHashMap<>();
```

10.4.6 处理 @Bean 注解

接下来将对 Bean 注解进行分析，负责处理的代码如下。

```
//处理 Bean 注解
Set<MethodMetadata> beanMethods = retrieveBeanMethodMetadata(sourceClass);
for (MethodMetadata methodMetadata : beanMethods) {
    configClass.addBeanMethod(new BeanMethod(methodMetadata, configClass));
}

//处理 Bean Method
processInterfaces(configClass, sourceClass);
```

在这段代码中有两个处理方法，第一个是 retrieveBeanMethodMetadata，第二个是 processInterfaces。首先对第二个方法进行分析，具体代码如下。

```
private void processInterfaces(ConfigurationClass configClass, SourceClass sourceClass)
        throws IOException {
    //获取接口
    for (SourceClass ifc : sourceClass.getInterfaces()) {
        //接口上获取方法元数据
        Set<MethodMetadata> beanMethods = retrieveBeanMethodMetadata(ifc);
        for (MethodMetadata methodMetadata : beanMethods) {
            if (!methodMetadata.isAbstract()) {
                configClass.addBeanMethod(new BeanMethod(methodMetadata, configClass));
            }
        }
        processInterfaces(configClass, ifc);
    }
}
```

在这段代码中可以看到它调用了第一个方法 retrieveBeanMethodMetadata。在这段代码中有下面两个核心操作：提取类中的方法元数据，将方法元数据经过处理转换成 BeanMethod 放入配置类中。下面回过来看 retrieveBeanMethodMetadata 方法，具体代码如下。

```
private Set<MethodMetadata> retrieveBeanMethodMetadata(SourceClass sourceClass) {
    //提取元数据
    AnnotationMetadata original = sourceClass.getMetadata();
    //提取带有 bean 标签的数据
    Set<MethodMetadata> beanMethods =
original.getAnnotatedMethods(Bean.class.getName());

    if (beanMethods.size() > 1 && original instanceof StandardAnnotationMetadata) {
        try {
            //类的注解元数据
            AnnotationMetadata asm =
```

```
            this.metadataReaderFactory.getMetadataReader(original.getClassName()).
getAnnotationMetadata();
            //提取带有 Bean 注解的方法元数据
            Set < MethodMetadata > asmMethods =
asm.getAnnotatedMethods(Bean.class.getName());
            if (asmMethods.size() >= beanMethods.size()) {

                //选中的方法元数据
                Set < MethodMetadata > selectedMethods =
new LinkedHashSet<>(asmMethods.size());
                //循环两个方法元数据进行 methodName 比较,如果相同会被加入选中集合
                for (MethodMetadata asmMethod : asmMethods) {
                    for (MethodMetadata beanMethod : beanMethods) {
                        if (beanMethod.getMethodName().equals(asmMethod.getMethodName())) {
                            selectedMethods.add(beanMethod);
                            break;
                        }
                    }
                }
                if (selectedMethods.size() == beanMethods.size()) {
                    beanMethods = selectedMethods;
                }
            }
        }
        catch (IOException ex) {
            logger.debug("Failed to read class file via ASM for determining @Bean method order", ex);
        }
    }
    return beanMethods;
}
```

在这段代码中,有以下三个步骤。

(1) 提取类当中的注解元数据。

(2) 提取带有 Bean 注解的数据,这里提取的都是方法上的注解数据,变量名称为 beanMethods。

(3) 返回值处理。在返回值处理中有以下两种情况。

① 当步骤(2)得到的数据数量大于 1 并且类的元数据类型是 StandardAnnotationMetadata 时,就提取注解元数据,然后提取注解元数据中方法被 Bean 注解标记的方法元数据,变量名称为 asmMethods。接着筛选方法元数据,通过筛选条件则加入候选集合。筛选条件为:对比 asmMethods 中的数据和 beanMethods 中的数据方法名称是否相同,相同则符合条件的。

② 不符合情况则返回步骤(2)得到的数据。

接下来对上述流程进行测试用例的编写。首先需要编写一个接口 InterA,具体代码如下。

```
public interface InterA {
    @Bean
    default AnnPeople annp() {
        AnnPeople annPeople = new AnnPeople();
        annPeople.setName("InterPeop");
```

```
        return annPeople;
    }
}
```

在这个接口中定义了一个默认函数(JDK 8 特性)并将该对象作为一个 Bean 进行修饰，下面对这个接口进行实现，实现代码如下。

```
@Configuration
public class InterAImpl implements InterA {

    @Bean
    public AnnPeople annPeople() {
        return new AnnPeople();
    }

    @Bean
    public AnnPeople annPeople2() {
        AnnPeople annPeople = new AnnPeople();
        annPeople.setName("InterPeop2");
        return annPeople;
    }

}
```

在这个实现类中需要再定义一个 Bean，这里是为了覆盖外层的解析。写出两个 Bean 是为了测试 beanMethods.size() > 1 && original instanceof StandardAnnotationMetadata 条件，最后编写一个测试方法，相关代码如下。

```
@Test
void testInterface(){
    AnnotationConfigApplicationContext context
        = new AnnotationConfigApplicationContext(InterAImpl.class);

    InterA bean = context.getBean(InterA.class);
}
```

测试用例准备完毕进入调试阶段，首先关注 beanMethods 的数据信息，具体数据如图 10.8 所示。

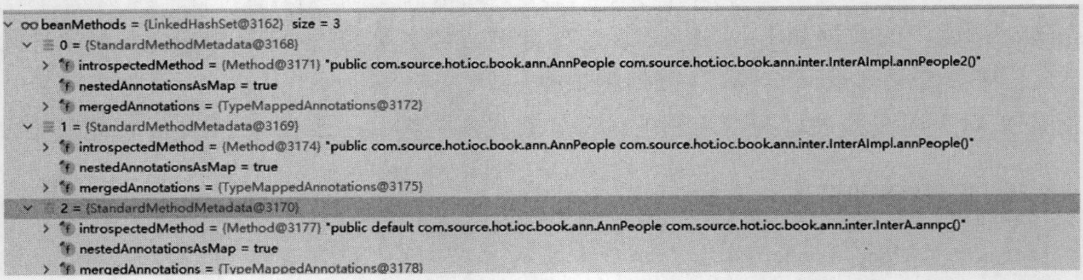

图 10.8　beanMethods 未经过 processInterfaces 方法处理

在这里可以看到三个 BeanMethod,最后一个元素是通过接口获取的,如果不需要这一个对象可以将 default 关键字删除,修改后代码如下。

```java
public interface InterA {
    @Bean
    AnnPeople annp();
}
```

但是这样编写就不会得到这个 Bean,它不符合 processInterfaces 中 !methodMetadata.isAbstract()条件。为了做出一个完整的测试,将代码进行修改变成如下内容。

```java
public interface InterA {
    @Bean
    default AnnPeople annpc() {
        AnnPeople annPeople = new AnnPeople();
        annPeople.setName("interface bean 1");
        return annPeople;
    }


    @Bean
    abstract AnnPeople annp();
}
```

修改实现类代码,改进后结果如下。

```java
@Configuration
public class InterAImpl implements InterA {

    @Bean
    public AnnPeople annPeople() {
        AnnPeople annPeople = new AnnPeople();
        annPeople.setName("bean1");
        return annPeople;
    }

    @Bean
    public AnnPeople annPeople2() {
        AnnPeople annPeople = new AnnPeople();
        annPeople.setName("bean2");
        return annPeople;
    }

    public AnnPeople annp() {
        AnnPeople annPeople = new AnnPeople();
        annPeople.setName("interface implements bean");
        return annPeople;
    }
}
```

此时去容器中根据类型来获取 Bean,查看现在在容器中类型是 AnnPeople 的内容,数据如图 10.9 所示。

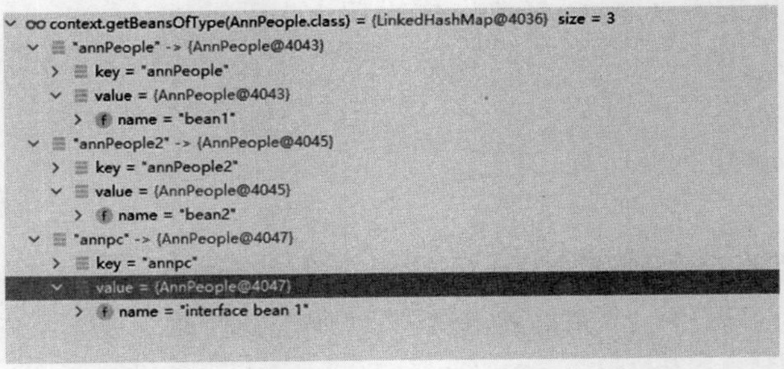

图 10.9 容器中的 AnnPeople 信息

10.5 处理父类配置

最后是关于父类配置的处理，相关代码如下。

```
if (sourceClass.getMetadata().hasSuperClass()) {
    String superclass = sourceClass.getMetadata().getSuperClassName();
    if (superclass != null && !superclass.startsWith("java") &&
            !this.knownSuperclasses.containsKey(superclass)) {
        this.knownSuperclasses.put(superclass, configClass);
        return sourceClass.getSuperClass();
    }
}
```

这段代码主要目的是向 knownSuperclasses 插入数据，插入数据的前提是当前正在处理的类存在父类，父类名称是以 java 开头，同时 knownSuperclasses 不存在。下面对父类配置进行相关测试代码编写，首先需要编写一个父类，类名为 SupperConfiguration，具体代码如下。

```
@Configuration
public class SupperConfiguration {

}
```

其次编写一个子类，类名为 ExtendConfiguration，具体代码如下。

```
@Configuration
public class ExtendConfiguration extends SupperConfiguration {

}
```

完成子类编写后编写一个应用上下文，具体代码如下。

```
@Configuration
@Import(ExtendConfiguration.class)
public class ThreeVarBeans {
}
```

最后编写一个测试方法，具体代码如下。

```
@Test
void testThreeVar() {
    AnnotationConfigApplicationContext context =
 new AnnotationConfigApplicationContext(ThreeVarBeans.class);
}
```

对于这段代码的分析可以参考 processConfigurationClass 的分析。

小结

本章对于 Spring 注解模式下配置类的解析做了分析，主要对各类注解进行了测试用例的编写和源码分析，了解这些注解的处理方式对于后续的 SpringBoot 及其他通过 Spring 注解模式启动的项目会有帮助。

第11章

ConfigurationClassPostProcessor 分析

在第 10 章介绍了 Spring 配置类的解析过程，本章将围绕 Spring 配置类解析后的处理行为进行分析，这个过程会引出 Spring 中的 ConfigurationClassPostProcessor 类，它是负责配置类解析后的处理行为，本章就将对这个类进行详细分析，包含测试用例搭建和 ConfigurationClassPostProcessor 的核心方法分析。

11.1 初识 ConfigurationClassPostProcessor

首先查看 ConfigurationClassPostProcessor 类图，如图 11.1 所示。

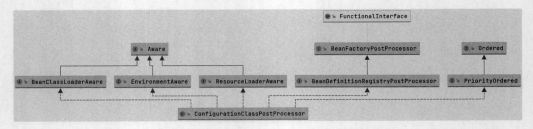

图 11.1 ConfigurationClassPostProcessor 类图

在这个类图中需要重点关注的是 BeanDefinitionRegistryPostProcessor 和 BeanFactoryPostProcessor。BeanFactoryPostProcessor 接口的作用是定制 BeanFactory 对象，BeanDefinitionRegistryPostProcessor 接口的作用是注册 BeanDefinition。

11.2 ConfigurationClassPostProcessor 测试用例搭建

接下来将对 ConfigurationClassPostProcessor 的测试搭建测试环境,首先需要编写一个 JavaBean 对象,类名为 AnnPeople,具体代码如下。

```java
public class AnnPeople {
    private String name;

    public String getName() {
        return name;
    }

    public void setName(String name) {
        this.name = name;
    }
}
```

接下来创建一个 Spring Component Bean,类名为 AnnBeans,具体代码如下。

```java
@Component
@Scope
public class AnnBeans {
    @Bean(name = "abc")
    public AnnPeople annPeople() {
        AnnPeople annPeople = new AnnPeople();
        annPeople.setName("people");
        return annPeople;
    }

    class InnerClass {

    }
}
```

完成 Spring Component Bean 编写后编写测试方法,具体代码如下。

```java
public class AnnotationContextTest {
    @Test
    void testBasePackages(){
        AnnotationConfigApplicationContext context =
                new AnnotationConfigApplicationContext("com.source.hot.ioc.book.ann");
        AnnPeople bean = context.getBean(AnnPeople.class);
        assert bean.getName().equals("people");
    }
}
```

11.3 postProcessBeanDefinitionRegistry 方法分析

接下来将对 postProcessBeanDefinitionRegistry 方法进行分析,具体处理代码如下。

```java
@Override
```

```java
public void postProcessBeanDefinitionRegistry(BeanDefinitionRegistry registry) {
    int registryId = System.identityHashCode(registry);
    if (this.registriesPostProcessed.contains(registryId)) {
        throw new IllegalStateException(
                "postProcessBeanDefinitionRegistry already called on this post - processor against " + registry);
    }
    if (this.factoriesPostProcessed.contains(registryId)) {
        throw new IllegalStateException(
                "postProcessBeanFactory already called on this post - processor against " + registry);
    }
    this.registriesPostProcessed.add(registryId);

    //注册 Configuration 相关注解的 Bean 定义
    processConfigBeanDefinitions(registry);
}
```

在这段代码中需要重点关注的方法是 processConfigBeanDefinitions，在这个方法中有很多细节可以讨论，具体的处理代码如下。

```java
public void processConfigBeanDefinitions(BeanDefinitionRegistry registry) {
    List<BeanDefinitionHolder> configCandidates = new ArrayList<>();
    //容器中已存在的 BeanName
    String[] candidateNames = registry.getBeanDefinitionNames();

    for (String beanName : candidateNames) {
        BeanDefinition beanDef = registry.getBeanDefinition(beanName);
        //BeanDefinition 中获取属性 CONFIGURATION_CLASS_ATTRIBUTE 是否为空
        if (beanDef.getAttribute(ConfigurationClassUtils.CONFIGURATION_CLASS_ATTRIBUTE) != null) {
            if (logger.isDebugEnabled()) {
                logger.debug("Bean definition has already been processed as a configuration class: " + beanDef);
            }
        }
        //判断当前的 BeanDefinition 是否属于候选对象
        else if(ConfigurationClassUtils.checkConfigurationClassCandidate(beanDef,this.metadataReaderFactory)) {
            configCandidates.add(new BeanDefinitionHolder(beanDef, beanName));
        }
    }

    //如果存有 @Configuration 的对象为空就返回,注意此时@Compoment 注解也算@Configuration
    //注解的子集,处理方法是 checkConfigurationClassCandidate
    if (configCandidates.isEmpty()) {
        return;
    }

    //将找到的 Configuration Bean 进行排序
    configCandidates.sort((bd1, bd2) -> {
```

```java
        int i1 = ConfigurationClassUtils.getOrder(bd1.getBeanDefinition());
        int i2 = ConfigurationClassUtils.getOrder(bd2.getBeanDefinition());
        return Integer.compare(i1, i2);
});

//创建两个 BeanName 生成器
//1. 组件扫描的 BeanName 生成器
//2. 导入的 BeanName 生成器
SingletonBeanRegistry sbr = null;
if (registry instanceof SingletonBeanRegistry) {
    sbr = (SingletonBeanRegistry) registry;
    if (!this.localBeanNameGeneratorSet) {
        BeanNameGenerator generator = (BeanNameGenerator) sbr.getSingleton(
            AnnotationConfigUtils.CONFIGURATION_BEAN_NAME_GENERATOR);
        if (generator != null) {
            this.componentScanBeanNameGenerator = generator;
            this.importBeanNameGenerator = generator;
        }
    }
}

//环境信息
if (this.environment == null) {
    this.environment = new StandardEnvironment();
}

//解析阶段
//创建 Configuration 解析对象
ConfigurationClassParser parser = new ConfigurationClassParser(
        this.metadataReaderFactory, this.problemReporter, this.environment,
        this.resourceLoader, this.componentScanBeanNameGenerator, registry);

//候选的需要解析的 BeanDefinition 容器
Set<BeanDefinitionHolder> candidates = new LinkedHashSet<>(configCandidates);
//已经完成解析的 Spring Configuration Bean
Set<ConfigurationClass> alreadyParsed = new HashSet<>(configCandidates.size());
do {
    //解析候选 Bean Definition Holder 集合
    parser.parse(candidates);
    //解析器中的验证
    parser.validate();

    //配置类
    Set<ConfigurationClass> configClasses =
new LinkedHashSet<>(parser.getConfigurationClasses());
    configClasses.removeAll(alreadyParsed);

    if (this.reader == null) {
        this.reader = new ConfigurationClassBeanDefinitionReader(
            registry, this.sourceExtractor, this.resourceLoader, this.environment,
```

```java
                    this.importBeanNameGenerator, parser.getImportRegistry());
        }
        //解析 Spring Configuration Class 的主要目的是提取其中的 Spring 注解并将其
        //转换成 BeanDefinition
        this.reader.loadBeanDefinitions(configClasses);
        alreadyParsed.addAll(configClasses);

        candidates.clear();
        if (registry.getBeanDefinitionCount() > candidateNames.length) {
            //新的候选名称
            String[] newCandidateNames = registry.getBeanDefinitionNames();
            //历史的候选名称
            Set<String> oldCandidateNames =
new HashSet<>(Arrays.asList(candidateNames));
            //已经完成解析的类名
            Set<String> alreadyParsedClasses = new HashSet<>();
            //加入解析类名
            for (ConfigurationClass configurationClass : alreadyParsed) {
                alreadyParsedClasses.add(configurationClass.getMetadata().getClassName());
            }
            //新的候选名称和对应的 BeanDefinition 添加到候选容器中
            for (String candidateName : newCandidateNames) {
                if (!oldCandidateNames.contains(candidateName)) {
                    BeanDefinition bd = registry.getBeanDefinition(candidateName);
                    if (ConfigurationClassUtils.checkConfigurationClassCandidate(bd, this.
metadataReaderFactory) && !alreadyParsedClasses.contains(bd.getBeanClassName())) {
                        candidates.add(new BeanDefinitionHolder(bd, candidateName));
                    }
                }
            }
            candidateNames = newCandidateNames;
        }
    }
    while (!candidates.isEmpty());

    //注册 Import 相关 Bean
    if (sbr != null && !sbr.containsSingleton(IMPORT_REGISTRY_BEAN_NAME)) {
        sbr.registerSingleton(IMPORT_REGISTRY_BEAN_NAME, parser.getImportRegistry());
    }

    if (this.metadataReaderFactory instanceof CachingMetadataReaderFactory) {
        ((CachingMetadataReaderFactory) this.metadataReaderFactory).clearCache();
    }
}
```

在这个方法中有以下六个处理步骤。

（1）容器内已存在的 Bean 进行候选分类。

（2）候选 Bean Definition Holder 的排序。

(3) BeanName 生成器的创建。
(4) 初始化基本环境信息。
(5) 解析候选 Bean。
(6) 注册 Import Bean 和清理数据。

11.3.1　容器内已存在的 Bean 进行候选分类

首先对容器内已存在的 Bean 进行候选分类,对应的处理代码如下。

```
List<BeanDefinitionHolder> configCandidates = new ArrayList<>();
//容器中已存在的 BeanName
String[] candidateNames = registry.getBeanDefinitionNames();

for (String beanName : candidateNames) {
    BeanDefinition beanDef = registry.getBeanDefinition(beanName);
    //BeanDefinition 中获取属性 CONFIGURATION_CLASS_ATTRIBUTE 是否为空
    if (beanDef.getAttribute(ConfigurationClassUtils.CONFIGURATION_CLASS_ATTRIBUTE) != null)
{
        if (logger.isDebugEnabled()) {
            logger.debug("Bean definition has already been processed as a configuration class: " +
beanDef);
        }
    }
    //判断当前的 BeanDefinition 是否属于候选对象
    else if
(ConfigurationClassUtils.checkConfigurationClassCandidate(beanDef, this.metadataReaderFactory)) {
        configCandidates.add(new BeanDefinitionHolder(beanDef, beanName));
    }
}
```

在这段代码中可以看到存储分类后的容器是 configCandidates,分类的依据是通过 ConfigurationClassUtils.checkConfigurationClassCandidate 方法进行,分类的依据(候选标准)有下面五种情况。

(1) 判断 BeanDefinition 中 className 和 FactoryMethodName 是否存在,只要有一个不存在就不符合候选标准。

(2) 判断 BeanDefinition 是否属于 AbstractBeanDefinition 类型,并且存在 className,与此同时,如果 className 是属于 BeanFactoryPostProcessor、BeanPostProcessor、AopInfrastructureBean 和 EventListenerFactory 中的某一个,那么这个 BeanDefiniton 不符合候选标准。

(3) 提取注解元数据失败的 BeanDefinition 不符合候选标准。

(4) 提取注解源数据中有关 Configuration 注解的数据后判断 proxyBeanMethods 属性是否为 false,如果为 true 该 BeanDefinition 不符合候选标准。

(5) 判断注解元信息中是否存在 Bean Component、ComponentScan、Import 和 ImportResource,如果不存在该 BeanDefinition 不符合候选标准。

在 checkConfigurationClassCandidate 方法中除了上述五种候选对象判断外还有三个

属性设置，这三个属性分别是 CONFIGURATION_CLASS_ATTRIBUTE、CONFIGURATION_CLASS_ATTRIBUTE 和 ORDER_ATTRIBUTE。

在测试用例中 AnnBeans 对象符合第五种候选条件，AnnBeans 存有 @Component 注解。接下来进入调试阶段查看 configCandidates 对象，具体数据如图 11.2 所示。

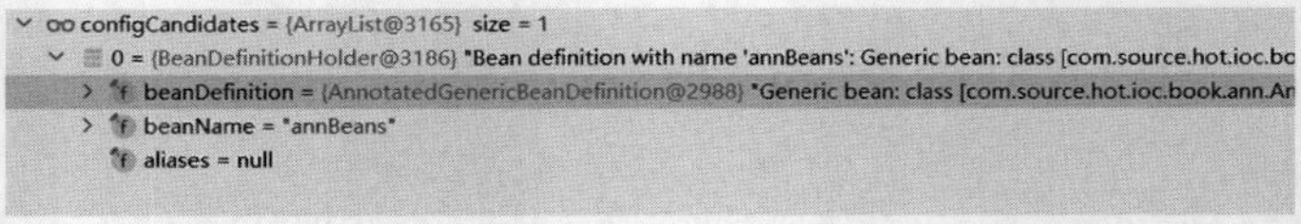

图 11.2　configCandidates 数据信息

通过图 11.2 可以看到，此时数据已经被加入容器中等待后续操作。

11.3.2　候选 BeanDefinition Holder 的排序

通过前文的候选操作此时拥有了 configCandidates 数据集合，Spring 会对这个集合进行排序操作，具体排序操作代码如下。

```
configCandidates.sort((bd1, bd2) -> {
    int i1 = ConfigurationClassUtils.getOrder(bd1.getBeanDefinition());
    int i2 = ConfigurationClassUtils.getOrder(bd2.getBeanDefinition());
    return Integer.compare(i1, i2);
});
```

这里的排序操作其本质是获取 order 的数据然后进行排序操作，负责获取 order 数据的代码如下。

```
public static int getOrder(BeanDefinition beanDef) {
    Integer order = (Integer) beanDef.getAttribute(ORDER_ATTRIBUTE);
    return (order != null ? order : Ordered.LOWEST_PRECEDENCE);
}
```

这段代码中 Spring 会从 BeanDefinition 中获取 ORDER_ATTRIBUTE 属性，如果数据不存在直接返回最大序号。

11.3.3　BeanName 生成器的创建

接下来将进行 BeanName 生成器的创建，具体创建代码如下。

```
//创建两个 BeanName 生成器
//1. 组件扫描的 BeanName 生成器
//2. 导入的 BeanName 生成器
SingletonBeanRegistry sbr = null;
if (registry instanceof SingletonBeanRegistry) {
    sbr = (SingletonBeanRegistry) registry;
    if (!this.localBeanNameGeneratorSet) {
        BeanNameGenerator generator = (BeanNameGenerator) sbr.getSingleton(
            AnnotationConfigUtils.CONFIGURATION_BEAN_NAME_GENERATOR);
```

```
            if (generator != null) {
                this.componentScanBeanNameGenerator = generator;
                this.importBeanNameGenerator = generator;
            }
        }
    }
```

从这段代码上可以看到会创建两个 BeanName 生成器，这两个 BeanName 生成器的本质是同一个，BeanName 生成器是 org. springframework. context. annotation. internal-ConfigurationBeanNameGenerator。

下面对两个 BeanNameGenerator 进行说明。

（1）componentScanBeanNameGenerator 是处理组件扫描注册的 BeanNameGenerator，默认采用短类名，默认实现是 AnnotationBeanNameGenerator. INSTANCE。

（2）importBeanNameGenerator 是用来处理导入的 BeanNameGenerator，默认采用全类名，默认实现是 FullyQualifiedAnnotationBeanNameGenerator。

11.3.4　初始化基本环境信息

接下来是关于基本环境信息的初始化，对应处理代码如下。

```
if (this.environment == null) {
    this.environment = new StandardEnvironment();
}
```

这段代码直接将 StandardEnvironment 对象创建出来作为基本环境信息对象。

11.3.5　解析候选 Bean

接下来将对每个候选 Bean 进行处理，具体的处理代码如下。

```
ConfigurationClassParser parser = new ConfigurationClassParser(
        this.metadataReaderFactory, this.problemReporter, this.environment,
        this.resourceLoader, this.componentScanBeanNameGenerator, registry);

//解析后的 BeanDefinition 容器
Set < BeanDefinitionHolder > candidates = new LinkedHashSet <>(configCandidates);
//需要解析的类
Set < ConfigurationClass > alreadyParsed = new HashSet <>(configCandidates.size());
do {
    //解析候选 Bean Definition Holder 集合
    parser.parse(candidates);
    //解析器中的验证
    parser.validate();

    //配置类
    Set < ConfigurationClass > configClasses =
new LinkedHashSet <>(parser.getConfigurationClasses());
```

```
            configClasses.removeAll(alreadyParsed);

            if (this.reader == null) {
                this.reader = new ConfigurationClassBeanDefinitionReader(
                        registry, this.sourceExtractor, this.resourceLoader, this.environment,
                        this.importBeanNameGenerator, parser.getImportRegistry());
            }
            //解析 Spring Configuration Class 的主要目的是提取其中的 Spring 注解并将其转换成
            //BeanDefinition
            this.reader.loadBeanDefinitions(configClasses);
            alreadyParsed.addAll(configClasses);

            candidates.clear();
            if (registry.getBeanDefinitionCount() > candidateNames.length) {
                //新的候选名称
                String[] newCandidateNames = registry.getBeanDefinitionNames();
                //历史的候选名称
                Set<String> oldCandidateNames = new HashSet<>(Arrays.asList(candidateNames));
                //已经完成解析的类名
                Set<String> alreadyParsedClasses = new HashSet<>();
                //加入解析类名
                for (ConfigurationClass configurationClass : alreadyParsed) {
                    alreadyParsedClasses.add(configurationClass.getMetadata().getClassName());
                }
                //新的候选名称和对应的 BeanDefinition 添加到候选容器中
                for (String candidateName : newCandidateNames) {
                    if (!oldCandidateNames.contains(candidateName)) {
                        BeanDefinition bd = registry.getBeanDefinition(candidateName);
                        if(ConfigurationClassUtils.checkConfigurationClassCandidate(bd,this.
metadataReaderFactory) &&
                                !alreadyParsedClasses.contains(bd.getBeanClassName())) {
                            candidates.add(new BeanDefinitionHolder(bd, candidateName));
                        }
                    }
                }
                candidateNames = newCandidateNames;
            }
        }
        while (!candidates.isEmpty());
```

从这段代码中可以确定负责解析的是 ConfigurationClassParser 类,在这个方法中会创建 ConfigurationClassParser 对象,通过 ConfigurationClassParser 对象的 parse 方法进行单个候选对象的处理,具体处理策略有以下三个。

(1) BeanDefinition 类型是 AnnotatedBeanDefinition。

(2) BeanDefinition 类型是 AbstractBeanDefinition 并且拥有 className 属性。

(3) 策略一和策略二以外的处理。

上述三个策略的处理方法是 processConfigurationClass,具体代码如下。

```
protected void processConfigurationClass(ConfigurationClass configClass) throws IOException {
```

```java
        if (this.conditionEvaluator.shouldSkip(configClass.getMetadata(), ConfigurationPhase.
PARSE_CONFIGURATION)) {
            return;
        }

        ConfigurationClass existingClass = this.configurationClasses.get(configClass);
        if (existingClass != null) {
            if (configClass.isImported()) {
                if (existingClass.isImported()) {
                    existingClass.mergeImportedBy(configClass);
                }
                return;
            }
            else {
                this.configurationClasses.remove(configClass);
                this.knownSuperclasses.values().removeIf(configClass::equals);
            }
        }

        SourceClass sourceClass = asSourceClass(configClass);
        do {
            sourceClass = doProcessConfigurationClass(configClass, sourceClass);
        }
        while (sourceClass != null);

        this.configurationClasses.put(configClass, configClass);
    }
```

在 processConfigurationClass 处理过程中建立了 configurationClasses 缓存和 importStack 数据，执行前数据如图 11.3 所示，执行后数据如图 11.4 所示。

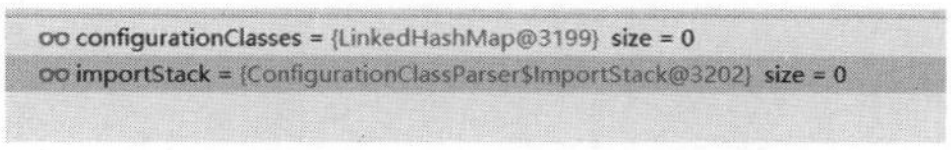

图 11.3 processConfigurationClass 执行前

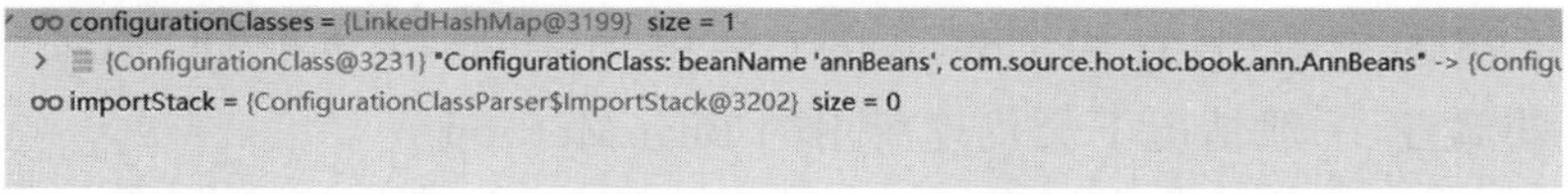

图 11.4 processConfigurationClass 执行后

经过 parse 方法处理候选项后会进入验证方法，具体验证代码如下。

```java
public void validate() {
    for (ConfigurationClass configClass : this.configurationClasses.keySet()) {
        configClass.validate(this.problemReporter);
    }
}


public void validate(ProblemReporter problemReporter) {
```

```
        Map<String, Object> attributes
= this.metadata.getAnnotationAttributes(Configuration.class.getName());
    if (attributes != null && (Boolean) attributes.get("proxyBeanMethods")) {
        if (this.metadata.isFinal()) {
            problemReporter.error(new FinalConfigurationProblem());
        }
        for (BeanMethod beanMethod : this.beanMethods) {
            beanMethod.validate(problemReporter);
        }
    }
}
```

在这个验证方法中会对所有候选对象进行验证,具体验证规则如下。

(1) 判断当前 BeanDefinition 的注解元数据是否存在 Configuration 注解属性。

(2) 判断注解元数据中 proxyBeanMethods 属性是否为 true,该属性从 Configuration 注解中获取。

(3) 判断注解元数据是否是 final 修饰的基本类。

(4) 通过 BeanMethod 来验证当前的数据,验证是否是 static 修饰,验证是否存在 Configuration 注解,以及是否存在需要重写的方法。

接下来将对在验证过程中出现的 Configuration 注解进行分析,Configuration 具体代码如下。

```
@Target(ElementType.TYPE)
@Retention(RetentionPolicy.RUNTIME)
@Documented
@Component
public @interface Configuration {

    @AliasFor(annotation = Component.class)
    String value() default "";

    boolean proxyBeanMethods() default true;

}
```

在这个注解的定义中它和 Component 注解具有类似功能,同时有个是否代理 BeanMethods 的属性,一般情况下会代理所用的 BeanMethod。

在处理完成验证相关的内容后会进行单个配置类解析,具体处理代码如下。

```
//配置类
Set<ConfigurationClass> configClasses =
new LinkedHashSet<>(parser.getConfigurationClasses());
configClasses.removeAll(alreadyParsed);

if (this.reader == null) {
    this.reader = new ConfigurationClassBeanDefinitionReader(
        registry, this.sourceExtractor, this.resourceLoader, this.environment,
        this.importBeanNameGenerator, parser.getImportRegistry());
}
```

```
//解析 Spring Configuration Class 的主要目的是提取其中的 Spring 注解并将其转换成
//BeanDefinition
this.reader.loadBeanDefinitions(configClasses);
alreadyParsed.addAll(configClasses);
```

在这段代码中第一个操作就是将 ConfigurationClassParser 的数据提取出来,该数据是 Spring Configuration Bean,这段代码中的后续操作 this.reader.loadBeanDefinitions 就是为了从中读取 BeanDefinition。在这里可以看到核心能力提供者是 Configuration-ClassBeanDefinitionReader,这个对象可以将 Spring Configuration Bean 中的 BeanDefinition 读取并加入容器即可。

当执行完成 this.reader.loadBeanDefinitions(configClasses)方法后开发者自定义的 Spring Bean 将会被加载到 Spring 容器中。当解析完成同一批 configClasses 后会将这批 configClasses 对象放入 alreadyParsed,同时将 candidates 清空,清空的目的是存放后续需要处理的数据,具体处理操作代码如下。

```
candidates.clear();
if (registry.getBeanDefinitionCount() > candidateNames.length) {
    //新的候选名称
    String[] newCandidateNames = registry.getBeanDefinitionNames();
    //历史的候选名称
    Set<String> oldCandidateNames = new HashSet<>(Arrays.asList(candidateNames));
    //已经完成解析的类名
    Set<String> alreadyParsedClasses = new HashSet<>();
    //加入解析类名
    for (ConfigurationClass configurationClass : alreadyParsed) {
        alreadyParsedClasses.add(configurationClass.getMetadata().getClassName());
    }
    //新的候选名称和对应的 BeanDefinition 添加到候选容器中
    for (String candidateName : newCandidateNames) {
        if (!oldCandidateNames.contains(candidateName)) {
            BeanDefinition bd = registry.getBeanDefinition(candidateName);
            if(ConfigurationClassUtils.checkConfigurationClassCandidate(bd, this.metadataReaderFactory) &&
                    !alreadyParsedClasses.contains(bd.getBeanClassName())) {
                candidates.add(new BeanDefinitionHolder(bd, candidateName));
            }
        }
    }
    candidateNames = newCandidateNames;
}
```

在这段代码中主要是一些集合对象的操作,重点关注执行的条件,当前注册其中 BeanDefinition 的数量大于方法最开始入口的候选 BeanDefinition 名称列表的数量就会进行该段代码处理。另外需要关注两个变量:newCandidateNames 和 oldCandidateNames。这两个变量的数据是存在差异的,具体差异的起源是因为 reader.loadBeanDefinitions (configClasses)方法的执行。oldCandidateNames 存储的是 loadBeanDefinitions 方法执行之前容器中的 BeanDefinitionName 列表，newCandidateNames 存储的是 loadBeanDefinitions 方法执行之后容器中的 BeanDefinitionName 列表。alreadyParsedClasses 变量也值得一看，alreadyParsedClasses 中的数据是从 alreadyParsed 对象中获取，alreadyParsed 中存储了已经解析完成的配置类（ConfigurationClass），进一步进行数据提取会将 ConfigurationClass 对应的类全路径放入 alreadyParsedClasses 中，它表示已经进行过解析的类。接下来将对添加到候选容器的操作做流程整理。

（1）newCandidateNames 中的 BeanName 是否在 oldCandidateNames 中存在，存在不会加入到容器，不存在就进入下一步判断。

（2）提取 New BeanName 对应的 BeanDefinition 进行两层验证。

① 验证当前的 BeanDefinition 是否符合候选类，这里的处理方法是 checkConfigurationClassCandidate。

② alreadyParsedClasses 中是否存在当前 BeanDefinition 的类名，不存在会被加入到容器。

符合上述逻辑验证的 BeanDefinition 会被添加到容器中以此完成处理操作。

11.3.6　注册 Import Bean 和清理数据

接下来将对注册 Import Bean 和清理数据相关内容进行分析，具体处理代码如下。

```
//注册 Import 相关 Bean
if (sbr != null && !sbr.containsSingleton(IMPORT_REGISTRY_BEAN_NAME)) {
    sbr.registerSingleton(IMPORT_REGISTRY_BEAN_NAME, parser.getImportRegistry());
}

if (this.metadataReaderFactory instanceof CachingMetadataReaderFactory) {
    ((CachingMetadataReaderFactory) this.metadataReaderFactory).clearCache();
}
```

这段代码处理以下两个事项。

（1）注册 Import Bean，将 Import Bean 注册到 ConfigurationClassPostProcessor.class.getName() + ".importRegistry" 名称对应的数据中。

（2）清理数据，将 metadataReaderCache 缓存清空。

11.4　postProcessBeanFactory 方法分析

接下来将对 postProcessBeanFactory 方法进行分析，具体处理代码如下。

```
@Override
public void postProcessBeanFactory(ConfigurableListableBeanFactory beanFactory) {
    int factoryId = System.identityHashCode(beanFactory);
    if (this.factoriesPostProcessed.contains(factoryId)) {
        throw new IllegalStateException(
                "postProcessBeanFactory already called on this post - processor against " +
beanFactory);
```

```java
        }
        this.factoriesPostProcessed.add(factoryId);
        if (!this.registriesPostProcessed.contains(factoryId)) {
            processConfigBeanDefinitions((BeanDefinitionRegistry) beanFactory);
        }

        enhanceConfigurationClasses(beanFactory);
        beanFactory.addBeanPostProcessor(new ImportAwareBeanPostProcessor(beanFactory));
}
```

在这段代码中主要对 enhanceConfigurationClasses 方法进行分析,该方法是一个增强方法,主要用来增强 Spring 配置类及代理 Spring 配置类。具体处理代码如下。

```java
public void enhanceConfigurationClasses(ConfigurableListableBeanFactory beanFactory) {
    //存储 Spring Configuration Bean Definiton
    Map<String, AbstractBeanDefinition> configBeanDefs = new LinkedHashMap<>();
    //容器中将符合条件的对象放入配置 Bean 容器中
    for (String beanName : beanFactory.getBeanDefinitionNames()) {
        BeanDefinition beanDef = beanFactory.getBeanDefinition(beanName);
        Object configClassAttr = beanDef.getAttribute(ConfigurationClassUtils.CONFIGURATION_CLASS_ATTRIBUTE);
        MethodMetadata methodMetadata = null;
        if (beanDef instanceof AnnotatedBeanDefinition) {
            methodMetadata = ((AnnotatedBeanDefinition) beanDef).getFactoryMethodMetadata();
        }
        if ((configClassAttr != null || methodMetadata != null) && beanDef instanceof AbstractBeanDefinition) {
            AbstractBeanDefinition abd = (AbstractBeanDefinition) beanDef;
            if (!abd.hasBeanClass()) {
                try {
                    abd.resolveBeanClass(this.beanClassLoader);
                }
                catch (Throwable ex) {
                    throw new IllegalStateException(
                            "Cannot load configuration class: " + beanDef.getBeanClassName(), ex);
                }
            }
        }
        if (ConfigurationClassUtils.CONFIGURATION_CLASS_FULL.equals(configClassAttr)) {
            if (!(beanDef instanceof AbstractBeanDefinition)) {
                throw new BeanDefinitionStoreException("Cannot enhance @Configuration bean definition '" + beanName +
                        "' since it is not stored in an AbstractBeanDefinition subclass");
            }
            else if (logger.isInfoEnabled() && beanFactory.containsSingleton(beanName)) {
                logger.info("Cannot enhance @Configuration bean definition '" + beanName
                        + "' since its singleton instance has been created too early. The typical cause " + "is a non-"
                        + "static @Bean method with a BeanDefinitionRegistryPostProcessor " +
                        "return type: Consider declaring such methods as 'static'.");
```

```
                    }
                    configBeanDefs.put(beanName, (AbstractBeanDefinition) beanDef);
                }
            }
            //Spring Configuration Bean Definition 不存在
            if (configBeanDefs.isEmpty()) {
                return;
            }

            //对 Spring Configuration Bean Definition 进行增强处理
            ConfigurationClassEnhancer enhancer = new ConfigurationClassEnhancer();
            for (Map.Entry<String, AbstractBeanDefinition> entry : configBeanDefs.entrySet()) {
                AbstractBeanDefinition beanDef = entry.getValue();
                beanDef.setAttribute(AutoProxyUtils.PRESERVE_TARGET_CLASS_ATTRIBUTE, Boolean.TRUE);
                Class<?> configClass = beanDef.getBeanClass();
                Class<?> enhancedClass = enhancer.enhance(configClass, this.beanClassLoader);
                if (configClass != enhancedClass) {
                    if (logger.isTraceEnabled()) {
                        logger.trace(String.format("Replacing bean definition '%s' existing class '%s' with " + "enhanced class '%s'", entry.getKey(), configClass.getName(), enhancedClass.getName()));
                    }
                    beanDef.setBeanClass(enhancedClass);
                }
            }
        }
```

在这个增强方法中做了以下两件事。

（1）将容器中的 BeanDefinition 过滤得到 Spring Configuration Bean。

（2）对得到的 Spring Configuration Bean 进行增强处理。

在处理第一件事的时候又分为三个步骤，具体步骤如下。

（1）提取三个属性，属性一根据 BeanName 在容器中找到对应的 BeanDefintion（beanDef）。属性二从 BeanDefintion 中提取属性 ConfigurationClassUtils.CONFIGURATION_CLASS_ATTRIBUTE（configClassAttr）。属性三如果 BeanDefintion 是 AnnotatedBeanDefinition 类型则提取 MethodMetadata（methodMetadata）。

（2）BeanDefinition 处理 BeanClass(abd.resolveBeanClass(this.beanClassLoader))。

（3）判断是否可以加入 configBeanDefs 中。判断依据有两层，第一层是当前 BeanDefintion 中的 ConfigurationClassUtils.CONFIGURATION_CLASS_ATTRIBUTE 属性是 full；第二层是当前 BeanDefintion 不是 AbstractBeanDefinition 抛出异常。

在本文的测试用例中是不符合这里的处理条件的。不符合的原因是 ConfigurationClassUtils.CONFIGURATION_CLASS_FULL.equals(configClassAttr)代码，有关 CONFIGURATION_CLASS_ATTRIBUTE 属性的设置在 checkConfigurationClassCandidate 方法中有所处理，具体处理代码如下。

```
Map<String, Object> config = metadata.getAnnotationAttributes(Configuration.class.getName());
if (config != null && !Boolean.FALSE.equals(config.get("proxyBeanMethods"))) {
```

```
   beanDef.setAttribute(CONFIGURATION_CLASS_ATTRIBUTE, CONFIGURATION_CLASS_FULL);
}
else if (config != null || isConfigurationCandidate(metadata)) {
   beanDef.setAttribute(CONFIGURATION_CLASS_ATTRIBUTE, CONFIGURATION_CLASS_LITE);
}
```

当前测试用例中 AnnBeans 的数据信息如图 11.5 所示。

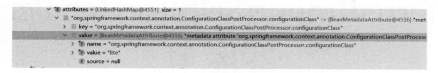

图 11.5　AnnBeans 中的数据信息

从图 11.5 中可以发现它不符合增强的要求。具体的增强处理逻辑是将原始的 Class 进行替换，从原来的 BeanClass 转换为 ConfigurationClassEnhancer 对象，具体操作有以下四步。

（1）设置 PRESERVE_TARGET_CLASS_ATTRIBUTE 属性为 true，表示这是一个增强类。

（2）提取 BeanDefintion 中原始的 Class。

（3）通过 ConfigurationClassEnhancer 对原始的 Class 进行增强。

（4）将增强后的 Class 填写到 BeanDefinition 的 BeanClass 中。

经过上述四步操作即可以得到一个增强的 BeanClass 对象以此完成增强操作。

小结

本章主要围绕 ConfigurationClassPostProcessor 对象的两个接口进行分析，ConfigurationClassPostProcessor 的主要作用是进行配置类的处理调度，本身不会处理关于配置类的操作，主要负责整体流程上的控制。

第12章 DeferredImportSelectorHandler 分析

本章将对 Spring 注解模式中的 DeferredImportSelectorHandler 类进行分析,本章包含创建 DeferredImportSelectorHandler 对象分析的测试环境和 DeferredImportSelectorHandler 相关源码分析。

12.1 初识 DeferredImportSelectorHandler

接下来将简单认识 DeferredImportSelectorHandler 对象,在这个对象中主要关注 parse 方法,具体处理代码如下。

```
public void parse(Set < BeanDefinitionHolder > configCandidates) {
    for (BeanDefinitionHolder holder : configCandidates) {
        BeanDefinition bd = holder.getBeanDefinition();
        try {
            if (bd instanceof AnnotatedBeanDefinition) {
                parse(((AnnotatedBeanDefinition) bd).getMetadata(), holder.getBeanName());
            }
             else if (bd instanceof AbstractBeanDefinition && ((AbstractBeanDefinition) bd).hasBeanClass()) {
                parse(((AbstractBeanDefinition) bd).getBeanClass(), holder.getBeanName());
            }
            else {
                parse(bd.getBeanClassName(), holder.getBeanName());
            }
        }
        catch (BeanDefinitionStoreException ex) {
```

```
            throw ex;
        }
        catch (Throwable ex) {
            throw new BeanDefinitionStoreException(
                    "Failed to parse configuration class [" + bd.getBeanClassName() + "]", ex);
        }
    }

    this.deferredImportSelectorHandler.process();
}
```

在这段代码中引出需要分析的变量是 deferredImportSelectorHandler，具体类型是 DeferredImportSelectorHandler，它能够对 ImportSelector 相关接口进行处理。接下来对这个对象的成员变量进行说明，它只有一个成员变量 deferredImportSelectors，它能够存储 DeferredImportSelectorHolder 对象，DeferredImportSelectorHolder 对象中包含配置类和接口 DeferredImportSelector，有关 deferredImportSelectors 的定义代码如下。

```
private class DeferredImportSelectorHandler {

    @Nullable
    private List<DeferredImportSelectorHolder> deferredImportSelectors =
new ArrayList<>();
}
```

有关 DeferredImportSelectorHolder 的定义代码如下。

```
private static class DeferredImportSelectorHolder {

    private final ConfigurationClass configurationClass;

    private final DeferredImportSelector importSelector;
}
```

12.2 DeferredImportSelectorHandler 测试环境搭建

接下来将创建一个用来对 DeferredImportSelectorHandler 类测试的环境，首先编写 DeferredImportSelector 的实现类，类名为 MyDeferredImportSelector，具体代码如下。

```
public class MyDeferredImportSelector implements DeferredImportSelector {

    @Override
    public String[] selectImports(AnnotationMetadata importingClassMetadata) {
        System.out.println("in MyDeferredImportSelector");
        return new String[0];
    }
}
```

完成 MyDeferredImportSelector 编写后继续编写 ImportBeanDefinitionRegistrar 的实现类，类名为 MyImportBeanDefinitionRegistrar，具体代码如下。

```java
public class MyImportBeanDefinitionRegistrar implements ImportBeanDefinitionRegistrar {
    @Override
    public void registerBeanDefinitions(AnnotationMetadata importingClassMetadata,
BeanDefinitionRegistry registry, BeanNameGenerator importBeanNameGenerator) {

    }

    @Override
    public void registerBeanDefinitions(AnnotationMetadata importingClassMetadata,
BeanDefinitionRegistry registry) {

    }
}
```

完成 MyImportBeanDefinitionRegistrar 编写后继续编写 ImportSelector 的实现类,类名为 MyImportSelector,具体代码如下。

```java
public class MyImportSelector implements ImportSelector {
    @Override
    public String[] selectImports(AnnotationMetadata importingClassMetadata) {
        System.out.println("in MyImportSelector");
        return new String[0];
    }
}
```

完成基本实现类的编写后编写 Spring 配置类,类名为 ImportSelectorBeans,具体代码如下。

```java
@Component
@Import(
        value = {
                MyDeferredImportSelector.class,
                MyImportBeanDefinitionRegistrar.class,
                MyImportSelector.class
        }
)
@Configuration
public class ImportSelectorBeans {

}
```

最后编写测试用例,具体代码如下。

```java
public class ImportSelectTest {
    @Test
    void testImportSelect() {
        AnnotationConfigApplicationContext context =
                new AnnotationConfigApplicationContext(ImportSelectorBeans.class);
    }
}
```

执行测试方法后会输出下面内容。

in MyDeferredImportSelector

12.3 handler 方法分析

接下来将对 handler 方法进行分析，handler 相关代码如下。

```
public void handle(ConfigurationClass configClass, DeferredImportSelector importSelector)
{
    DeferredImportSelectorHolder holder = new DeferredImportSelectorHolder(
        configClass, importSelector);
    if (this.deferredImportSelectors == null) {
        DeferredImportSelectorGroupingHandler handler =
new DeferredImportSelectorGroupingHandler();
        handler.register(holder);
        handler.processGroupImports();
    }
    else {
        this.deferredImportSelectors.add(holder);
    }
}
```

上述代码本身的操作复杂度不高，在这段代码中需要引入另一个对象 DeferredImportSelectorGroupingHandler。在引入这个对象前需要知道的是 deferredImportSelectors 为空才会创建 DeferredImportSelectorGroupingHandler 对象，否则就会直接在 deferredImportSelectors 中将数据加入进去。

在这个 handler 方法中需要关注第二个参数，注意只有类型是 DeferredImportSelector 的才会进行处理，在测试用例中符合该条件的对象是 MyDeferredImportSelector，目前容器中的情况如图 12.1 所示。

图 12.1　MyDeferredImportSelector 加入容器

12.4　DeferredImportSelectorGroupingHandler 分析

当 MyDeferredImportSelector 数据加入容器之后会进行 process 方法的调度，具体调度代码如下。

```
public void process() {
        List < DeferredImportSelectorHolder > deferredImports =
this.deferredImportSelectors;
        this.deferredImportSelectors = null;
        try {
```

```
                if (deferredImports != null) {
                    DeferredImportSelectorGroupingHandler handler =
new DeferredImportSelectorGroupingHandler();
                    deferredImports.sort(DEFERRED_IMPORT_COMPARATOR);
                    deferredImports.forEach(handler::register);
                    handler.processGroupImports();
                }
            }
            finally {
                this.deferredImportSelectors = new ArrayList<>();
            }
        }
```

在这个调度方法中最终处理的对象是 DeferredImportSelectorGroupingHandler，首先认识这个对象的成员变量，成员变量定义代码如下。

```
private final Map< Object, DeferredImportSelectorGrouping > groupings =
new LinkedHashMap<>();

private final Map < AnnotationMetadata, ConfigurationClass > configurationClasses = new HashMap<>();
```

下面对 groupings 变量进行分析，该变量是一个存储结构，存储的 key 有下面两种情况。

（1）Class<? extends Group>：该数据从 DeferredImportSelector 接口的 getImportGroup 方法中获取。

（2）DeferredImportSelectorHolder：该数据就是 DeferredImportSelectorHolder。

在这个存储结构中 value 是一个明确的数据类型，具体类型是 DeferredImportSelectorGrouping。下面对 DeferredImportSelectorGrouping 对象进行分析，它的成员变量有两个，第一个变量 group，表示了一个分组，具体是对 DeferredImportSelectorHolder 对象进行分组；第二个变量是 deferredImports，它存储了 DeferredImportSelectorHolder 对象的数据集合。DeferredImportSelectorGrouping 成员变量定义代码如下。

```
private static class DeferredImportSelectorGrouping {

    private final DeferredImportSelector.Group group;

    private final List< DeferredImportSelectorHolder > deferredImports = new ArrayList<>();
}
```

在 DeferredImportSelectorGroupingHandler 类中存放了两个方法，第一个是注册方法（register），它能够将 DeferredImportSelectorHolder 对象进行注册转换成 DeferredImportSelectorGrouping 对象；第二个方法是 processGroupImports，它能够处理 ImportSelector 接口。

12.5 processImports 方法分析

接下来将对 processImports 方法进行分析，在这个方法中具体处理的代码如下面这一段。

```java
private void processImports(ConfigurationClass configClass,
SourceClass currentSourceClass,
        Collection<SourceClass> importCandidates, boolean checkForCircularImports) {

    //判断是否存在需要处理的 ImportSelector 集合，如果不需要则不处理直接返回
    if (importCandidates.isEmpty()) {
        return;
    }

    //判断是否需要进行 import 循环检查
    //判断当前配置类是否在 importStack 中
    if (checkForCircularImports && isChainedImportOnStack(configClass)) {
        this.problemReporter.error(new
CircularImportProblem(configClass, this.importStack));
    }
    else {
        //向 importStack 中加入当前正在处理的配置类
        this.importStack.push(configClass);
        try {
            //循环处理参数传递过来的 importSelector
            for (SourceClass candidate : importCandidates) {

                //判断是否是 ImportSelector 类型
                if (candidate.isAssignable(ImportSelector.class)) {
                    Class<?> candidateClass = candidate.loadClass();
                    //将 class 转换成对象
                    ImportSelector selector
 = ParserStrategyUtils.instantiateClass(candidateClass, ImportSelector.class,
                            this.environment, this.resourceLoader, this.registry);
                    //如果类型是 DeferredImportSelector
                    if (selector instanceof DeferredImportSelector) {
                        this.deferredImportSelectorHandler.handle(configClass, (DeferredImportSelector)
selector);
                    }
                    else {
                        //获取需要导入的类
                        String[] importClassNames
 = selector.selectImports(currentSourceClass.getMetadata());
                        //将需要导入的类从字符串转换成 SourceClass
                        Collection<SourceClass> importSourceClasses
 = asSourceClasses(importClassNames);
                        //递归处理需要导入的类
                        processImports(configClass, currentSourceClass, importSourceClasses, false);
                    }
                }

                //处理类型是 ImportBeanDefinitionRegistrar 的情况
                else if (candidate.isAssignable(ImportBeanDefinitionRegistrar.class)) {
                    Class<?> candidateClass = candidate.loadClass();
                    ImportBeanDefinitionRegistrar registrar =
                            ParserStrategyUtils.instantiateClass(candidateClass,
```

```
                    ImportBeanDefinitionRegistrar.class,
                                this.environment, this.resourceLoader, this.registry);
                        configClass.addImportBeanDefinitionRegistrar(registrar, currentSourceClass.
getMetadata());
                    }
                    //其他情况
                    else {
                        this.importStack.registerImport(
                                currentSourceClass.getMetadata(), candidate.getMetadata().
getClassName());
                        processConfigurationClass(candidate.asConfigClass(configClass));
                    }
                }
            }
            catch (BeanDefinitionStoreException ex) {
                throw ex;
            }
            catch (Throwable ex) {
                throw new BeanDefinitionStoreException(
                        "Failed to process import candidates for configuration class [" +
                        configClass.getMetadata().getClassName() + "]", ex);
            }
            finally {
                //从 importStack 删除配置类
                this.importStack.pop();
            }
        }
    }
```

在这段代码中的处理细节如下。

（1）判断参数 importCandidates 是否存在数据，如果不存在数据就不进行后续处理。

（2）判断参数 checkForCircularImports 是否为 true，同时验证 importStack 中是否存在当前参数 configClass。如果 checkForCircularImports 为 true 同时验证通过就会抛出异常。

（3）处理 importCandidates 集合中的每个 ImportSelector 对象，处理方式如下三点。

① 类型是 ImportSelector，将 SourceClass 转换成 ImportSelector 对象，得到 ImportSelector 后分为两种情况处理。第一种，类型是 DeferredImportSelector，重复 handler 方法进行处理。第二种，类型不是 DeferredImportSelector，调用 ImportSelector 中 selectImports 方法得到需要导入的类，将类名转换成 SourceClass，最后解析每个 SourceClass 处理方法是本方法（processImports），递归调用。

② 类型是 ImportBeanDefinitionRegistrar，将 SourceClass 转换成 ImportBeanDefinitionRegistrar 实例，将这个实例放入 importBeanDefinitionRegistrars 容器。

③ 类型既不是 ImportSelector 也不是 ImportBeanDefinitionRegistrar，将这个对象当作配置类进行处理。

小结

关于 DeferredImportSelectorHandler 对象它主要是为了三个接口的处理而存在的对象,它对 DeferredImportSelector 接口、ImportBeanDefinitionRegistrar 接口和 ImportSelector 接口进行了统一的处理流程约束,具体的处理方法是 processImports,整体的处理流程可以简单理解为获取三个接口中对应的实现类再进行接口实现方法的调用。

第13章

ConfigurationClassBeanDefinitionReader 分析

在第 10 章中讲述了 Spring 配置类解析的过程,本章将对解析过程中的 ConfigurationClassBeanDefinitionReader 对象单独提取出来进行分析,主要分析目标是对配置类中 Bean 的加载进行分析。

13.1 ConfigurationClassBeanDefinitionReader 测试环境搭建

在开始分析 ConfigurationClassBeanDefinitionReader 对象之前需要建立一个测试环境,这个测试环境没有特殊要求只需要写一个 Spring 注解模式的启动代码即可。首先定义一个 JavaBean,类名为 AnnPeople,具体代码如下。

```
public class AnnPeople {
    private String name;

    public String getName() {
        return name;
    }

    public void setName(String name) {
        this.name = name;
    }
}
```

完成 JavaBean 编写后编写 Spring 注解模式的配置类,类名为 AnnBeans,具体代码如下。

```
@Component
@Scope
public class AnnBeans {
    @Bean(name = "abc")
    public AnnPeople annPeople() {
        AnnPeople annPeople = new AnnPeople();
        annPeople.setName("people");
        return annPeople;
    }
}
```

完成 Spring 注解模式中配置类的编写后编写一个测试方法,具体代码如下。

```
@Test
void testComponentClasses(){
    AnnotationConfigApplicationContext context =
            new AnnotationConfigApplicationContext(AnnBeans.class);
    AnnPeople bean = context.getBean(AnnPeople.class);
    assert bean.getName().equals("people");
}
```

这一次需要分析的目标对象是在分析 ConfigurationClassPostProcessor 类的时候遇到的一个对象,当时做了一个简单的说明:ConfigurationClassBeanDefinitionReader 可以读取 Spring Configuration Bean 中的 Bean 信息将其注册到 Spring 容器中,本文将对整个处理过程做详细的分析。分析前需要找到入口,入口在 processConfigBeanDefinitions 方法中,具体代码如下。

```
if (this.reader == null) {
    this.reader = new ConfigurationClassBeanDefinitionReader(
            registry, this.sourceExtractor, this.resourceLoader, this.environment,
            this.importBeanNameGenerator, parser.getImportRegistry());
}
//解析 Spring Configuration Class 的主要目的是提取其中的 Spring 注解并将其转换成
//BeanDefinition
this.reader.loadBeanDefinitions(configClasses);
```

13.2 ConfigurationClassBeanDefinitionReader 构造函数

在执行 loadBeanDefinitions 方法之前需要先创建 ConfigurationClassBeanDefinitionReader 对象,创建方法是构造函数,构造函数代码如下。

```
ConfigurationClassBeanDefinitionReader (BeanDefinitionRegistry registry, SourceExtractor
sourceExtractor,ResourceLoader resourceLoader, Environment environment, BeanNameGenerator
importBeanNameGenerator,
        ImportRegistry importRegistry) {

        this.registry = registry;
```

```
            this.sourceExtractor = sourceExtractor;
            this.resourceLoader = resourceLoader;
            this.environment = environment;
            this.importBeanNameGenerator = importBeanNameGenerator;
            this.importRegistry = importRegistry;
            this.conditionEvaluator =
    new ConditionEvaluator(registry, environment, resourceLoader);
    }
```

这个构造函数涉及多个对象,这些对象的含义见表 13.1。

表 13.1 ConfigurationClassBeanDefinitionReader 构造函数参数

参 数 名 称	参 数 类 型	参 数 说 明
registry	BeanDefinitionRegistry	BeanDefinition 注册器
sourceExtractor	SourceExtractor	元数据解析器
resourceLoader	ResourceLoader	资源读取器
environment	Environment	配置信息接口
importBeanNameGenerator	BeanNameGenerator	BeanName 生成接口
importRegistry	ImportRegistry	import 注册器

在 ConfigurationClassBeanDefinitionReader 的构造函数中主要目的是将各类资源进行初始化等待后续使用,在构造参数列表中重点可以关注 registry,该变量在后续进行 Bean 定义注册时会产生作用。

13.3 loadBeanDefinitions 方法分析

接下来将对 loadBeanDefinitions 方法进行分析,处理代码如下。

```
public void loadBeanDefinitions(Set<ConfigurationClass> configurationModel) {
    TrackedConditionEvaluator trackedConditionEvaluator =
new TrackedConditionEvaluator();
    for (ConfigurationClass configClass : configurationModel) {
        loadBeanDefinitionsForConfigurationClass(configClass, trackedConditionEvaluator);
    }
}
```

在这段代码中重点关注一个类 TrackedConditionEvaluator 和一个方法 loadBeanDefinitionsForConfigurationClass。方法 loadBeanDefinitions 的处理围绕这两个内容做了调用。前者主要负责存储条件解析结果的缓存,后者主要负责从 Spring 配置类中解析 BeanDefinition 对象。

13.4 TrackedConditionEvaluator 分析

接下来对 TrackedConditionEvaluator 进行分析,前文简单描述了这是一个负责存储条件解析结果缓存的工具,具体的存储结构是 Map,key 存储的是配置类,value 存储的是条件解析结果,具体定义信息如下。

```java
private final Map<ConfigurationClass, Boolean> skipped = new HashMap<>();
```

有关单个配置类的数据处理方法如下。

```java
public boolean shouldSkip(ConfigurationClass configClass) {
    Boolean skip = this.skipped.get(configClass);
    if (skip == null) {
        if (configClass.isImported()) {
            boolean allSkipped = true;
            for (ConfigurationClass importedBy : configClass.getImportedBy()) {
                if (!shouldSkip(importedBy)) {
                    allSkipped = false;
                    break;
                }
            }
            if (allSkipped) {
                skip = true;
            }
        }
        if (skip == null) {
            skip = conditionEvaluator.shouldSkip(configClass.getMetadata(), ConfigurationPhase.REGISTER_BEAN);
        }
        this.skipped.put(configClass, skip);
    }
    return skip;
}
```

在这个方法中会使用到第 21 章中关于条件注解相关的技术。该方法会解析配置类上的条件注解（@Conditional）的信息，将配置类和解析结果放在缓存容器中。

13.5　loadBeanDefinitionsForConfigurationClass 方法分析

接下来将对 loadBeanDefinitionsForConfigurationClass 方法进行分析，下面是该方法的代码。

```java
private void loadBeanDefinitionsForConfigurationClass(
        ConfigurationClass configClass,
        TrackedConditionEvaluator trackedConditionEvaluator) {

    if (trackedConditionEvaluator.shouldSkip(configClass)) {
        //BeanName 提取
        String beanName = configClass.getBeanName();
        //BeanName 是否存在
        //注册容器中是否存在当前处理的 BeanName
        if (StringUtils.hasLength(beanName)
                && this.registry.containsBeanDefinition(beanName)) {
            //注册容器中移除 BeanName 对应的数据
            this.registry.removeBeanDefinition(beanName);
```

```
    }
    //import 注册器移除当前的类名
    this.importRegistry.removeImportingClass(configClass.getMetadata().getClassName());
    return;
}

//判断当前处理的配置类是不是 import 导入的对象
if (configClass.isImported()) {
    registerBeanDefinitionForImportedConfigurationClass(configClass);
}
//获取配置类中的 BeanMethod 列表
for (BeanMethod beanMethod : configClass.getBeanMethods()) {
    //处理单个 BeanMethod
    loadBeanDefinitionsForBeanMethod(beanMethod);
}

//处理 ImportedResoruce 数据
loadBeanDefinitionsFromImportedResources(configClass.getImportedResources());
//处理 ImportBeanDefinitionRegistrar
loadBeanDefinitionsFromRegistrars(configClass.getImportBeanDefinitionRegistrars());
}
```

该方法的主要处理流程如下。

（1）对当前需要处理的配置类进行条件注解验证，如果通过验证，则进行下面的处理。

① 提取配置类的 BeanName。

② BeanName 为空并且当前容器中存在 BeanName。如果符合这个条件，此时需要将容器中 BeanName 对应的数据删除。

③ 在 importRegistry 容器中移除当前类名。

（2）判断当前配置类是否是通过 import 导入的对象。如果是，则进入 registerBean-DefinitionForImportedConfigurationClass 进行处理。

（3）获取配置类中的 BeanMethod 列表，让每个 BeanMethod 都经过 loadBeanDefinitionsForBeanMethod 方法处理。

（4）处理 ImportedResoruce 相关内容。

（5）处理 ImportBeanDefinitionRegistrar 相关内容。

这个处理过程中出现的各个方法会在接下来的内容中做出具体分析。

13.6　loadBeanDefinitionsForBeanMethod 方法分析

接下来将对 loadBeanDefinitionsForBeanMethod 方法进行分析。该方法的作用是从 BeanMethod 对象中提取 BeanDefinition 对象。在这个提取过程中，BeanMethod 对象的内容很关键。BeanMethod 继承自 ConfigurationMethod 类，在源代码中 BeanMethod 没有提供成员变量，其成员变量都是由 ConfigurationMethod 提供，具体成员变量见表 13.2。

第13章 ConfigurationClassBeanDefinitionReader 分析

表 13.2 ConfigurationMethod 成员变量

变量名称	变量类型	变量说明
metadata	MethodMetadata	方法元数据
configurationClass	ConfigurationClass	配置类对象

测试用例中使用的配置类是 AnnBeans，它的 BeanMethod 信息如图 13.1 所示。

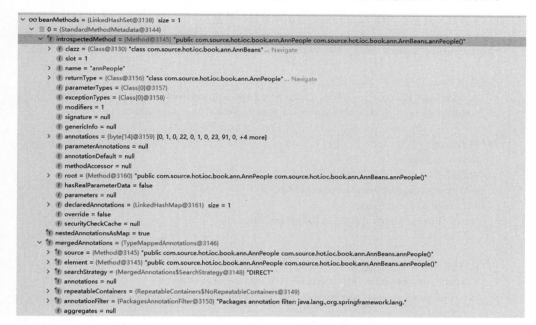

图 13.1　AnnBeans 对象中的 BeanMethod

在图 13.1 中可以看到，AnnBeans 对象中标记@Bean 注解的方法被解析成了 BeanMethod 对象，同时不难发现，在这个 BeanMethod 中还存有了关于注解相关的信息，具体存储是在 mergedAnnotations 变量中。

在得到 BeanMethod 对象后需要将其转换为 BeanDefinition 对象才可供开发者使用，具体处理代码如下。

```
@SuppressWarnings("deprecation")
//for RequiredAnnotationBeanPostProcessor.SKIP_REQUIRED_CHECK_ATTRIBUTE
private void loadBeanDefinitionsForBeanMethod(BeanMethod beanMethod) {

    //提取配置类
    ConfigurationClass configClass = beanMethod.getConfigurationClass();
    //提取方法元数据
    MethodMetadata metadata = beanMethod.getMetadata();
    //提取方法名称
    String methodName = metadata.getMethodName();

    //进行 Condition 注解处理
    if
```

```java
    (this.conditionEvaluator.shouldSkip(metadata, ConfigurationPhase.REGISTER_BEAN)) {
        configClass.skippedBeanMethods.add(methodName);
        return;
    }
    if (configClass.skippedBeanMethods.contains(methodName)) {
        return;
    }

    //从注解元数据中提取 Bean 注解相关的注解属性对象
    AnnotationAttributes bean = AnnotationConfigUtils.attributesFor(metadata, Bean.class);
    Assert.state(bean != null, "No @Bean annotation attributes");

    //提取 Bean 属性中的 name 值
    List<String> names = new ArrayList<>(Arrays.asList(bean.getStringArray("name")));
    String beanName = (!names.isEmpty() ? names.remove(0) : methodName);

    //别名处理
    //真名 = 方法名称
    //别名等于 @Bean 中的 name 属性
    //Register aliases even when overridden
    for (String alias : names) {
        this.registry.registerAlias(beanName, alias);
    }

    //检查是否存在覆盖的定义
    if (isOverriddenByExistingDefinition(beanMethod, beanName)) {
        if (beanName.equals(beanMethod.getConfigurationClass().getBeanName())) {
            throw
new BeanDefinitionStoreException(beanMethod.getConfigurationClass().getResource().
getDescription(), beanName, "Bean name derived from @Bean method '"
+ beanMethod.getMetadata().getMethodName() +
                "' clashes with bean name for containing configuration class; please make those names unique!");
        }
        return;
    }

    //BeanDefinition 的创建和属性设置
    ConfigurationClassBeanDefinition beanDef =
new ConfigurationClassBeanDefinition(configClass, metadata);
    beanDef.setResource(configClass.getResource());
    beanDef.setSource(this.sourceExtractor.extractSource(metadata, configClass.getResource()));

    if (metadata.isStatic()) {
        if (configClass.getMetadata() instanceof StandardAnnotationMetadata) {
            beanDef.setBeanClass(((StandardAnnotationMetadata) configClass.getMetadata()).
getIntrospectedClass());
        }
        else {
            beanDef.setBeanClassName(configClass.getMetadata().getClassName());
        }
```

```java
        beanDef.setUniqueFactoryMethodName(methodName);
    }
    else {
        beanDef.setFactoryBeanName(configClass.getBeanName());
        beanDef.setUniqueFactoryMethodName(methodName);
    }

    if (metadata instanceof StandardMethodMetadata) {
        beanDef.setResolvedFactoryMethod(((StandardMethodMetadata) metadata).
getIntrospectedMethod());
    }

    beanDef.setAutowireMode(AbstractBeanDefinition.AUTOWIRE_CONSTRUCTOR);

beanDef.setAttribute(org.springframework.beans.factory.annotation.
RequiredAnnotationBeanPostProcessor.SKIP_REQUIRED_CHECK_ATTRIBUTE, Boolean.TRUE);

    //Lazy Primary DependsOn Role Description 注解数据设置
    AnnotationConfigUtils.processCommonDefinitionAnnotations(beanDef, metadata);

    Autowire autowire = bean.getEnum("autowire");
    if (autowire.isAutowire()) {
        beanDef.setAutowireMode(autowire.value());
    }

    boolean autowireCandidate = bean.getBoolean("autowireCandidate");
    if (!autowireCandidate) {
        beanDef.setAutowireCandidate(false);
    }

    String initMethodName = bean.getString("initMethod");
    if (StringUtils.hasText(initMethodName)) {
        beanDef.setInitMethodName(initMethodName);
    }

    String destroyMethodName = bean.getString("destroyMethod");
    beanDef.setDestroyMethodName(destroyMethodName);

    //Scope 处理
    ScopedProxyMode proxyMode = ScopedProxyMode.NO;
    AnnotationAttributes attributes =
AnnotationConfigUtils.attributesFor(metadata, Scope.class);
    if (attributes != null) {
        beanDef.setScope(attributes.getString("value"));
        proxyMode = attributes.getEnum("proxyMode");
        if (proxyMode == ScopedProxyMode.DEFAULT) {
            proxyMode = ScopedProxyMode.NO;
        }
    }

    BeanDefinition beanDefToRegister = beanDef;
```

```
            if (proxyMode != ScopedProxyMode.NO) {
                BeanDefinitionHolder proxyDef = ScopedProxyCreator.createScopedProxy(
                        new BeanDefinitionHolder(beanDef, beanName), this.registry,
                        proxyMode == ScopedProxyMode.TARGET_CLASS);
                beanDefToRegister = new ConfigurationClassBeanDefinition(
                        (RootBeanDefinition) proxyDef.getBeanDefinition(), configClass, metadata);
            }

            if (logger.isTraceEnabled()) {
                logger.trace(String.format("Registering bean definition for @Bean method %s.%s()",
                        configClass.getMetadata().getClassName(), beanName));
            }
            //注册 BeanDefinition
            this.registry.registerBeanDefinition(beanName, beanDefToRegister);
        }
```

上述代码的整体处理过程简单地说就是将 BeanMethod 中的各个数据提取出来并赋值给 BeanDefinition，在这个操作过程中用到的 BeanDefinition 对象是 ConfigurationClassBeanDefinition 类型，整体的数据处理过程有下面 13 个。

（1）从 BeanMethod 中提取配置。
（2）从 BeanMethod 中提取方法元数据。
（3）从方法元数据中提取方法名称。
（4）在方法元数据中进行条件注解处理。
（5）在配置类中进行条件注解处理。
（6）从方法元数据中提取 Bean 注解的数据。
（7）提取注解属性中 name 属性列表。
（8）确定 BeanName，如果注解属性中的 name 列表不为空，则取第一个作为 BeanName，否则将使用方法名称作为 BeanName。
（9）别名注册，真名是上一步得到的 BeanName，别名是 name 属性列表中的其他名称。
（10）检查是否存在覆盖的 BeanDefinition。如果存在，就会抛出异常 BeanDefinition-StoreException。
（11）创建 BeanDefinition，实际对象是 ConfigurationClassBeanDefinition。
（12）解析和设置 BeanDefinition 中的属性。
（13）进行 BeanDefinition 注册。

上述 13 个操作过程就是 loadBeanDefinitionsForBeanMethod 方法中各类处理的总结。

13.7 registerBeanDefinitionForImportedConfigurationClass 方法分析

接下来将对 registerBeanDefinitionForImportedConfigurationClass 方法进行分析。当一个类是被导入的（具体表现形式为通过 @Import 注解注入一个类）会执行该方法，具体代码如下。

```
        if (configClass.isImported()) {
```

```
    //导入
    registerBeanDefinitionForImportedConfigurationClass(configClass);
}
```

根据表结构可以编写下面这些代码,首先需要编写一个需要导入的类,类名为 ConfigurationA,代码信息如下。

```
public class ConfigurationA {
    @Bean(name = "ConfigurationA.annPeople")
    public AnnPeople annPeople() {
        return new AnnPeople();
    }
}
```

完成需要注入的类编写后进一步编写 Spring 配置类,类名为 ImportedBeans,具体代码如下。

```
@Configuration
@Import(ConfigurationA.class)
public class ImportedBeans {
}
```

完成 Spring 配置类编写后编写一个测试方法,具体代码如下。

```
@Test
void testAnnImportConfiguration() {
    AnnotationConfigApplicationContext context
            = new AnnotationConfigApplicationContext(ImportedBeans.class);
    AnnPeople bean = context.getBean("ConfigurationA.annPeople", AnnPeople.class);
    assert bean != null;
}
```

基本测试环境准备完成后,将断点放在 registerBeanDefinitionForImportedConfigurationClass 方法上进行调试。首先关注 configClass 对象,它是 com.source.hot.ioc.book.ann.imported. ConfigurationA 类,完整信息如图 13.2 所示。

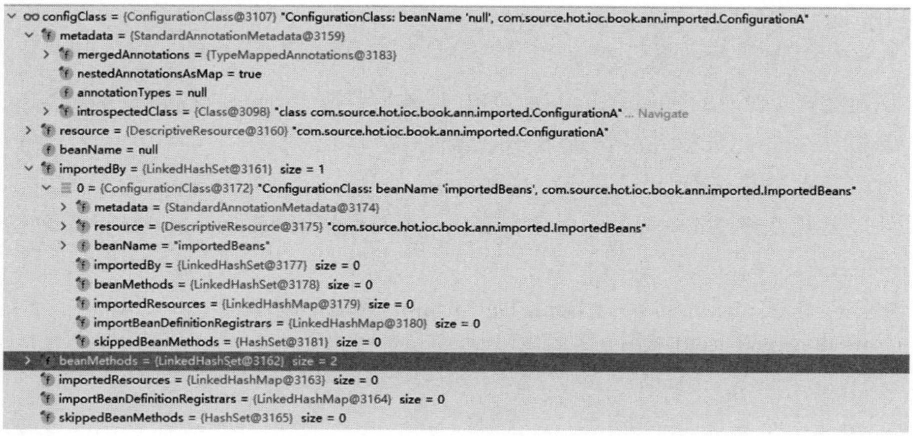

图 13.2 configClass 完整信息

拥有 configClass 数据对象后进入 registerBeanDefinitionForImportedConfigurationClass 方法，具体处理代码如下。

```java
private void registerBeanDefinitionForImportedConfigurationClass(ConfigurationClass configClass) {
    //提取注解元数据
    AnnotationMetadata metadata = configClass.getMetadata();
    //将注解元数据转换成 BeanDefinition
    AnnotatedGenericBeanDefinition configBeanDef =
            new AnnotatedGenericBeanDefinition(metadata);

    //解析 Scope 元数据
    ScopeMetadata scopeMetadata =
            scopeMetadataResolver.resolveScopeMetadata(configBeanDef);
    configBeanDef.setScope(scopeMetadata.getScopeName());
    //处理 BeanName
    String configBeanName =
            this.importBeanNameGenerator.generateBeanName(configBeanDef, this.registry);
    //处理常规注解
    AnnotationConfigUtils.processCommonDefinitionAnnotations(configBeanDef, metadata);

    //BeanDefinition 转换成 Bean Definition Holder
    BeanDefinitionHolder definitionHolder =
            new BeanDefinitionHolder(configBeanDef, configBeanName);
    //应用 Scope 代理
    definitionHolder =
            AnnotationConfigUtils.applyScopedProxyMode(scopeMetadata,
                    definitionHolder, this.registry);
    //注册 BeanDefinition
    this.registry.registerBeanDefinition(definitionHolder.getBeanName(), definitionHolder.getBeanDefinition());
    configClass.setBeanName(configBeanName);

    if (logger.isTraceEnabled()) {
        logger.trace("Registered bean definition for imported class '" + configBeanName + "'");
    }
}
```

这段代码的处理过程有以下 8 步。

（1）从配置类中提取注解元数据。

（2）创建配置类对应的 BeanDefinition 对象，具体类型是 AnnotatedGenericBeanDefinition。

（3）解析 Scope 元数据并且设置 Scope 属性。

（4）对当前配置类进行 BeanName 的初始化。

（5）处理常规注解 Lazy、Primary、DependsOn、Role 和 Description 并将其数据赋值给 BeanDefinition。

（6）将 BeanDefinition 转换为 Bean Definition Holder 对象。

（7）处理 Scope 代理相关内容。

（8）BeanDefinition 注册。

完成上述 8 个步骤后有关 @Import 的注入就结束了，数据都被放在 Spring 容器中等待后续使用。

13.8 loadBeanDefinitionsFromImportedResources 方法分析

接下来将分析 loadBeanDefinitionsFromImportedResources 方法。该方法是围绕注解 @ImportedResources 进行处理。首先编写一个测试用例,第一步编写一个 SpringXML 配置文件,文件名为 first-ioc.xml,具体内容如下。

```xml
<?xml version = "1.0" encoding = "UTF-8"?>
<beans xmlns:xsi = "http://www.w3.org/2001/XMLSchema-instance"
       xmlns = "http://www.springframework.org/schema/beans"
       xsi:schemaLocation = "http://www.springframework.org/schema/beans http://www.springframework.org/schema/beans/spring-beans.xsd">


    <bean id = "people" class = "com.source.hot.ioc.book.pojo.PeopleBean">
        <property name = "name" value = "zhangsan"/>
    </bean>
</beans>
```

完成 SpringXML 配置文件的编写后,将这个配置类和 @ImportedResources 注解配合使用编写 ImportResourceBeansTest 类,具体代码如下。

```java
@ImportResource(locations = {
        "classpath:/META-INF/first-ioc.xml"
})
@Configuration
public class ImportResourceBeansTest {
    @Test
    void testImportResource(){
        AnnotationConfigApplicationContext context =
new AnnotationConfigApplicationContext(ImportResourceBeansTest.class);
        PeopleBean bean = context.getBean(PeopleBean.class);
        assert bean != null;
    }
}
```

注解 @ImportedResources 的作用其实是将一个 SpringXML 文件导入注解模式的 Spring 中。

测试用例准备完成后,进入调试阶段查看 importedResources 的数据信息,具体数据如图 13.3 所示。

得到 ImportedResources 的数据信息后就可以进入 loadBeanDefinitionsFromImportedResources 方法进行处理,具体处理代码如下。

```java
private void loadBeanDefinitionsFromImportedResources(
        Map<String, Class<? extends BeanDefinitionReader>> importedResources) {

    Map<Class<?>, BeanDefinitionReader> readerInstanceCache = new HashMap<>();

    //循环处理 importedResources
```

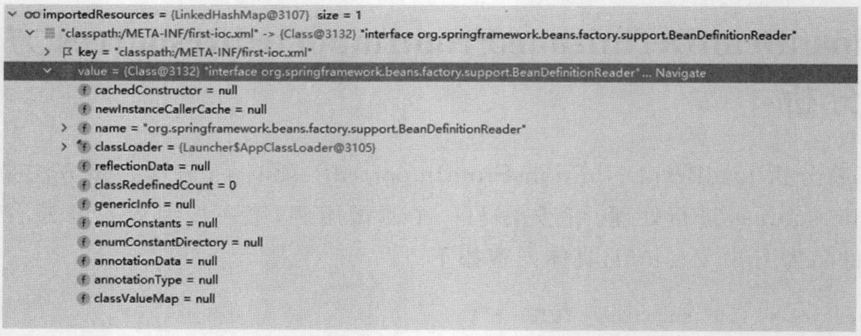

图 13.3 importedResources 数据信息

```
importedResources.forEach((resource, readerClass) -> {
    //判断 Bean Defintion Reader 是不是以 BeanDefinitionReader
    if (BeanDefinitionReader.class == readerClass) {
        //判断文件地址是不是以.groovy 结尾
        if (StringUtils.endsWithIgnoreCase(resource, ".groovy")) {
            readerClass = GroovyBeanDefinitionReader.class;
        }
        //其他情况都当作 XML 处理
        else {
            readerClass = XmlBeanDefinitionReader.class;
        }
    }

    //尝试从缓存中获取 BeanDefinitionReader
    BeanDefinitionReader reader = readerInstanceCache.get(readerClass);
    //如果不存在则创建一个新的 BeanDefinitionReader
    if (reader == null) {
        try {
            //Instantiate the specified BeanDefinitionReader
            reader =
    readerClass.getConstructor(BeanDefinitionRegistry.class).newInstance(this.registry);
            if (reader instanceof AbstractBeanDefinitionReader) {
                AbstractBeanDefinitionReader abdr = ((AbstractBeanDefinitionReader) reader);
                abdr.setResourceLoader(this.resourceLoader);
                abdr.setEnvironment(this.environment);
            }
            readerInstanceCache.put(readerClass, reader);
        }
        catch (Throwable ex) {
            throw new IllegalStateException(
                    "Could not instantiate BeanDefinitionReader class [" +
    readerClass.getName() + "]");
        }
    }

    //Bean Defintion Reader 进行解析
    reader.loadBeanDefinitions(resource);
```

    });

}
```

这段代码的处理过程如下。

（1）创建一个用来存储 Class 和 BeanDefinitionReader 关系的容器，key 表示 BeanDefinition Reader 类型，value 表示 Bean Definition Reader 实例。

（2）处理 importedResources 对象。

① 判断读取器是不是 BeanDefinitionReader 类型。如果是 BeanDefinitionReader 类型进一步处理，如果资源文件是以 .groovy 结尾将读取器类型定义成 GroovyBeanDefinitionReader。

② 其他情况都将读取器类型定义为 XmlBeanDefinitionReader。

③ 从容器中获取当前读取器类型对应的读取器实例，如果不存在就创建一个实例（创建方式是反射创建），并将这个实例放入绑定关系中。

④ 解析 resource，这里的解析就是对 Spring XML 模式下配置文件的解析。

完成上述操作后有关注解 @ImportResource 的内容就处理完成。接下来将对注解 ImportResource 数据的初始化进行分析，有关注解数据提取的代码如下。

```
AnnotationAttributes importResource =
        AnnotationConfigUtils.attributesFor(sourceClass.getMetadata(), ImportResource.class);
if (importResource != null) {
   String[] resources = importResource.getStringArray("locations");
   Class<? extends BeanDefinitionReader> readerClass =
importResource.getClass("reader");
   for (String resource : resources) {
      String resolvedResource = this.environment.resolveRequiredPlaceholders(resource);
      configClass.addImportedResource(resolvedResource, readerClass);
   }
}
```

这段代码的主要作用是提取注解元数据中关于注解 ImportResource 的数据信息，在提取完成数据信息后会将数据放在配置类的 importedResource 属性中。

13.9 loadBeanDefinitionsFromRegistrars 方法分析

接下来将对 loadBeanDefinitionsFromRegistrars 方法进行分析，该方法对应的处理不是注解的处理而是对应接口的处理，对应接口是 ImportBeanDefinitionRegistrar，但是它需要通过注解 @Import 进行导入，否则不会被注解模式下的 Spring 识别。

首先需要编写一个测试用例，先实现 ImportBeanDefinitionRegistrar 接口，实现类为 MyBeanRegistrar，具体代码如下。

```
public class MyBeanRegistrar implements ImportBeanDefinitionRegistrar {

   @Override
   public void registerBeanDefinitions(AnnotationMetadata importingClassMetadata,
         BeanDefinitionRegistry registry) {
```
```

```
        GenericBeanDefinition gbd = new GenericBeanDefinition();
        gbd.setBeanClass(AnnPeople.class);
        gbd.getPropertyValues().addPropertyValue("name", "name");
        registry.registerBeanDefinition("abn", gbd);
    }
}
```

编写完成实现类后编写 Spring 配置类,类名为 ImportBeanDefinitionRegistrarBeans,具体代码如下。

```
@Configuration
@Import(MyBeanRegistrar.class)
public class ImportBeanDefinitionRegistrarBeans {
}
```

完成 Spring 配置类编写后进行测试方法编写,具体代码如下。

```
@Test
void testImportBeanDefinitionRegistrar() {
    AnnotationConfigApplicationContext context =
    new AnnotationConfigApplicationContext(ImportBeanDefinitionRegistrarBeans.class);

    AnnPeople bean = context.getBean("abn", AnnPeople.class);
    assert bean != null;
}
```

测试用例编写完成后进入调试阶段查看容器中 importBeanDefinitionRegistrars 对象信息,具体数据如图 13.4 所示。

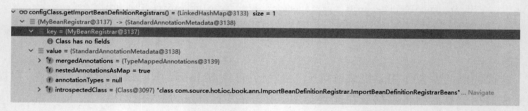

图 13.4　importBeanDefinitionRegistrars 数据信息

得到数据信息后进入 loadBeanDefinitionsFromRegistrars 方法,具体代码如下。

```
private void
loadBeanDefinitionsFromRegistrars(Map < ImportBeanDefinitionRegistrar, AnnotationMetadata >
registrars) {
    registrars.forEach((registrar, metadata) ->
        registrar.registerBeanDefinitions(metadata, this.registry, this.importBeanNameGenerator));
}
```

在这段代码中会执行每个 ImportBeanDefinitionRegistrar 接口实现类的 registerBeanDefinitions 方法。

小结

本章围绕 ConfigurationClassBeanDefinitionReader 类进行分析，分析了以下四种处理模式。

（1）@Import 中 value 填写的是一个 Spring Configuration Bean，对其进行解析，处理方法是 registerBeanDefinitionForImportedConfigurationClass。

（2）当前配置类中 BeanMethod 的解析，处理方法是 loadBeanDefinitionsForBeanMethod。

（3）处理 @ImportedResoruce 注解相关内容，该处理操作是读取 XML 或者 groovy 类型的 Spring 配置文件，将其中定义的 Bean 注册到 Spring 中，处理方法是 loadBeanDefinitionsFromImportedResources。

（4）@Import 中 value 填写的是一个 ImportBeanDefinitionRegistrar 实现类，处理方法是 loadBeanDefinitionsFromRegistrars。

第14章

Spring元数据

第9~13章一直在探讨 Spring 注解模式相关内容，但是一直未对注解本身做出相关说明，本章将对 Spring 注解模式下常见的注解元数据进行介绍。

14.1 认识 MetadataReaderFactory

接口 MetadataReaderFactory 作为一个工厂类，它的主要目的是创建对象，提供了关于 MetadataReader 对象的创建方式，接口定义如下。

```
public interface MetadataReaderFactory {

    MetadataReader getMetadataReader(String className) throws IOException;

    MetadataReader getMetadataReader(Resource resource) throws IOException;

}
```

MetadataReaderFactory 中提供了以下两种创建方式。

（1）通过 className 创建 MetaDataReader。
（2）通过 Resource 创建 MetaDataReader。

Spring 中关于 MetadataReaderFactory 的实现有两个：SimpleMetadataReaderFactory 和 CachingMetadataReaderFactory。下面将对这两项内容进行分析。

14.2 SimpleMetadataReaderFactory 分析

在 SimpleMetadataReaderFactory 中存在一个成员变量，该成员变量是 ResourceLoader，默认为 DefaultResourceLoader，该对象的主要作用是进行资源加载，它的主要目的是将

className 转换成 Resource 对象。接下来对 getMetadataReader 方法进行分析，这个方法是能够读取类元数据，具体处理代码如下。

```
@Override
public MetadataReader getMetadataReader(String className) throws IOException {
    try {
        String resourcePath = ResourceLoader.CLASSPATH_URL_PREFIX +
                ClassUtils.convertClassNameToResourcePath(className) +
                ClassUtils.CLASS_FILE_SUFFIX;
        Resource resource = this.resourceLoader.getResource(resourcePath);
        return getMetadataReader(resource);
    }
    catch (FileNotFoundException ex) {
        int lastDotIndex = className.lastIndexOf('.');
        if (lastDotIndex != -1) {
            String innerClassName =
                    className.substring(0, lastDotIndex) + '$'
                    + className.substring(lastDotIndex + 1);
            String innerClassResourcePath = ResourceLoader.CLASSPATH_URL_PREFIX +
                    ClassUtils.convertClassNameToResourcePath(innerClassName) +
                    ClassUtils.CLASS_FILE_SUFFIX;
            Resource innerClassResource =
                    this.resourceLoader.getResource(innerClassResourcePath);
            if (innerClassResource.exists()) {
                return getMetadataReader(innerClassResource);
            }
        }
        throw ex;
    }
}
```

这段代码经过整理可以分为下面三个处理步骤。

(1) 将 className 进行转换，转换成内部类的类名（从代码上表现为文件找不到的异常处理）或者独立类的类名。

(2) 通过资源加载器（ResourceLoader）将第一项中得到的数据转换成 Resource。

(3) 将 Resource 转换成 MetadataReader。

步骤(1)的处理过程详情是类名的转换，具体处理代码是：

```
String resourcePath = ResourceLoader.CLASSPATH_URL_PREFIX +
        ClassUtils.convertClassNameToResourcePath(className) +
        ClassUtils.CLASS_FILE_SUFFIX;
```

这个转换过程的细节遵守这个公式：classpath：+ className.replace('.','/') +.class(className 是参数)。Java 中关于类还有内部类这种情况，对于内部类的处理代码如下。

```
int lastDotIndex = className.lastIndexOf('.');
String innerClassName =
```

```
            className.substring(0, lastDotIndex) + '$' + className.substring(lastDotIndex + 1);
String innerClassResourcePath = ResourceLoader.CLASSPATH_URL_PREFIX +
        ClassUtils.convertClassNameToResourcePath(innerClassName) +
 ClassUtils.CLASS_FILE_SUFFIX;
```

假设现在有一个内部类 com. source. hot. ioc. book. ioc. MetadataReaderFactoryTest ＄A,在这个类名上会做一次字符串拆分再重组。简而言之,将最后一个.替换为＄,解析后 innerClassName 的结果是 com. source. hot. ioc. book. ioc ＄MetadataReaderFactoryTest ＄A。 innerClassName 解析完成后还会继续经过以下两个操作。

（1）classpath: + innerClassName.replace('.','/') +.class。

（2）innerClassResourcePath: classpath:com/source/hot/ioc/book/ioc ＄MetadataReaderFactoryTest ＄A.class。

通过前文的操作得到了创建 SimpleMetadataReader 对象的基本信息即可进行对象创建,创建代码如下。

```
@Override
public MetadataReader getMetadataReader(Resource resource) throws IOException {
    return new SimpleMetadataReader(resource, this.resourceLoader.getClassLoader());
}
```

14.3 CachingMetadataReaderFactory 分析

接下来将对 CachingMetadataReaderFactory 类进行分析。CachingMetadataReaderFactory 是 SimpleMetadataReaderFactory 对象的子类,CachingMetadataReaderFactory 和父类比较增加了缓存相关的处理,具体信息可以通过成员变量查看,成员变量如下。

```
public static final int DEFAULT_CACHE_LIMIT = 256;

@Nullable
private Map<Resource, MetadataReader> metadataReaderCache;
```

变量 DEFAULT_CACHE_LIMIT 表示缓存容器的初始化空间,具体空间是 256 个。

变量 metadataReaderCache 表示缓存容器,key 表示资源,value 表示读取器。

关于读取器的获取和初始化,可以查看下面的代码。

```
@Override
public MetadataReader getMetadataReader(Resource resource) throws IOException {
    if (this.metadataReaderCache instanceof ConcurrentMap) {
        MetadataReader metadataReader = this.metadataReaderCache.get(resource);
        if (metadataReader == null) {
            metadataReader = super.getMetadataReader(resource);
            this.metadataReaderCache.put(resource, metadataReader);
        }
        return metadataReader;
    }
    else if (this.metadataReaderCache != null) {
```

```
            synchronized (this.metadataReaderCache) {
                MetadataReader metadataReader = this.metadataReaderCache.get(resource);
                if (metadataReader == null) {
                    metadataReader = super.getMetadataReader(resource);
                    this.metadataReaderCache.put(resource, metadataReader);
                }
                return metadataReader;
            }
        }
        else {
            return super.getMetadataReader(resource);
        }
    }
```

从这里可以看到它会从 metadataReaderCache 中获取数据,但是 metadataReaderCache 的类型存在多样性(在 CachingMetadataReaderFactory 中有 LocalResourceCache,该类会作为缓存的存储类的类型)。获取方式遵循下面的操作步骤。

(1)从缓存中获取。
(2)从父类中获取。
(3)父类中获取后设置到缓存。

14.4　注解元数据读取工厂总结

注解元数据读取工厂的分析围绕 MetadataReaderFactory 接口出发了解了它的两个实现类:SimpleMetadataReaderFactory 和 CachingMetadataReaderFactory。在分析过程中比较了它们两者的差异,后者具有缓存能力,前者提供基本的工厂功能。

14.5　初识 MetadataReader

MetadataReader 接口是用来读取元数据的接口,它可以读取这两类元数据:ClassMetadata 类元数据,AnnotationMetadata 注解元数据。MetadataReader 在 Spring 中只有一个实现类 SimpleMetadataReader,它是主要分析目标。在 SimpleMetadataReader 对象中有三个成员变量,这三个成员变量是主要的返回对象,成员变量详细代码如下。

```
private static final int PARSING_OPTIONS = ClassReader.SKIP_DEBUG
        | ClassReader.SKIP_CODE | ClassReader.SKIP_FRAMES;

private final Resource resource;

private final AnnotationMetadata annotationMetadata;
```

三个变量的含义如下:PARSING_OPTIONS 表示解析标记,resource 表示资源对象,annotationMetadata 表示注解元数据。

14.6 MetadataReader 接口实现类说明

14.6.1 SimpleAnnotationMetadataReadingVisitor 成员变量

SimpleMetadataReader 的构造函数中引用了 SimpleAnnotationMetadataReadingVisitor 对象，这个对象的重点信息都在成员变量中，具体信息查看表 14.1。

表 14.1 SimpleAnnotationMetadataReadingVisitor 成员变量

变量名称	变量类型	变量说明
classLoader	ClassLoader	类加载器
className	String	类名
access	int	
superClassName	String	父类名称
interfaceNames	String[]	实现的接口名称列表
enclosingClassName	String	外部类名称
independentInnerClass	boolean	是否是内部独立类
memberClassNames	Set < String >	成员类列表，内部类名称列表
annotations	List < MergedAnnotation <?>>	合并后的注解数据列表
annotatedMethods	List < SimpleMethodMetadata >	方法元数据列表
metadata	SimpleAnnotationMetadata	注解元数据
source	Source	源对象，存储 className

SimpleAnnotationMetadataReadingVisitor 的成员变量表主要对 memberClassNames 进行了说明。在 Java 开发中一个类就是一个简单的类，并不会在类中包含一个类（也有包含的情况），如果一个类没有包含一个类，那么 memberClassNames 数据将不存在。成员变量 annotations 表示的信息是类上的注解数据，并非所有的注解数据，多个注解合并的本质是分组存放数据，分组依据是注解类型。

14.6.2 SimpleAnnotationMetadata 成员变量

本节将介绍 SimpleAnnotationMetadata 成员变量，具体信息查看表 14.2。

表 14.2 SimpleAnnotationMetadata 成员变量

变量名称	变量类型	变量说明
className	String	类名
access	int	
enclosingClassName	String	外部类名称
superClassName	String	父类名称
independentInnerClass	boolean	是否是内部独立类
interfaceNames	String[]	实现的接口名称列表
memberClassNames	String[]	成员类列表，内部类名称列表

续表

变量名称	变量类型	变量说明
annotatedMethods	MethodMetadata[]	方法元数据
annotations	MergedAnnotations	合并的注解接口
annotationTypes	Set < String >	注解类型

在 SimpleAnnotationMetadata 成员变量中对 interfaceNames 变量进行了说明。在 Java 中一个类可以继承（关键字 extends）类或实现（implements）类，interfaceNames 表示的就是通过关键字 implements 实现的接口名称列表，需要使用这个数据的原因是在 Spring 中定义了大量接口，可以方便地判断是否实现某个接口，进一步可以直接进行反射生成对象。

14.6.3 SimpleMethodMetadata 成员变量

本节将介绍 SimpleMethodMetadata 成员变量，具体信息查看表 14.3。

表 14.3 SimpleMethodMetadata 成员变量

变量名称	变量类型	变量说明
methodName	String	方法名称
access	int	
declaringClassName	String	类名
returnTypeName	String	返回值类型，返回值类全名称
annotations	MergedAnnotations	合并的注解

在 SimpleMethodMetadata 成员变量中需要关注 declaringClassName 属性和 returnTypeName 属性。前者表示了当前方法所在的类，通过该属性可以确定一个方法的唯一路径，后者表示返回值类型名称（全类名），通过该属性可以对比方法执行结果（反射执行）是否和原有方法类型匹配。

14.6.4 MergedAnnotationsCollection 成员变量

本节将介绍 MergedAnnotationsCollection 成员变量，具体信息查看表 14.4。

表 14.4 MergedAnnotationsCollection 成员变量

变量名称	变量类型	变量说明
annotations	MergedAnnotation<?>[]	合并的注解列表
mappings	AnnotationTypeMappings[]	注解和类型的映射关系列表

在 MergedAnnotationsCollection 成员变量中的数据内容都是以数组的形式出现，在其中所蕴含的数据内容都是组合形式，该对象主要存储集合元素。

14.6.5 TypeMappedAnnotation 成员变量

本节将介绍 TypeMappedAnnotation 成员变量，具体信息查看表 14.5。

表 14.5 TypeMappedAnnotation 成员变量

变量名称	变量类型	变量说明
mapping	AnnotationTypeMapping	注解和类型的映射关系
classLoader	ClassLoader	类加载器
source	Object	元数据
rootAttributes	Object	根属性，注解的属性表
valueExtractor	BiFunction < Method, Object, Object >	解析值函数
aggregateIndex	int	参数索引
useMergedValues	boolean	是否需要合并属性
attributeFilter	Predicate < String >	属性过滤器
resolvedRootMirrors	int[]	
resolvedMirrors	int[]	
string	String	注解的全量数据

在 TypeMappedAnnotation 成员变量中对 attributeFilter 进行说明，该变量的数据类型是 Predicate，核心目的是过滤无效属性值。此外还有 useMergedValues 和 string 变量，前者表示的是是否可以进行合并，如果不能进行合并数据会被单独存储，如果合并会将数据进行整合，通常使用数组进行保存。后者表示的是注解的全量数据，注解的全量数据表示在 Java 源代码中所编写的注解将其完全转换为字符串，例如，在源代码中使用 Value 注解：@Value(value="ad")，此时 string 所对应的数据就是 @Value(value="ad")，如果存在一些默认值的设置，string 会随之增加。假设 Value 注解中包含 bol，默认值为 true，此时 string 的数据应该是 @Value(value="ad",bol=true)。

14.6.6 AnnotationTypeMappings 成员变量

本节将介绍 AnnotationTypeMappings 成员变量，具体信息查看表 14.6。

表 14.6 AnnotationTypeMappings 成员变量

变量名称	变量类型	变量说明
repeatableContainers	RepeatableContainers	可重复容器
filter	AnnotationFilter	注解过滤器
mappings	List < AnnotationTypeMapping >	注解和类型的映射集合

在 AnnotationTypeMappings 成员变量表中主要关注 RepeatableContainers。这个对象和 Java 中的 Repeatable 注解有一定关联，一般情况下，默认支持的就是 Repeatable 注解，该注解可重复注释类型的包含注释类型。RepeatableContainers 对象中提供了查找可重复注解的方法和建立可重复注解和可包含注解之间关系的方法。成员变量 filter 的数据类型

是 AnnotationFilter,它是一个函数式接口,主要用于注解过滤。

14.6.7 ClassMetadataReadingVisitor 成员变量

本节将介绍 ClassMetadataReadingVisitor 成员变量,具体信息查看表 14.7。

表 14.7 ClassMetadataReadingVisitor 成员变量

变量名称	变量类型	变量说明
className	String	类名
isInterface	boolean	是不是接口
isAnnotation	boolean	是否是注解
isAbstract	boolean	是否是 abstract 修饰
isFinal	boolean	是否是 final 修饰
enclosingClassName	String	外部类名称
independentInnerClass	boolean	是否是内部独立类
superClassName	String	父类名称
interfaces	String[]	实现的接口名称列表
memberClassNames	Set< String >	内部类列表

在 ClassMetadataReadingVisitor 中所提供的成员变量都可以通过 Class 中提供的方法进行获取,常见的内容有 className、isInterface、isAnnotation、isAbstract 和 isFinal。这些内容在 Java 反射中十分重要,这些数据不仅是在 Spring 中有较高的价值,在其他开源项目中也有较高的价值。

14.6.8 AnnotationMetadataReadingVisitor 成员变量

本节将介绍 AnnotationMetadataReadingVisitor 成员变量,具体信息查看表 14.8。

表 14.8 AnnotationMetadataReadingVisitor 成员变量

变量名称	变量类型	变量说明
classLoader	ClassLoader	类加载器
annotationSet	Set< String >	注解名称列表
metaAnnotationMap	Map< String, Set< String >>	key:注解名称 value:注解属性集合
attributesMap	LinkedMultiValueMap< String, AnnotationAttributes >	key:注解名称 value:注解属性对象
methodMetadataSet	Set< MethodMetadata >	方法元数据

在 AnnotationMetadataReadingVisitor 的成员变量中可以关注 metaAnnotationMap 变量,该变量存储的信息有注解名称和元注解属性表,它保存了顶层注解 Annotation ClassName 和它的所有元注解(即根据递归查找找到的 Annotation ClassName)。另一个变量是 attributesMap,它所存储的内容是顶层注解与所有元注解与属性值的对应关系,注意它不会进行递归搜索,只保留顶层。

14.7 类元数据接口说明

ClassMetadata 作为一个接口,它所定义的方法十分重要,具体说明查看表 14.9。

表 14.9 ClassMetadata 方法

方法名称	返回值类型	方法说明
getClassName	String	获取类名称
isInterface	boolean	判断是不是接口
isAnnotation	boolean	判断是不是注解
isAbstract	boolean	判断是否 abstract 修饰
isConcrete	boolean	是否允许实例化
isFinal	boolean	是否是 final 修饰
isIndependent	boolean	是否是一个独立类
hasEnclosingClass	boolean	是否存在封闭类类名
getEnclosingClassName	String	获取封闭类类名
hasSuperClass	boolean	是否存在父类
getSuperClassName	String	获取父类名称
getInterfaceNames	String[]	获取实现的接口名称
getMemberClassNames	String[]	获取内部类名称

接下来对 getEnclosingClassName 方法进行说明,编写一个测试类,类名为 ClassMetaTest,具体代码如下。

```java
public class ClassMetaTest {
    @Test
    void testEnclosingClassName() {
        Class<A> aClass = A.class;
        Class<?> enclosingClass = aClass.getEnclosingClass();
        System.out.println(enclosingClass);

        Class<B> bClass = B.class;
        Class<?> enclosingClass1 = bClass.getEnclosingClass();
        System.out.println(enclosingClass1);

    }

    public static class B {

    }

    public class A {

    }
}
```

该代码输出结果如下。

```
class com.source.hot.ioc.book.ioc.ClassMetaTest
class com.source.hot.ioc.book.ioc.ClassMetaTest
```

接下来设计一个类放在独立的包路径下，具体代码如下。

```
package com.source.hot.ioc.book.ioc;

public class CTest {
}
```

对应的测试方法如下。

```
@Test
void testCtest(){
    System.out.println(CTest.class.getEnclosingClass());
}
```

输出结果为 null。

方法 getEnclosingClass 的作用是：如果当前类是一个内部类，那么这个方法的执行结果将是当前类包裹住的类。

14.8 StandardClassMetadata 对象分析

接下来将对 StandardClassMetadata 方法进行分析，该方法的实现都是基于 JDK Class 对象进行的，各类方法都是 Class 方法的调用。首先查看 StandardClassMetadata 的构造方法，具体代码如下。

```
@Deprecated
public StandardClassMetadata(Class<?> introspectedClass) {
    Assert.notNull(introspectedClass, "Class must not be null");
    this.introspectedClass = introspectedClass;
}
```

这个构造方法传递了一个 Class 对象，这个参数在整个方法实现中承担了所有的功能提供者，如 getClassName 的实现，具体处理代码如下。

```
@Override
public String getClassName() {
    return this.introspectedClass.getName();
}
```

在 StandardClassMetadata 的实现过程中大量地采用了 Class 所提供的方法。

14.9 注解元数据基础认识

对于注解元数据的基础认知主要以类图为主，AnnotationMetadata 类图如图 14.1 所示。

从这个类图中可以发现它依赖类元数据对象，此外

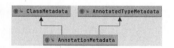

图 14.1 AnnotationMetadata 类图

还依赖注解类型元数据,注解元数据是它们两个接口的合集。

接下来将对 AnnotatedTypeMetadata 接口进行说明,具体方法查看表 14.10。

表 14.10 AnnotatedTypeMetadata 方法

方 法 名 称	方法返回值	方 法 参 数	方 法 作 用
getAnnotations	MergedAnnotations		提取注解集合
isAnnotated	boolean	annotationName:注解名称	判断是否存在该注解
getAnnotationAttributes	Map<String,Object>	annotationName:注解名称	获取指定注解的属性表

接下来对 AnnotationMetadata 接口进行说明,具体方法查看表 14.11。

表 14.11 AnnotationMetadata 方法

方 法 名 称	方法返回值	方 法 作 用
getAnnotationTypes	Set<String>	获取注解类名列表
getMetaAnnotationTypes	Set<String>	获取一个类型的注解名称列表
hasAnnotation	boolean	是否直接存在某个注解
hasMetaAnnotation	boolean	是否存在某个注解
hasAnnotatedMethods	boolean	是否存在某个注解
getAnnotatedMethods	Set<MethodMetadata>	获取注解属性
introspect	AnnotationMetadata	构造器提供 AnnotationMetadata 初始化方法

在这个接口中提供的 hasAnnotation 方法和 hasMetaAnnotation 方法存在差异,它们的差异是是否直接使用了注解。验证这个逻辑可以编写下面的测试环境,首先编写两个具有注解的类,具体代码如下。

```
@Component
public class A {

}

@Configuration
public class C {

}
```

对应的测试代码如下。

```
@Test
void testAnn() {
    AnnotationMetadata introspect = AnnotationMetadata.introspect(C.class);
    System.out.println(introspect.hasAnnotation(Component.class.getName()));
    System.out.println(introspect.hasMetaAnnotation(Component.class.getName()));
}
```

上述代码会输出下面的结果。

```
false
true
```

在这个例子中，Component 注解是间接被引用的，它在 Configuration 上被使用的具体代码如下。

```
@Target(ElementType.TYPE)
@Retention(RetentionPolicy.RUNTIME)
@Documented
@Component
public @interface Configuration {}
```

方法 hasMetaAnnotation 可以判断出是否间接使用了某个注解。

接下来将对 StandardAnnotationMetadata 对象进行分析，在这个对象中重点关注成员变量，具体信息见表 14.12。

表 14.12　StandardAnnotationMetadata 成员变量

变量名称	变量类型	变量说明
mergedAnnotations	MergedAnnotations	合并的注解属性
nestedAnnotationsAsMap	boolean	是否需要将注解属性作成嵌套的 map 对象
annotationTypes	Set<String>	注解类型列表

在这三个成员变量中最关键的变量是 mergedAnnotations，它为后续的注解方法使用提供了基础数据内容。具体处理代码如下。

```
@Nullable
default Map<String, Object> getAnnotationAttributes(String annotationName,
        boolean classValuesAsString) {

    MergedAnnotation<Annotation> annotation = getAnnotations().get(annotationName,
            null, MergedAnnotationSelectors.firstDirectlyDeclared());
    if (!annotation.isPresent()) {
        return null;
    }
    return annotation.asAnnotationAttributes(Adapt.values(classValuesAsString, true));
}
```

在这个提取数据的方法中主要有下面两个操作。

（1）提取 mergedAnnotations 中指定注解的注解信息。
（2）将注解信息转换成 Map 对象。

14.10　Java 中注解数据获取

接下来将使用 Java 来提取注解中的数据信息，通过这个方式来对 Spring 中的提取方式有一个简单认识。

第一步需要创建一个对象,对象名为 A,具体代码如下。

```java
@Component(value = "abcd")
public class A {

}
```

第二步编写提取注解数据的代码。

```java
@Test
void annAttribute() throws InvocationTargetException, IllegalAccessException {
   A a = new A();
   Class<? extends A> aClass = a.getClass();
   Annotation[] annotations = aClass.getAnnotations();

   Map<String, Map<String, Object>> annotationAttributes = new HashMap<>();

   for (Annotation annotation : annotations) {
      Class<? extends Annotation> annClass = annotation.annotationType();

      String name = annClass.getName();
      Method[] declaredMethods = annClass.getDeclaredMethods();

      Map<String, Object> oneAnnotationAttributes = new HashMap<>();

      for (Method declaredMethod : declaredMethods) {
         String annAtrrName = declaredMethod.getName();
         Object invoke = declaredMethod.invoke(annotation);
         oneAnnotationAttributes.put(annAtrrName, invoke);

      }
      annotationAttributes.put(name, oneAnnotationAttributes);
   }

   System.out.println();
}
```

在这个方法中使用 annotationAttributes 作为存储结构,第一层 key 表示注解名称,value 表示注解数据;第二层 key 表示注解属性名称,value 表示注解属性值。

14.11　ScopeMetadataResolver 分析

对 ScopeMetadataResolver 接口的作用及其源码进行分析,对于一个接口而言需要关注的就是它的定义,具体定义如下。

```java
@FunctionalInterface
public interface ScopeMetadataResolver {

   ScopeMetadata resolveScopeMetadata(BeanDefinition definition);

}
```

从这段代码中可以发现它的作用是从 BeanDefinition 对象中提取 ScopeMetadata 对象。

14.11.1 ScopeMetadata 分析

ScopeMetadata 是一个简单的 Java 对象，不具备特殊的处理方法，ScopeMetadata 的定义变量如下。

```
private String scopeName = BeanDefinition.SCOPE_SINGLETON;

private ScopedProxyMode scopedProxyMode = ScopedProxyMode.NO;
```

scopeName 表示作用域名称，scopedProxyMode 表示代理类型。在 ScopeMetadata 中不存在有参构造，其他方法都是 getter 和 setter 方法，因此可以进一步理解 ScopeMetadataResolver 的作用，解析 BeanDefiniton 中的属性然后创建 ScopeMetadata。

14.11.2 AnnotationScopeMetadataResolver 分析

接下来将对注解环境中的 ScopeMetadata 进行解析，具体代码在 AnnotationScopeMetadataResolver 中，相关实现如下。

```java
@Override
public ScopeMetadata resolveScopeMetadata(BeanDefinition definition) {
    ScopeMetadata metadata = new ScopeMetadata();
    if (definition instanceof AnnotatedBeanDefinition) {
        AnnotatedBeanDefinition annDef = (AnnotatedBeanDefinition) definition;
        AnnotationAttributes attributes = AnnotationConfigUtils.attributesFor(
            annDef.getMetadata(), this.scopeAnnotationType);
        if (attributes != null) {
            metadata.setScopeName(attributes.getString("value"));
            ScopedProxyMode proxyMode = attributes.getEnum("proxyMode");
            if (proxyMode == ScopedProxyMode.DEFAULT) {
                proxyMode = this.defaultProxyMode;
            }
            metadata.setScopedProxyMode(proxyMode);
        }
    }
    return metadata;
}
```

这里这个方法处理过程中需要关注一个注解 @Scope，只有这个注解中的数据才是需要的数据，换句话说，Scope 注解的属性对应的实体类是 ScopeMetadata，下面是 Scope 注解相关的定义代码。

```
@Target({ElementType.TYPE, ElementType.METHOD})
@Retention(RetentionPolicy.RUNTIME)
@Documented
```

```
public @interface Scope {

    @AliasFor("scopeName")
    String value() default "";

    @AliasFor("value")
    String scopeName() default "";

    ScopedProxyMode proxyMode() default ScopedProxyMode.DEFAULT;

}
```

在处理过程中通过 resolveScopeMetadata 方法提取的数据对应了 Scope 注解中的数据,具体处理代码如下。

```
AnnotationAttributes attributes = AnnotationConfigUtils.attributesFor(
    annDef.getMetadata(), this.scopeAnnotationType);
```

理解这段代码需要理解 scopeAnnotationType 变量,该变量表示了 Scope 的类,为了对这段代码进行测试可以编写一个测试类,类名为 AnnBeans,具体代码如下。

```
@Component
@Scope
public class AnnBeans {
}
```

在这个 Bean 定义中可以看到它拥有了两个注解,这两个注解的数据会被放在 metadata 中,通过 AnnotationConfigUtils.attributesFor 方法可以单独提取需要的注解属性值。接下来进入调试阶段查看 metadata 数据,详细信息如图 14.2 所示。

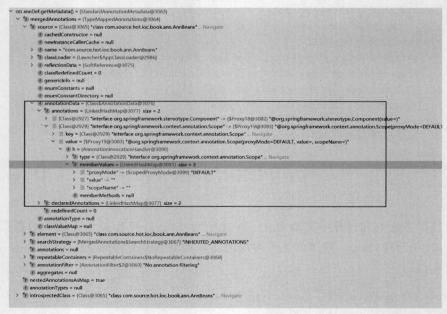

图 14.2 metadata 数据信息

经过提取 Scope 后 attributes 的数据信息如图 14.3 所示。

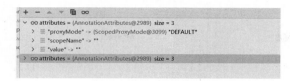

图 14.3　attributes 数据信息

在得到 attributes 数据信息后即可对 BeanDefinition 进行数据设置。

小结

本章主要对 Scope 元数据解析进行相关分析，在处理这个注解时有下面这样一个比较通用的处理逻辑。

（1）提取 BeanDefinition 中的注解元数据。

（2）从注解元数据中提取需要处理的注解。

（3）将提取得到的注解进行数据填充（给 BeanDefinition 进行赋值）。

本章对 Spring 中常用的注解、类的元数据存储对象进行说明，这些存储对象的核心是各个字段的存储内容。此外，对元数据的读取工具做了分析。

第15章

Spring事件

从本章开始将介绍、分析 Spring IoC 中的辅助工具。本章将对 Spring IoC 中的事件相关内容进行分析，包括 Spring 事件的使用和 Spring 事件的处理流程分析。

15.1 Spring 事件测试环境搭建

在 Spring 的启动阶段有下面这段代码：AbstractApplicationContext#finishRefresh，在这个方法中有一个关于事件推送的方法：publishEvent，本章的测试用例就围绕这个方法开始，下面是 publishEvent 方法的调用代码。

```
publishEvent(new ContextRefreshedEvent(this));
```

在 Spring 的启动流程中还有这些地方用到了 Spring 提供的基本事件。
（1）start 方法中的推送：

```
publishEvent(new ContextStartedEvent(this));
```

（2）stop 方法中的推送：

```
publishEvent(new ContextStoppedEvent(this));
```

测试用例将围绕 ContextRefreshedEvent、ContextStartedEvent 和 ContextStoppedEvent 三个事件对象编写。首先编写 ContextRefreshedEvent 的处理器，类名为 ContextRefreshedEvent，该处理器将在发布 ContextRefreshedEvent 事件时调用，具体代码如下。

```
public class CRefreshEventHandler
    implements ApplicationListener<ContextRefreshedEvent>{

    public void onApplicationEvent(ContextRefreshedEvent event) {
        System.out.println("ContextRefreshedEvent Received");
```

其次编写 ContextStartedEvent 的处理器,类名为 CStartEventHandler,该处理器将在发布 ContextStartedEvent 事件时调用,具体代码如下。

```java
public class CStartEventHandler
    implements ApplicationListener<ContextStartedEvent>{

    public void onApplicationEvent(ContextStartedEvent event) {
        System.out.println("ContextStartedEvent Received");
    }
}
```

最后编写 ContextStoppedEvent 的处理器,类名为 CStopEventHandler,该处理器将在发布 ContextStoppedEvent 事件时调用,具体代码如下。

```java
public class CStopEventHandler
    implements ApplicationListener<ContextStoppedEvent>{

    public void onApplicationEvent(ContextStoppedEvent event) {
        System.out.println("ContextStoppedEvent Received");
    }
}
```

完成各个事件处理器代码编写后将这些处理器都放到 Spring 容器中,编写一个 SpringXML 配置文件,文件名为 spring-event.xml,文件内容如下。

```xml
<?xml version="1.0" encoding="UTF-8"?>
<beans xmlns:xsi="http://www.w3.org/2001/XMLSchema-instance"
    xmlns="http://www.springframework.org/schema/beans"
    xsi:schemaLocation="http://www.springframework.org/schema/beans http://www.springframework.org/schema/beans/spring-beans.xsd">

    <bean id="cStartEventHandler" class="com.source.hot.ioc.book.event.CStartEventHandler"/>
    <bean id="cStopEventHandler" class="com.source.hot.ioc.book.event.CStopEventHandler"/>
    <bean class="com.source.hot.ioc.book.event.CRefreshEventHandler"/>
</beans>
```

完成 SpringXML 配置文件编写后制作一个测试类,测试类名为 EventTest,具体代码如下。

```java
public class EventTest {

    @Test
    void testEvent() {
        ClassPathXmlApplicationContext context =
            new ClassPathXmlApplicationContext("META-INF/spring-event.xml");
        context.start();
```

```
            context.stop();
            context.close();
        }
}
```

测试方法 testEvent 的执行结果如下。

```
ContextRefreshedEvent Received
ContextStartedEvent Received
ContextStoppedEvent Received
```

15.2 Spring 事件处理器注册

事件处理器要发挥能力前需要先注入 Spring 容器中，负责自定义事件处理器注册的方法是 beanFactory.preInstantiateSingletons()。这个方法就是执行事件处理器注册的方法。它可以作为事件处理器注册的入口方法，通过入口进一步往下追踪源码可以在 AbstractAutowireCapableBeanFactory ♯ initializeBean（java.lang.String，java.lang.Object，org.springframework.beans.factory.support.RootBeanDefinition）中找到 org.springframework.beans.factory.support.AbstractAutowireCapableBeanFactory ♯ applyBeanPostProcessorsAfterInitialization 方法，这个方法中主要做的事情是 BeanPostProcessor 的调用。Spring 中 BeanPostProcessor 的实现类有很多，重点关注 ApplicationListenerDetector 对象。在 ApplicationListenerDetector ♯ postProcessAfterInitialization 方法中就有将事件处理器加入容器中的具体代码，整个事件处理器注册的调用链路如下。

（1）org.springframework.context.support.AbstractApplicationContext ♯ finishBeanFactoryInitialization。

（2）org.springframework.context.support.AbstractApplicationContext ♯ finishBeanFactoryInitialization。

（3）org.springframework.beans.factory.support.AbstractAutowireCapableBeanFactory ♯ initializeBean(java.lang.String，java.lang.Object，org.springframework.beans.factory.support.RootBeanDefinition)。

（4）org.springframework.beans.factory.support.AbstractAutowireCapableBeanFactory ♯ applyBeanPostProcessorsAfterInitialization。

（5）org.springframework.context.support.ApplicationListenerDetector ♯ postProcessAfterInitialization。

首先查看 ApplicationListenerDetector 对象的类图，如图 15.1 所示。

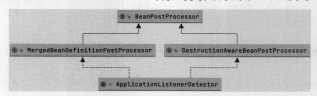

图 15.1 ApplicationListenerDetector 类图

在类图中可以看到它间接实现了 BeanPostProcessor 接口，出现这个接口可以按照下面两个内容作为分析，第一个是实例创建之后需要做什么，第二个是实例摧毁之前需要做什么。

15.2.1　事件处理器实例创建后

接下来将对事件处理器实例创建后做的事项进行分析，下面是处理代码。

```
@Override
public Object postProcessAfterInitialization(Object bean, String beanName) {
    if (bean instanceof ApplicationListener) {
        Boolean flag = this.singletonNames.get(beanName);
        if (Boolean.TRUE.equals(flag)) {
            this.applicationContext.addApplicationListener((ApplicationListener<?>) bean);
        }
        else if (Boolean.FALSE.equals(flag)) {
            if (logger.isWarnEnabled() && !this.applicationContext.containsBean()) {
                logger.warn("Inner bean '" + beanName + "' implements ApplicationListener " +
                    "interface " + "but is not reachable for event multicasting by its containing ApplicationContext " +
                    "because it does not have singleton scope. Only top-level listener beans are allowed " + "to " +
                    "be of non-singleton scope.");
            }
            this.singletonNames.remove(beanName);
        }
    }
    return bean;
}
```

在这段代码中首先需要了解 singletonNames 对象的存储结构，key 存储的是 BeanName，value 存储的是布尔值，表示是否单例，具体定义代码如下。

```
private final transient Map<String, Boolean> singletonNames =
    new ConcurrentHashMap<>(256);
```

该对象的设置代码如下。

```
@Override
public void postProcessMergedBeanDefinition(RootBeanDefinition beanDefinition, Class<?> beanType, String beanName) {
    if (ApplicationListener.class.isAssignableFrom(beanType)) {
        this.singletonNames.put(beanName, beanDefinition.isSingleton());
    }
}
```

在这段代码中可以进一步知道是否单例的数据来源是在 BeanDefinition 中。有关 postProcessAfterInitialization 的处理流程如图 15.2 所示。

```
                    ┌──────────┐
                    │ BeanName │
                    └────┬─────┘
                         ↓
              ┌────────────────────────┐
              │ 从singletonNames取出标志 │
              └──────────┬─────────────┘
                         ↓
                    ╱ 是否单例? ╲
                Y ╱             ╲ N
                 ↓               ↓
    ┌──────────────────────┐  ┌──────────────────────────────┐
    │ 将Bean加入事件处理器列表 │  │ 将BeanName从singletonNames中删除 │
    └──────────────────────┘  └──────────────────────────────┘
```

图 15.2 postProcessAfterInitialization 处理流程

经过上述代码处理会将开发者自定义的各类事件处理器（事件监听器）放入到 Spring 的应用上下文中，具体数据如图 15.3 所示。

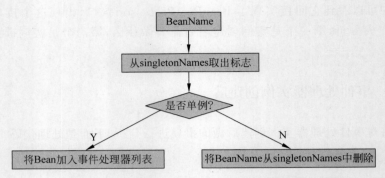

图 15.3 应用上下文中的事件处理器列表

15.2.2 事件处理器实例摧毁前

接下来将分析事件处理器摧毁前做的事项，具体处理代码如下。

```
@Override
public void postProcessBeforeDestruction(Object bean, String beanName) {
    if (bean instanceof ApplicationListener) {
        try {
            ApplicationEventMulticaster multicaster
                = this.applicationContext.getApplicationEventMulticaster();
            multicaster.removeApplicationListener((ApplicationListener<?>) bean);
            multicaster.removeApplicationListenerBean(beanName);
        }
        catch (IllegalStateException ex) {
            //ApplicationEventMulticaster not initialized yet - no need to remove a listener
        }
    }
}
```

在这段代码中可以看到具体处理都是关于移除事件处理器的，在这里体现出来以下两种移除方式。

（1）根据 Bean 实例进行移除。
（2）根据 BeanName 进行移除。

在这段代码中移除的数据对象是 multicaster 中 defaultRetriever 的事件监听器实例列表（applicationListeners）和事件监听器名称列表（applicationListenerBeans）。移除一个数据前数据如图 15.4 所示，移除一个数据后数据如图 15.5 所示。

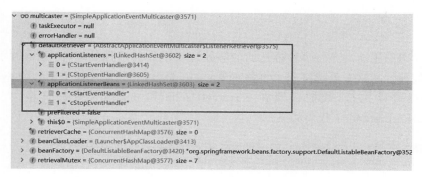

图 15.4　移除数据信息前

图 15.5　移除单个数据后

当所有事件处理器（事件监听器）都移除了，defaultRetriever 中的数据内容将会为空。在 multicaster 中还有一个变量是 retrieverCache，它是一个数据缓存，重点关注以下一些数据信息。

（1）key 中的 eventType#type：该信息表示需要处理的事件类型。

（2）value 中的 applicationListeners：该信息表示提供处理的处理接口。

这两项信息在事件处理阶段会产生作用，将 applicationListeners 循环并将事件对象放入执行每个 applicationLisenter。

15.3　Spring 事件推送和处理

接下来将对事件推送相关内容进行分析，首先需要找到能够进行事件推送的对象或者接口。在 org. springframework. context. support. AbstractApplicationContext ♯ publishEvent(java. lang. Object, org. springframework. core. ResolvableType) 方法中可以在其中发现这样一段代码 getApplicationEventMulticaster(). multicastEvent(applicationEvent, eventType)，这段代码就是进行事件推送（事件广播）的核心了。在这里需要进一步查找可以发现真正处理的接口是 ApplicationEventMulticaster，在 Spring 中可以找到的唯一的一个处理事件推送的对象 SimpleApplicationEventMulticaster。接下来就对

SimpleApplicationEventMulticaster 类中的发布事件方法做分析,发布事件的代码内容如下。

```java
@Override
public void multicastEvent(final ApplicationEvent event, @Nullable ResolvableType eventType) {
    ResolvableType type = (eventType != null ? eventType : resolveDefaultEventType(event));
    Executor executor = getTaskExecutor();
    for (ApplicationListener<?> listener : getApplicationListeners(event, type)) {
        if (executor != null) {
            executor.execute(() -> invokeListener(listener, event));
        }
        else {
            invokeListener(listener, event);
        }
    }
}
```

这段代码的用途就是将事件推送给能够处理该事件的事件处理器,并通过事件处理器进行事件处理。在这个处理过程中事件对象 event 是已知的,通过这个已知的 event 找到事件处理器就可以执行,数据存储在 retrieverCache 和 defaultRetriever 中,处理方法是 getApplicationListeners,代码如下。

```java
protected Collection<ApplicationListener<?>> getApplicationListeners(
        ApplicationEvent event, ResolvableType eventType) {

    //获取事件源信息
    Object source = event.getSource();
    //事件源信息类型
    Class<?> sourceType = (source != null ? source.getClass() : null);
    //缓存对象 key 的创建
    ListenerCacheKey cacheKey = new ListenerCacheKey(eventType, sourceType);

    //从容器中获取
    //Quick check for existing entry on ConcurrentHashMap...
    ListenerRetriever retriever = this.retrieverCache.get(cacheKey);
    //不为空,直接从缓存值中获取
    if (retriever != null) {
        //从缓存值中获取应用监听器列表
        return retriever.getApplicationListeners();
    }

    if (this.beanClassLoader == null ||
        (ClassUtils.isCacheSafe(event.getClass(), this.beanClassLoader) &&
            (sourceType == null ||
                ClassUtils.isCacheSafe(sourceType, this.beanClassLoader)))) {
        synchronized (this.retrievalMutex) {
            retriever = this.retrieverCache.get(cacheKey);
            if (retriever != null) {
```

```
                return retriever.getApplicationListeners();
            }
            retriever = new ListenerRetriever(true);
            //从 defaultRetriever 推算 ApplicationListener 对象
            Collection<ApplicationListener<?>> listeners =
                retrieveApplicationListeners(eventType, sourceType, retriever);
            this.retrieverCache.put(cacheKey, retriever);
            return listeners;
        }
    }
    else {
        return retrieveApplicationListeners(eventType, sourceType, null);
    }
}
```

这个方法中主要操作逻辑如下。

（1）从 retrieverCache 中读取数据，注意第一次读取的事件类型（cacheKey）是不存在的，Spring 启动完成后 retrieverCache 是一个空数据对象。

（2）从 retrieverCache 中搜索不到数据后在 defaultRetriever 中进行推算 ApplicationLisenter，这里的推算围绕以下两种情况。

① 从 ApplicationListener 实例列表中进行搜索。

② 从 ApplicationListener 实例名称列表中搜索。

这两种情况的本质都是用来判断实现了 ApplicationListener 的对象是否支持当前传递的事件类型，如果支持就会添加到返回值中。

经过上述操作，ApplicationListener 数据对象已经获取成功，接下来需要将事件对象放入到事件处理器中进行处理，具体处理代码如下。

```
protected void invokeListener(ApplicationListener<?> listener, ApplicationEvent event) {
    ErrorHandler errorHandler = getErrorHandler();
    if (errorHandler != null) {
        try {
            doInvokeListener(listener, event);
        }
        catch (Throwable err) {
            errorHandler.handleError(err);
        }
    }
    else {
        doInvokeListener(listener, event);
    }
}


@SuppressWarnings({"rawtypes", "unchecked"})
private void doInvokeListener(ApplicationListener listener, ApplicationEvent event) {
    try {
        listener.onApplicationEvent(event);
    }
    catch (ClassCastException ex) {
```

```
            String msg = ex.getMessage();
            if (msg == null || matchesClassCastMessage(msg, event.getClass())) {
                Log logger = LogFactory.getLog(getClass());
                if (logger.isTraceEnabled()) {
                    logger.trace("Non-matching event type for listener: " + listener, ex);
                }
            }
            else {
                throw ex;
            }
        }
    }
```

在这段方法中主要关注 doInvokeListener 方法,在这个方法中可以看到直接进行 onApplicationEvent 方法调用。注意:在调用 onApplicationEvent 方法时会有一个独立的线程进行处理。

小结

本章主要围绕 Spring 启动和停止阶段中的三个事件进行拓展,在这三个事件基础上编写了三个对应的事件处理器(事件监听器)。根据三个事件处理器进入到处理器的源码分析,对事件处理器的注册、事件推送和事件处理进行了源码分析。

第16章

占位符解析

本章将对 Spring IoC 辅助工具中的占位符解析进行源码分析,本章包含占位符解析的基本环境搭建、占位符解析算法分析。

16.1 基本环节搭建

在 Spring 开发过程中,当开发者使用 Spring JDBC 相关技术点的时候对占位符的使用有一个比较直观的感受,下面将围绕这个场景创建一个测试工程(模拟 JDBC 的使用,并非去搭建一个 JDBC 使用案例)。第一步需要编写一个 Java 对象用来存储数据信息,该对象名为 PropertyBean,具体代码如下。

```
public class PropertyBean {
    private String name;
    //省略 getter & setter
}
```

第二步需要编写一个外部配置文件,文件名为 property.properties,文件内容如下。

name = zhangsan

第三步需要编写 SpringXML 配置文件,文件名为 PropertyResolution.xml,文件内容如下。

```
<?xml version = "1.0" encoding = "UTF-8"?>
< beans xmlns:xsi = "http://www.w3.org/2001/XMLSchema - instance"
       xmlns:context = "http://www.springframework.org/schema/context" xmlns = "http://www.springframework.org/schema/beans"
       xsi:schemaLocation = "http://www.springframework.org/schema/beans http://www.springframework.org/schema/beans/spring - beans.xsd
```

```
http://www.springframework.org/schema/context http://www.springframework.org/schema/
context/spring-context.xsd">
    <context:property-placeholder location="classpath:property.properties"/>

    <bean class="com.source.hot.ioc.book.pojo.PropertyBean">
        <property name="name" value="${name}"/>
    </bean>
</beans>
```

第四步编写测试方法,方法名为 testPropertyPlaceholder,具体代码如下。

```
void testPropertyPlaceholder() {
    ClassPathXmlApplicationContext context =
 new ClassPathXmlApplicationContext("META-INF/PropertyResolution.xml");
    PropertyBean bean = context.getBean(PropertyBean.class);
    assumeTrue(bean.getName().equals("zhangsan"));
}
```

在这个测试用例中希望执行的结果是 zhangsan,通过外部配置＋占位符解析的模式来得到外部配置文件中具体的配置内容从而达到可配置的需求。

16.2　XML 的解析

首先需要对 SpringXML 配置文件进行分析。从配置文件中可以看到<context:property-placeholder location="classpath:property.properties"/>配置,这个配置的作用是将外部数据导入 Spring 中。在这里需要重点关注 context 标签的内容,通过第 4 章的内容可以知道,这是一个自定义标签,对于自定义标签的分析需要通过 spring.handlers 来找到对应的处理方法,context 标签对应的 spring.handlers 文件位于 spring-context\src\main\resources\META-INF\spring.handlers,该文件的内容如下。

```
http\://www.springframework.org/schema/context=org.springframework.context.config.
ContextNamespaceHandler
http\://www.springframework.org/schema/jee=org.springframework.ejb.config.
JeeNamespaceHandler
http\://www.springframework.org/schema/lang=org.springframework.scripting.config.
LangNamespaceHandler
http\://www.springframework.org/schema/task=org.springframework.scheduling.config.
TaskNamespaceHandler
http\://www.springframework.org/schema/cache=org.springframework.cache.config.
CacheNamespaceHandler
```

在这个文件中可以找到 context 对应的处理类是 org.springframework.context.config.ContextNamespaceHandler,在 Spring 工程中找到这个类,阅读它的 init 方法,具体代码如下。

```
public class ContextNamespaceHandler extends NamespaceHandlerSupport {

    @Override
```

```
    public void init() {
        registerBeanDefinitionParser("property-placeholder",
new PropertyPlaceholderBeanDefinitionParser());
        registerBeanDefinitionParser("property-override",
new PropertyOverrideBeanDefinitionParser());
        registerBeanDefinitionParser("annotation-config",
new AnnotationConfigBeanDefinitionParser());
        registerBeanDefinitionParser("component-scan",
new ComponentScanBeanDefinitionParser());
        registerBeanDefinitionParser("load-time-weaver",
new LoadTimeWeaverBeanDefinitionParser());
        registerBeanDefinitionParser("spring-configured",
new SpringConfiguredBeanDefinitionParser());
        registerBeanDefinitionParser("mbean-export",
new MBeanExportBeanDefinitionParser());
        registerBeanDefinitionParser("mbean-server",
new MBeanServerBeanDefinitionParser());
    }

}
```

在 init 方法中定义了具体标签对应的具体解析类,这里需要重点关注的是 property-placeholder 对应的解析类 PropertyPlaceholderBeanDefinitionParser。具体的处理过程可以结合第 4 章中的内容进行阅读,对于这个方法的处理重点是关注数据信息的存储,它会被存储在 BeanDefinitionBuilder 中,具体信息如图 16.1 所示。

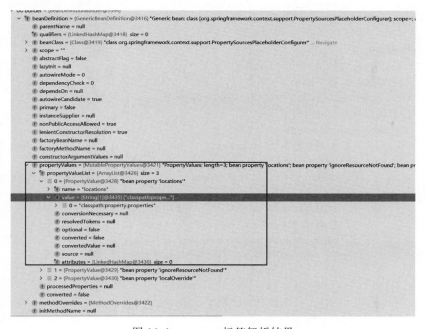

图 16.1　context 标签解析结果

从图 16.1 中可以发现,数据信息 classpath:property.properties 和 location 已经建立绑定关系。接下来需要做的事情就是将配置文件的数据解析到 Spring 中进行保存。

16.3 外部配置的读取

接下来将对外部配置读取进行分析,在前文得到了 classpath:property.properties 数据信息后需要将其读取并解析其中的数据信息,解析的处理对象是 PropertySources-PlaceholderConfigurer,类图如图 16.2 所示。

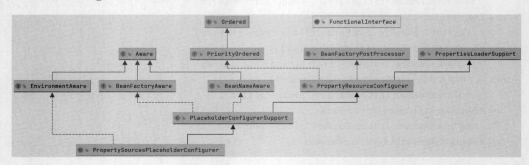

图 16.2 PropertySourcesPlaceholderConfigurer 类图

在整个类图中需要关注的接口是 BeanFactoryPostProcessor,在这个接口的方法中有关于 BeanFactory 初始化之前做的一些行为。PropertySourcesPlaceholderConfigurer 方法中的 postProcessBeanFactory 方法代码如下。

```java
@Override
public void postProcessBeanFactory(ConfigurableListableBeanFactory beanFactory) throws BeansException {

    //属性容器为空
    if (this.propertySources == null) {
        //创建属性容器
        this.propertySources = new MutablePropertySources();
        //环境属性存在的情况下
        if (this.environment != null) {
            //往属性容器中添加 environment
            this.propertySources.addLast(
                new PropertySource<Environment>(ENVIRONMENT_PROPERTIES_PROPERTY_SOURCE_NAME, this.environment) {
                    @Override
                    @Nullable
                    public String getProperty(String key) {
                        return this.source.getProperty(key);
                    }
                }
            );
        }
        try {
            //加载外部属性文件
            PropertySource<?> localPropertySource =
                new PropertiesPropertySource(LOCAL_PROPERTIES_PROPERTY_SOURCE_NAME,
```

```
            mergeProperties()));
                if (this.localOverride) {
                    this.propertySources.addFirst(localPropertySource);
                }
                else {
                    this.propertySources.addLast(localPropertySource);
                }
            }
            catch (IOException ex) {
                throw new BeanInitializationException("Could not load properties", ex);
            }
        }

        //处理占位符
        processProperties(beanFactory,
    new PropertySourcesPropertyResolver(this.propertySources));
        this.appliedPropertySources = this.propertySources;
}
```

这段代码的处理流程如图 16.3 所示。

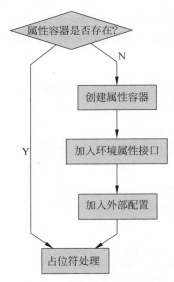

图 16.3 PropertySourcesPlaceholderConfigurer#postProcessBeanFactory 处理流程

在这段代码中需要重点关注下面两段代码。

```
PropertySource<?> localPropertySource =
      new PropertiesPropertySource(LOCAL_PROPERTIES_PROPERTY_SOURCE_NAME, mergeProperties());
processProperties(beanFactory, new PropertySourcesPropertyResolver(this.propertySources));
```

首先关注第一行代码，它是一个构造函数的使用，在这个构造函数中没有什么特别代码，重点关注构造函数调用之前的第二个参数，第二个参数是通过一个函数进行获取数据，函数是 mergeProperties，具体处理代码如下。

```
protected Properties mergeProperties() throws IOException {
```

```java
        Properties result = new Properties();

        if (this.localOverride) {
            loadProperties(result);
        }

        if (this.localProperties != null) {
            for (Properties localProp : this.localProperties) {
                CollectionUtils.mergePropertiesIntoMap(localProp, result);
            }
        }

        if (!this.localOverride) {
            loadProperties(result);
        }

        return result;
    }
```

在这段代码中需要关注两个变量,第一个变量是 localOverride,它表示是否需要重写本地数据,默认值为 false;第二个变量是 localProperties,它表示本地的属性数据表。在测试用例中并未对 localOverride 属性进行设置,接下来对 loadProperties 方法进行分析,具体代码如下。

```java
    protected void loadProperties(Properties props) throws IOException {
        if (this.locations != null) {
            for (Resource location : this.locations) {
                if (logger.isTraceEnabled()) {
                    logger.trace("Loading properties file from " + location);
                }
                try {
                    PropertiesLoaderUtils.fillProperties(
                            props, new EncodedResource(location,
this.fileEncoding), this.propertiesPersister);
                }
                catch (FileNotFoundException | UnknownHostException ex) {
                    if (this.ignoreResourceNotFound) {
                        if (logger.isDebugEnabled()) {
                            logger.debug("Properties resource not found: " + ex.getMessage());
                        }
                    }
                    else {
                        throw ex;
                    }
                }
            }
        }
    }
```

在 loadProperties 方法中需要理解一个属性 locations,这个属性存储了在 context:property-placeholder 中配置的 location 属性,得到单个 location 后 Spring 会去读取这个文

件再转换成 Properties 对象。locations 数据信息如图 16.4 所示。

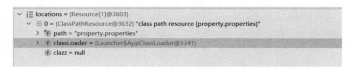

图 16.4　locations 数据信息

在 locations 中存储了外部资源配置的文件地址，通过循环处理每个外部资源配置地址得到文件中的数据，具体处理方法是下面这段代码。

```
PropertiesLoaderUtils.fillProperties(props, new
EncodedResource(location, this.fileEncoding), this.propertiesPersister)
```

这段代码可以简单理解为 Java 中 Properties 对象的资源读取操作，资源读取结果如图 16.5 所示。

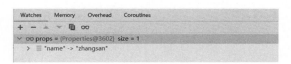

图 16.5　资源读取结果

在资源读取完成后会进行头插法或者尾插法的操作，操作对象是 propertySources，经过操作后的数据结果如图 16.6 所示。

图 16.6　propertySources 数据结果

可以从图 16.6 的结果发现这是一个尾插法，当数据被插入 propertySources 后就完成了外部配置的导入。

16.4　字符串占位符解析

接下来将进入占位符解析相关的内容，首先需要找到占位符解析的入口，Spring 中负责进行占位符解析的方法是 org.springframework.context.support.PropertySourcesPlaceholderConfigurer#processProperties(org.springframework.beans.factory.config.ConfigurableListableBeanFactory, org.springframework.core.env.ConfigurablePropertyResolver)，它的处理阶段是在外部资源解析之后，也就是 PropertySourcesPlaceholderConfigurer#postProcessBeanFactory 方法之后，PropertySourcesPlaceholderConfigurer#postProcessBeanFactory 中最后两行代码：

```
        processProperties(beanFactory,
new PropertySourcesPropertyResolver(this.propertySources));
        this.appliedPropertySources = this.propertySources;
```

第一行代码就是进行占位符解析,下面将对这个方法进行分析。首先需要关注 SpringXML 配置文件中的 bean 定义,具体定义信息如下。

```
<bean class="com.source.hot.ioc.book.pojo.PropertyBean">
    <property name="name" value="${name}"/>
</bean>
```

这个 bean 标签所对应的 BeanDefinition 信息如图 16.7 所示。

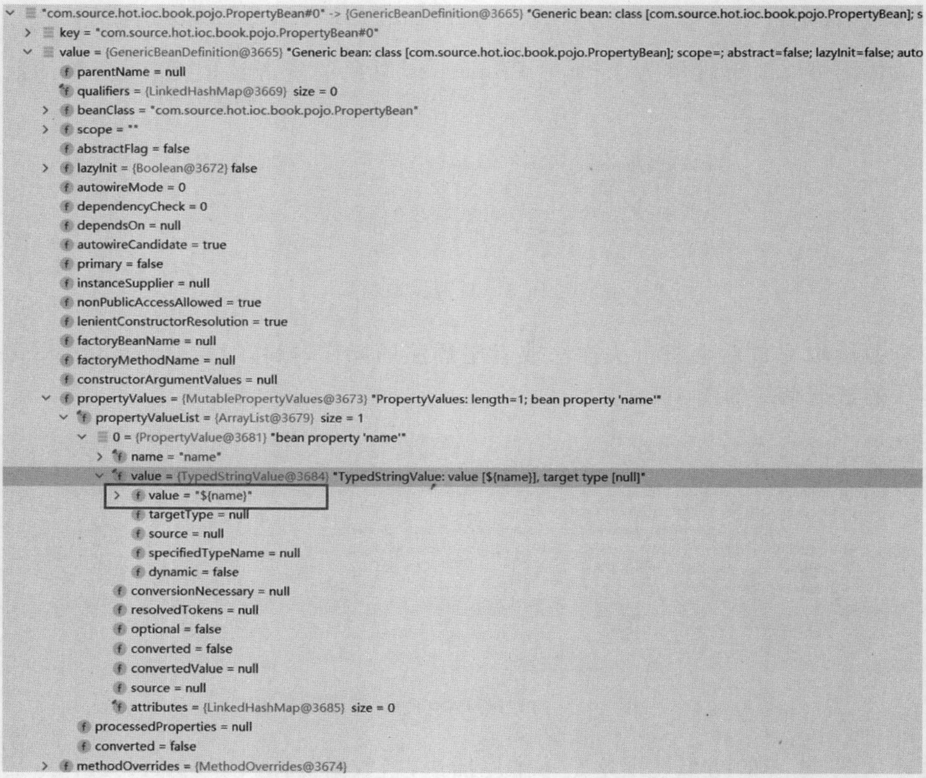

图 16.7 未进行占位符解析的 BeanDefinition 对象

从图 16.7 可以发现目前属性表中关于 name 对应的数据信息还是一个占位符的状态,具体信息是 ${name}。当经过占位符解析后会将 ${name} 解析成为 zhangsan,解析后结果如图 16.8 所示。

通过对比图 16.7 和图 16.8 可以发现,value 已经替换成外部配置文件中的数据。接下来对 processProperties 方法进行分析,下面是 processProperties 方法代码。

```
protected void processProperties(ConfigurableListableBeanFactory beanFactoryToProcess,
    final ConfigurablePropertyResolver propertyResolver) throws BeansException {

    //设置前置占位符
```

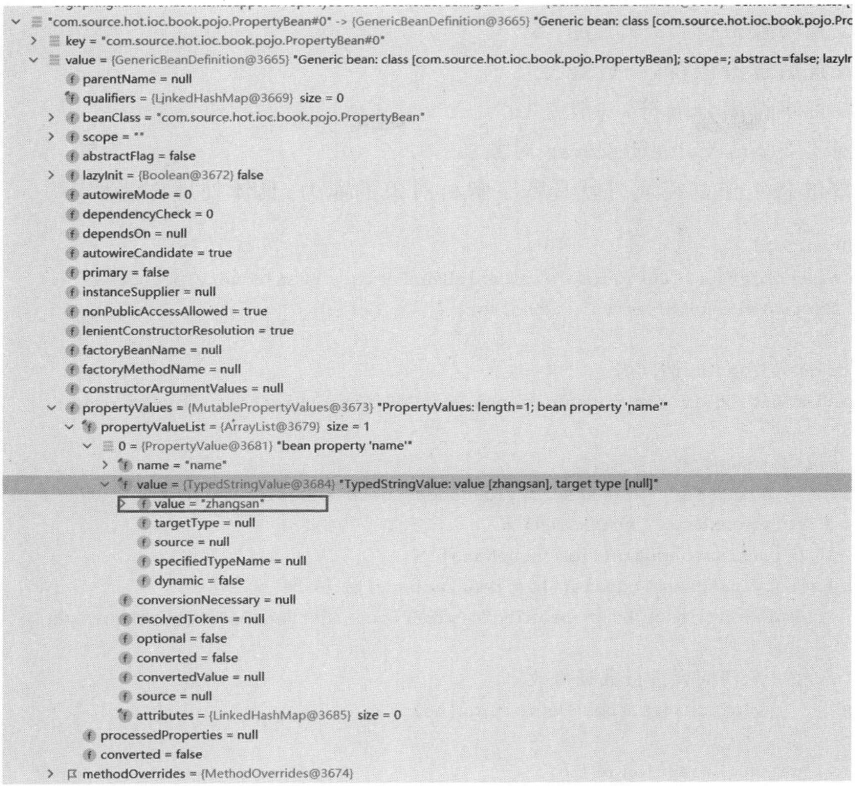

图 16.8　占位符解析后的 BeanDefinition 对象

```
propertyResolver.setPlaceholderPrefix(this.placeholderPrefix);
//设置后置占位符
propertyResolver.setPlaceholderSuffix(this.placeholderSuffix);
//设置字符分隔符
propertyResolver.setValueSeparator(this.valueSeparator);

//字符串解析器
StringValueResolver valueResolver = strVal -> {
    String resolved = (this.ignoreUnresolvablePlaceholders ?
        propertyResolver.resolvePlaceholders(strVal) :
        propertyResolver.resolveRequiredPlaceholders(strVal));
    if (this.trimValues) {
        resolved = resolved.trim();
    }
    return (resolved.equals(this.nullValue) ? null : resolved);
};

//真正的解析方法
doProcessProperties(beanFactoryToProcess, valueResolver);
}
```

在 processProperties 方法中除了最后一行代码以外其他的都是在为处理做准备，准备内容如下。

(1) 设置前置占位符,一般是"${"。
(2) 设置后置占位符,一般是"}"。
(3) 设置字符串分隔符,一般是";"。
(4) 创建 StringValueResolver 对象。

在这些准备工作都完成时就具备了解析对象的能力,具体处理方法如下。

```
protected void
doProcessProperties(ConfigurableListableBeanFactory beanFactoryToProcess,
      StringValueResolver valueResolver) {

   //beanDefinition 访问者
   BeanDefinitionVisitor visitor = new BeanDefinitionVisitor(valueResolver);

   //获取 BeanName 列表
   String[] beanNames = beanFactoryToProcess.getBeanDefinitionNames();
   for (String curName : beanNames) {
      if (!(curName.equals(this.beanName)
&& beanFactoryToProcess.equals(this.beanFactory))) {
         BeanDefinition bd = beanFactoryToProcess.getBeanDefinition(curName);
         try {
            //访问者进行数据修改
            visitor.visitBeanDefinition(bd);
         }
         catch (Exception ex) {
            throw new
BeanDefinitionStoreException(bd.getResourceDescription(), curName, ex.getMessage(), ex);
         }
      }
   }

   beanFactoryToProcess.resolveAliases(valueResolver);

   beanFactoryToProcess.addEmbeddedValueResolver(valueResolver);
}
```

在这段代码中最关键的对象是 BeanDefinitionVisitor,这个对象中的细节方法可以关注 visitBeanDefinition,具体处理代码如下。

```
public void visitBeanDefinition(BeanDefinition beanDefinition) {
   //Parent Name 解析
   visitParentName(beanDefinition);
   //Bean Class 解析
   visitBeanClassName(beanDefinition);
   //Factory Bean 解析
   visitFactoryBeanName(beanDefinition);
   //Factory Method 解析
   visitFactoryMethodName(beanDefinition);
   //Scope 解析
   visitScope(beanDefinition);
   if (beanDefinition.hasPropertyValues()) {
```

```
            //属性解析
            visitPropertyValues(beanDefinition.getPropertyValues());
        }
        if (beanDefinition.hasConstructorArgumentValues()) {
            //构造器解析
            ConstructorArgumentValues cas = beanDefinition.getConstructorArgumentValues();
            visitIndexedArgumentValues(cas.getIndexedArgumentValues());
            visitGenericArgumentValues(cas.getGenericArgumentValues());
        }
    }
```

在这个方法处理过程中有很多以 visit 开头的方法,这些方法中又会有不同的处理,但是其目的是将原数据替换成解析后的数据,它们之间对应的转换关系如下。

(1) resolveStringValue 方法提供给：visitParentName、visitBeanClassName、visitFactoryBeanName、visitFactoryMethodName、visitScope 使用。

(2) resolveValue 方法提供给：visitPropertyValues、visitIndexedArgumentValues、visitGenericArgumentValues 使用。

在上述两种转换关系下还有一层潜在关系,可以通过 visitParentName 方法找到潜在关系,下面是 visitParentName 方法的处理细节。

```
    protected void visitParentName(BeanDefinition beanDefinition) {
        String parentName = beanDefinition.getParentName();
        if (parentName != null) {
            String resolvedName = resolveStringValue(parentName);
            if (!parentName.equals(resolvedName)) {
                beanDefinition.setParentName(resolvedName);
            }
        }
    }
```

在这段代码中可以看到关键方法是 resolveStringValue。在前文提到的依赖 resolveStringValue 方法的使用者都遵守下面这个处理逻辑。

(1) 从 BeanDefinition 中获取需要解析的变量。

(2) 通过 resolveStringValue 方法解析对应数据。

(3) 将 resolveStringValue 解析得到的数据重新设置给 BeanDefinition。

继续观察另一个方法 visitPropertyValues,具体处理代码如下。

```
    protected void visitPropertyValues(MutablePropertyValues pvs) {
        PropertyValue[] pvArray = pvs.getPropertyValues();
        for (PropertyValue pv : pvArray) {
            Object newVal = resolveValue(pv.getValue());
            if (!ObjectUtils.nullSafeEquals(newVal, pv.getValue())) {
                pvs.add(pv.getName(), newVal);
            }
        }
    }
```

这段代码的关键处理是通过 resolveValue 方法进行的,在前文提到的依赖 resolveValue 方法的使用者都遵守下面这个处理逻辑。

(1) 循环数据列表中的 value 值。
(2) 通过 resolveValue 方法解析对应数据。
(3) 将 resolveValue 解析得到的数据重新设置给当前 value。

16.4.1 resolveStringValue 分析

接下来将对 resolveStringValue 方法进行分析，resolveStringValue 处理代码如下。

```
@Nullable
protected String resolveStringValue(String strVal) {
    if (this.valueResolver == null) {
        throw new IllegalStateException("No StringValueResolver specified - pass a resolver " +
                "object into the constructor or override the 'resolveStringValue' method");
    }
    String resolvedValue = this.valueResolver.resolveStringValue(strVal);
    return (strVal.equals(resolvedValue) ? strVal : resolvedValue);
}
```

在这段代码中主要关注的是 this.valueResolver.resolveStringValue(strVal) 方法，整个处理过程可以分为以下三步。

(1) 解析器是否存在，不存在则抛出异常(IllegalStateException)。
(2) 进行解析。
(3) 解析结果和原数据对比，如果不同就将解析结果作为返回值。

了解整体流程后关注 valueResolver 的一些信息。valueResolver 的类型是 StringValueResolver，这个接口的方法定义就是用来进行字符串解析的，在这里需要弄清楚 StringValueResolver 的实际类型。在进行占位符解析前存在具体的处理代码：

```
StringValueResolver valueResolver = strVal -> {
    String resolved = (this.ignoreUnresolvablePlaceholders ?
            propertyResolver.resolvePlaceholders(strVal) :
            propertyResolver.resolveRequiredPlaceholders(strVal));
    if (this.trimValues) {
        resolved = resolved.trim();
    }
    return (resolved.equals(this.nullValue) ? null : resolved);
};
```

从这段代码中可以发现 StringValueResolver 的内容。首先需要了解两个参数，第一个参数是 ignoreUnresolvablePlaceholders，该参数表示是否需要忽略不可解析的占位符，默认值为 false；第二个参数是 propertyResolver，该参数是接口 ConfigurablePropertyResolver，需要确认具体的类型，从调用方法中可以看到它是 PropertySourcesPropertyResolver 类型。

16.4.2 resolvePlaceholders 分析

接下来对 resolvePlaceholders 方法进行分析，下面是 resolvePlaceholders 代码详情。

```
@Override
public String resolvePlaceholders(String text) {
    if (this.nonStrictHelper == null) {
        this.nonStrictHelper = createPlaceholderHelper(true);
    }
    return doResolvePlaceholders(text, this.nonStrictHelper);
}
```

这段代码中首先关注 nonStrictHelper 的类型和这里的 createPlaceholderHelper(true) 方法，通过 createPlaceholderHelper(true) 方法会创建出 PropertyPlaceholderHelper 对象，这个对象就是能够提供解析能力的核心。了解 nonStrictHelper 后进一步对 doResolvePlaceholders 方法进行分析，doResolvePlaceholders 方法如下。

```
private String doResolvePlaceholders(String text, PropertyPlaceholderHelper helper) {
    return helper.replacePlaceholders(text, this::getPropertyAsRawString);
}
```

在这段代码中真正的处理能力是由前面准备的 nonStrictHelper 对象提供的，具体处理代码如下。

```
public String replacePlaceholders(String value, PlaceholderResolver placeholderResolver) {
    Assert.notNull(value, "'value' must not be null");
    return parseStringValue(value, placeholderResolver, null);
}
```

首先理解这个方法的两个参数，第一个参数是 String value：需要解析的字符串，可以带有占位符也可以不带有占位符；第二个参数是 PlaceholderResolver placeholderResolver：占位符解析器。

进入调试阶段确认这两个变量的数据。首先查看 value 变量数据，如图 16.9 所示；其次查看 placeholderResolver 变量，数据如图 16.10 所示。

图 16.9 value 数据信息

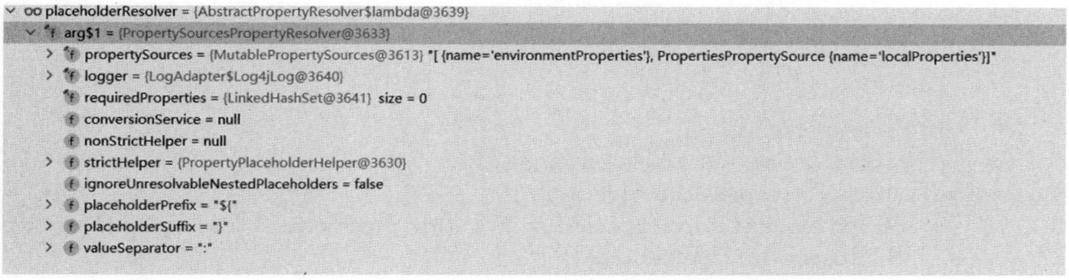

图 16.10 placeholderResolver 数据信息

了解了数据信息后就可以对 parseStringValue 方法进行分析，具体代码如下。

```java
protected String parseStringValue(
        String value, PlaceholderResolver placeholderResolver, @Nullable Set<String> visitedPlaceholders) {

    //占位符所在位置
    int startIndex = value.indexOf(this.placeholderPrefix);
    if (startIndex == -1) {
        return value;
    }

    //返回值
    StringBuilder result = new StringBuilder(value);
    while (startIndex != -1) {
        //寻找结尾占位符
        int endIndex = findPlaceholderEndIndex(result, startIndex);
        if (endIndex != -1) {
            //返回值切分留下中间内容
            String placeholder = result.substring(startIndex + this.placeholderPrefix.length(), endIndex);
            String originalPlaceholder = placeholder;
            if (visitedPlaceholders == null) {
                visitedPlaceholders = new HashSet<>(4);
            }
            if (!visitedPlaceholders.add(originalPlaceholder)) {
                throw new IllegalArgumentException(
                        "Circular placeholder reference '" + originalPlaceholder +
                "' in property definitions");
            }
            //Recursive invocation, parsing placeholders contained in the placeholder key.
            //递归获取占位符内容
            placeholder = parseStringValue(placeholder, placeholderResolver, visitedPlaceholders);
            String propVal = placeholderResolver.resolvePlaceholder(placeholder);
            if (propVal == null && this.valueSeparator != null) {
                int separatorIndex = placeholder.indexOf(this.valueSeparator);
                if (separatorIndex != -1) {
                    String actualPlaceholder = placeholder.substring(0, separatorIndex);
                    String defaultValue = placeholder.substring(separatorIndex + this.valueSeparator.length());
                    propVal = placeholderResolver.resolvePlaceholder(actualPlaceholder);
                    if (propVal == null) {
                        propVal = defaultValue;
                    }
                }
            }
            if (propVal != null) {
                propVal = parseStringValue(propVal, placeholderResolver, visitedPlaceholders);
                result.replace(startIndex, endIndex + this.placeholderSuffix.length(), propVal);
                if (logger.isTraceEnabled()) {
                    logger.trace("Resolved placeholder '" + placeholder + "'");
                }
                startIndex = result.indexOf(this.placeholderPrefix, startIndex + propVal.length());
```

```
            }
            else if (this.ignoreUnresolvablePlaceholders) {
                startIndex = result.indexOf(this.placeholderPrefix,
    endIndex + this.placeholderSuffix.length());
            }
            else {
                throw new IllegalArgumentException("Could not resolve placeholder '" +
                    placeholder + "'" + " in value \"" + value + "\"");
            }
            visitedPlaceholders.remove(originalPlaceholder);
        }
        else {
            startIndex = -1;
        }
    }
    return result.toString();
}
```

在这个方法中有以下三种处理情况。

(1) 需要解析的字符串没有占位符。

(2) 需要解析的字符串只有一对占位符。

(3) 需要解析的字符串存在一对及以上占位符。在这种情况下占位符有两种情况,第一种是平级占位符,第二种是嵌套占位符。

为了对上述三种处理情况(具体可以细分为四种,将第三种情况当作两个情况)进行分析,需要分别编写不同的测试方法。由于现在找到了核心处理对象 PropertyPlaceholderHelper,因此可以不通过编写 SpringXML 相关内容来进行测试用例的编写。有关占位符测试的代码信息如下。

```
public class PropertyPlaceholderHelperTest {
    Properties properties = new Properties();

    PropertyPlaceholderHelper propertyPlaceholderHelper;

    @NotNull
    private static
PropertyPlaceholderHelper.PlaceholderResolver
getPlaceholderResolver(Properties properties) {
        return new PropertyPlaceholderHelper.PlaceholderResolver() {
            @Override
            public String resolvePlaceholder(String placeholderName) {
                String value = properties.getProperty(placeholderName);
                return value;
            }
        };
    }


    @BeforeEach
    void initProperties() {
        properties.put("a", "1");
        properties.put("b", "2");
```

```java
            properties.put("c", "3");
            properties.put("a23", "abc");
            propertyPlaceholderHelper = new PropertyPlaceholderHelper("{", "}");
    }

    /**
     * 没有占位符
     */
    @Test
    void testNoPlaceholder() {
        String noPlaceholder = "a";
        String replacePlaceholders
                = propertyPlaceholderHelper.replacePlaceholders(noPlaceholder, this::getValue);
        assumeTrue(replacePlaceholders.equals("a"));
    }

    /**
     * 存在一个占位符
     */
    @Test
    void testOnePlaceholder() {
        String onePlaceholder = "{a}";
        String replacePlaceholders
                = propertyPlaceholderHelper.replacePlaceholders(onePlaceholder, this::getValue);
        assumeTrue(replacePlaceholders.equals("1"));

    }

    /**
     * 存在平级占位符
     */
    @Test
    void testSameLevelPlaceholder() {
        String sameLevelPlaceholder = "{a}{b}{c}";
        String replacePlaceholders
                = propertyPlaceholderHelper.replacePlaceholders(sameLevelPlaceholder, getPlaceholderResolver(properties));
        assumeTrue(replacePlaceholders.equals("163"));

    }

    /**
     * 存在嵌套占位符
     */
    @Test
    void testNestedPlaceholder() {
        String nestedPlaceholder = "{a{b}{c}}";
        String replacePlaceholders
                = propertyPlaceholderHelper.replacePlaceholders(nestedPlaceholder, getPlaceholderResolver(properties));
        assumeTrue(replacePlaceholders.equals("abc"));

    }
```

```
    private String getValue(String key) {
        return this.properties.getProperty(key);
    }
}
```

1. 没有占位符的分析

接下来将对没有占位符的情况进行分析，对应的测试方法是 testNoPlaceholder。关于没有占位符的处理具体代码如下。

```
int startIndex = value.indexOf(this.placeholderPrefix);
if (startIndex == -1) {
    return value;
}
```

这段代码的作用就是确认是否存在占位符，如果不存在就直接返回了。在测试用例中需要解析的字符串是"a"，前置占位符是"{"，很明显，前置占位符不存在于需要解析的字符串中，此时就将原始数据返回。

2. 只有一对占位符的分析

接下来将对只有一对占位符的情况进行分析，对应的测试方法是 testOnePlaceholder，关于只有一对占位符的处理流程如图 16.11 所示。

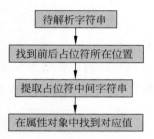

图 16.11 只有一对占位符的处理流程

下面对整个处理过程中的细节进行说明。

（1）通过 int startIndex = value.indexOf(this.placeholderPrefix); 找到前置占位符的索引。

（2）通过 int endIndex = findPlaceholderEndIndex(result,startIndex); 找到后置占位符的索引。

（3）通过 String placeholder = result.substring(startIndex + this.placeholderPrefix.length(),endIndex); 提取占位符中间的字符串。

（4）通过 String propVal = placeholderResolver.resolvePlaceholder(placeholder); 来获取占位符中间字符串对应的实际值。

对上述第四点进行说明，传递的是 this::getValue，此时进行的是 PlaceholderResolver 接口调用，其本质最终会调用 getValue 来得到数据。这里不太好理解为什么参数是接口传递的却是函数，这里主要是 @FunctionalInterface 在起作用，相关内容可以查看 JDK 8 的知识。

下面对现有内容进行整理，第一，占位符中间的字符串：a；第二，获取数据的方法：getValue；第三，数据存储容器：properties。通过这三点信息，占位符解析的信息可谓是呼之欲出。a 对应的数据是 1。

3. 平级占位符的分析

接下来将对平级占位符的情况进行分析。对应的测试方法是 testSameLevelPlaceholder，测试用例中编写了 {a}{b}{c} 这样一个字符串，这样的占位符不存在嵌套关系，都是一对一对包裹自己，不会出现一对占位符中还有占位符的情况，因此将其称为平级占位符。处理平级占位符的情况其实质就是多次执行一对占位符的逻辑,但其中也有一些差异,具体处理流程如图 16.12 所示。

接下来对{a}{b}{c}这个字符串的处理过程做每一步细节描述,省略提取占位符中的字符串和从属性表中获取的过程,处理流程如下。

(1) 输入变量{a}{b}{c}。
(2) 处理第一对占位符 {a}。
(3) 返回值标记为 1{b}{c}。
(4) 处理 1{b}{c} 中的 {b}。
(5) 返回值标记为 16{c}。
(6) 处理 16{c} 中的 {c}。
(7) 返回值标记为 163。

4. 嵌套占位符的分析

接下来将对嵌套占位符的情况进行分析。对应的测试方法是 testNestedPlaceholder,在测试用例中编写了 {a{b}{c}} 这样一个字符串,从这个字符串中的占位符表现形式上可以发现它出现了层级关系。对于这种层级关系,Spring 的处理流程如图 16.13 所示。

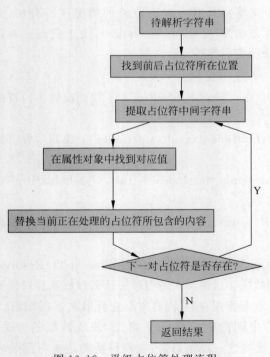
图 16.12　平级占位符处理流程

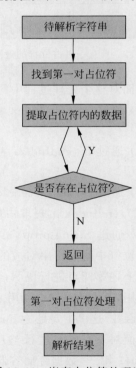
图 16.13　嵌套占位符处理流程

以测试用例中的{a{b}{c}}为例将整个转换过程中的一些核心步骤提取出来如下。

(1) 提取第一对占位符后的结果:a{b}{c}。
(2) 处理{b}得到第一个解析结果:a2{c}。
(3) 处理{c}得到第二个结果:a23。

(4)处理整个 a23 得到第三个结果：abc。

第(2)步和第(3)步可以看作是平级占位符的处理，第(4)步可以看作是只有一对占位符的处理。对于嵌套占位符的解析其本质上就是一个递归操作。

5．占位符解析小结

前文对占位符解析的四种情况进行了说明，下面将四种流程整合得到图 16.14。

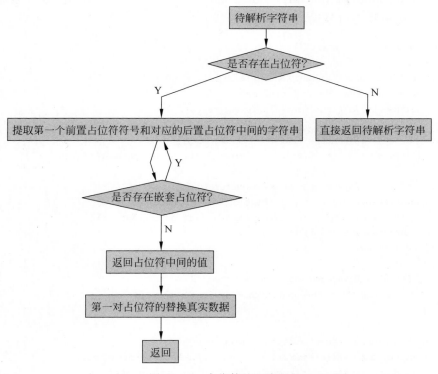

图 16.14　占位符处理流程

16.4.3　resolveRequiredPlaceholders 分析

接下来将对 resolveRequiredPlaceholders 方法进行分析，下面是具体处理代码。

```
@Override
public String resolveRequiredPlaceholders(String text) throws IllegalArgumentException {
    if (this.strictHelper == null) {
        this.strictHelper = createPlaceholderHelper(false);
    }
    return doResolvePlaceholders(text, this.strictHelper);
}
```

在这段代码中可以发现核心处理代码和 resolvePlaceholders 中的核心处理方法相同，都是 doResolvePlaceholders，有关分析可以查看前文内容。

16.4.4　BeanDefinitionVisitor#visitBeanDefinition 分析

现在回到 BeanDefinitionVisitor#visitBeanDefinition 方法中，在这个方法中使用到的技术就是前面提到的四种占位符处理。这样就可以理解 visitParentName、visitBeanClassName、visitFactoryBeanName、visitFactoryMethodName 和 visitScope 五个方法了。现在还需要对 resolveValue 方法进行分析，下面是具体代码。

```java
@SuppressWarnings("rawtypes")
@Nullable
protected Object resolveValue(@Nullable Object value) {
    if (value instanceof BeanDefinition) {
        visitBeanDefinition((BeanDefinition) value);
    }
    else if (value instanceof BeanDefinitionHolder) {
        visitBeanDefinition(((BeanDefinitionHolder) value).getBeanDefinition());
    }
    else if (value instanceof RuntimeBeanReference) {
        RuntimeBeanReference ref = (RuntimeBeanReference) value;
        String newBeanName = resolveStringValue(ref.getBeanName());
        if (newBeanName == null) {
            return null;
        }
        if (!newBeanName.equals(ref.getBeanName())) {
            return new RuntimeBeanReference(newBeanName);
        }
    }
    else if (value instanceof RuntimeBeanNameReference) {
        RuntimeBeanNameReference ref = (RuntimeBeanNameReference) value;
        String newBeanName = resolveStringValue(ref.getBeanName());
        if (newBeanName == null) {
            return null;
        }
        if (!newBeanName.equals(ref.getBeanName())) {
            return new RuntimeBeanNameReference(newBeanName);
        }
    }
    else if (value instanceof Object[]) {
        visitArray((Object[]) value);
    }
    else if (value instanceof List) {
        visitList((List) value);
    }
    else if (value instanceof Set) {
        visitSet((Set) value);
    }
    else if (value instanceof Map) {
        visitMap((Map) value);
    }
```

```
        else if (value instanceof TypedStringValue) {
            TypedStringValue typedStringValue = (TypedStringValue) value;
            String stringValue = typedStringValue.getValue();
            if (stringValue != null) {
                String visitedString = resolveStringValue(stringValue);
                typedStringValue.setValue(visitedString);
            }
        }
        else if (value instanceof String) {
            return resolveStringValue((String) value);
        }
        return value;
    }
```

在这里出现的以 visit 开头的方法都指向了 resolveStringValue 方法，它就是前文提到过的方法，具体分析可以查看前文内容。

小结

本章主要围绕 Spring 中的占位符解析讨论了下面 4 点。
（1）占位符的使用。
（2）外部资源的读取和加载。
（3）占位符解析对象的处理。
（4）占位符解析算法。

第17章

Spring中的转换服务

本章将对 Spring IoC 辅助工具中的转换服务进行使用说明和源码分析,本章还包含脱离 Spring 实现转换服务。

17.1 初识 Spring 转换服务

在整个 Spring 框架中有很多关于转换服务的使用场景,例如,在第 9 章获取 Bean 实例时需要进行转换。Spring 中提供了 Java 基础类型的转换和复杂对象的转换能力,下面将编写一个测试用来对 Spring 转换服务做一个简单的认知。

首先创建一个 Java 对象,类名为 Product,具体代码如下。

```
public class Product {
 private String name;

 private BigDecimal price;
}
```

完成 Java 对象编写后编写一个转换器,该转换器是将字符串转换成 Product 对象,转换器的代码如下。

```
public class StringToProduct implements Converter<String, Product> {
 @Override
 public Product convert(String source) {
    String[] split = source.split(";");
    return new Product(split[0],new BigDecimal(split[1]));
 }
}
```

完成转换器编写后再编写一个 SpringXML 配置文件,文件名为 convert.xml,文件内容如下。

```xml
<?xml version="1.0" encoding="UTF-8"?>
<beans xmlns:xsi="http://www.w3.org/2001/XMLSchema-instance"
    xmlns="http://www.springframework.org/schema/beans"
    xsi:schemaLocation="http://www.springframework.org/schema/beans http://www.springframework.org/schema/beans/spring-beans.xsd">


    <bean id="conversionService" class="org.springframework.context.support.ConversionServiceFactoryBean">
        <property name="converters">
            <list>
                <bean class="com.source.hot.ioc.book.convert.StringToProduct"/>
            </list>
        </property>
    </bean>
</beans>
```

完成 SpringXML 配置文件编写后编写单元测试,测试代码如下。

```
class ConvertTest {
 @Test
 void convertTest() {
    ClassPathXmlApplicationContext context =
new ClassPathXmlApplicationContext("/META-INF/convert.xml");
    ConversionService bean = context.getBean(ConversionService.class);
    Product convert = bean.convert("product;10.00", Product.class);
    System.out.println();
 }

}
```

在这个测试用例中可以通过调试发现 convert 符合转换规则,转换后数据如图 17.1 所示。

图 17.1 字符串转换 Product 对象结果

上述操作就是 Spring 中对于转换服务的一个简单使用案例。

17.2 ConversionServiceFactoryBean 对象的实例化

通过前文测试用例中的 SpringXML 配置文件可以发现,在配置阶段所使用的类是 org.springframework.context.support.ConversionServiceFactoryBean,并设置了 converters 属性。在 Java 中对于一个对象的属性设置需要创建对象后才可以进行,在 Spring 中也是一样的,必须要创建 ConversionServiceFactoryBean 对象,才可以进行数据操作。在 ConversionServiceFactoryBean 的代码中可以看到它实现了 InitializingBean 接口,实现了这个接口的方法很重要,同时 ConversionServiceFactoryBean 还实现了 FactoryBean 接口,它也很重要。下面介绍两个接口的关系,在 InitializingBean 接口方法中进行了数据的初始化,提供给 FactoryBean 使用,即前者完成数据准备,后者完成数据提供数据访问相关功能。

17.2.1 afterPropertiesSet 方法分析

接下来将对 afterPropertiesSet 方法进行详细说明,下面是 afterPropertiesSet 的具体代码。

```
@Override
public void afterPropertiesSet() {
    this.conversionService = createConversionService();
    ConversionServiceFactory.registerConverters(this.converters, this.conversionService);
}
```

这段代码总共两行,第一行代码通过 createConversionService 方法创建了 GenericConversionService 对象,实际对象是 DefaultConversionService;第二行代码作用是注册转换服务。

17.2.2 GenericConversionService 对象创建

在 Spring 中关于 GenericConversionService 对象的创建是调用 new 关键字进行创建,在 GenericConversionService 构造方法中还有一些其他操作,构造函数代码如下。

```
public DefaultConversionService() {
    addDefaultConverters(this);
}
```

前文提到的其他操作是 addDefaultConverters 代码,addDefaultConverters 方法会往 Spring 容器中进行注册默认的一些转换服务。该方法的提供者是 DefaultConversionService 接口,有关 DefaultConversionService 接口的方法说明见表 17.1。

表 17.1 DefaultConversionService 方法说明

方法代码	参数说明	方法作用
void addConverter(Converter<?,?> converter);	converter 实现了 Converter 接口的对象	注册转换服务
<S, T> void addConverter (Class<S> sourceType, Class<T> targetType, Converter<? super S,? extends T> converter);	sourceType:原始对象类型 targetType:目标对象类型 converter:提供转换服务的对象	注册转换服务
void addConverter(GenericConverter converter);	converter:实现了 GenericConverter 的对象	注册转换服务

续表

方法代码	参数说明	方法作用
void addConverterFactory（ConverterFactory <?,?> factory）；	factory：实现了 ConverterFactory 的对象	注册转换服务
void removeConvertible（Class <?> sourceType，Class <?> targetType）；	sourceType：原始对象类型 targetType：目标对象类型	根据原始对象类型和目标对象类型移除转换服务

17.2.3 注册转换服务

在 DefaultConversionService 的构造方法中可以看到提供注册能力的是 this，也就是 DefaultConversionService 本身，最终提供 ConverterRegistry 实现方法的其实是 DefaultConversionService 的父类 GenericConversionService，现在找到了分析目标。ConverterRegistry 的实现提供者是 GenericConversionService，接下来将对注册转换服务进行分析。

1. 转换器注册方式 1——Converter 直接注册

首先是对第一种注册方式——Converter 接口直接注册进行分析，对应的处理代码如下。

```
@Override
public void addConverter(Converter<?, ?> converter) {
    //获取解析类型
    ResolvableType[] typeInfo = getRequiredTypeInfo(converter.getClass(), Converter.class);
    if (typeInfo == null && converter instanceof DecoratingProxy) {
        typeInfo = getRequiredTypeInfo(((DecoratingProxy) converter).getDecoratedClass(), Converter.class);
    }
    if (typeInfo == null) {
        throw new IllegalArgumentException("Unable to determine source type <S> and target type <T> for your " + "Converter [" + converter.getClass().getName() + "]; does the class parameterize those types?");
    }
    //添加 converter
    addConverter(new ConverterAdapter(converter, typeInfo[0], typeInfo[1]));
}
```

在这个方法中需要理解两个内容，第一个是 ResolvableType 类，第二个是 addConverter 方法。ResolvableType 是 Spring 对 java.lang.reflect.Type 的一个封装，可以简单理解成这个变量中有反射相关的数据信息。在上述代码中可以发现 typeInfo 是一个数组，它存储了 Converter 接口的两个泛型数据。在前文的测试用例中，Converter < String, Product > 这里可以直接理解成数组变量存储了 String 类型和 Product 类型，具体信息如图 17.2 所示。

接下来对 addConverter 方法进行分析。在执行该方法的时候会将前置数据转换成 ConverterAdapter 对象。ConverterAdapter 对象是 GenericConverter 类型，可以发现在第

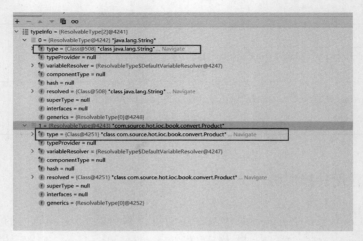

图 17.2　typeInfo 信息

三种处理方式上会使用 GenericConverter 对象进行注册,具体注册请看第三种注册方式的分析。了解 ConverterAdapter 对象需要从它的成员变量说起,它有三个成员变量,具体信息见表 17.2。

表 17.2　ConverterAdapter 成员变量

变量名称	变量类型	变量说明
converter	Converter<Object,Object>	转换接口提供者
typeInfo	ConvertiblePair	存储源数据类型和目标类型的对象
targetType	ResolvableType	转换目标的类型描述

2. 转换器注册方式 2——原始类型＋目标类型＋Converter 注册

下面介绍第二种注册方式：原始类型＋目标类型＋Converter 注册,对应处理代码如下。

```
@Override
public <S, T> void addConverter(Class<S> sourceType, Class<T> targetType, Converter<? super S, ? extends T> converter) {
    //添加 convert 的适配器对象
    addConverter(new ConverterAdapter(
        converter,
        ResolvableType.forClass(sourceType), ResolvableType.forClass(targetType)));
}
```

在这段代码中可以看到最终进行添加的对象依然是 ConverterAdapter,有关 ConverterAdapter 对象的注册可以查看第三种注册方式。除了 ConverterAdapter 对象以外还需要关注 ResolvableType 对象,该对象可以将一个 Class 对象转换成 ConverterAdapter 所需要的 ResolvableType 对象,具体代码如下。

```
private ResolvableType(@Nullable Class<?> clazz) {
    this.resolved = (clazz != null ? clazz : Object.class);
    this.type = this.resolved;
```

```
        this.typeProvider = null;
        this.variableResolver = null;
        this.componentType = null;
        this.hash = null;
}
```

在这段代码中关注 resolved 和 type 两个变量即可，这两个变量存储了原始的 Class 信息。

3. 转换器注册方式 3——GenericConverter 注册

接下来介绍第三种注册方式，通过 GenericConverter 进行注册，相关处理代码如下。

```
@Override
public void addConverter(GenericConverter converter) {
    //加入 convert 接口
    this.converters.add(converter);
    //缓存清除
    invalidateCache();
}
```

在这段代码中有两个操作，第一个操作是将参数转换器放入到转换器容器（converters）中，第二个操作是缓存的清理。首先需要关注存储容器 converters，它是一个对象，作为一个容器它的数据结构十分重要，具体存储结构可见表 17.3。

表 17.3　converters 的存储结构

变量名称	变量类型	变量说明
globalConverters	Set < GenericConverter >	存储 GenericConverter 的容器，全局共享的转换服务，没有类型约束
converters	Map < ConvertiblePair, ConvertersForPair >	key：存储原始类型和目标类型 value：存储多个 GenericConverter

了解成员变量后可以进一步查看 add 方法，add 方法详情如下。

```
public void add(GenericConverter converter) {
    //获取转换对象 ConvertiblePair
    Set < ConvertiblePair > convertibleTypes = converter.getConvertibleTypes();
    //判空
    if (convertibleTypes == null) {
        Assert.state(converter instanceof ConditionalConverter,
                "Only conditional converters may return null convertible types");
        this.globalConverters.add(converter);
    }
    else {
        for (ConvertiblePair convertiblePair : convertibleTypes) {
            //获取 ConvertersForPair 对象
            ConvertersForPair convertersForPair = getMatchableConverters(convertiblePair);
            convertersForPair.add(converter);
        }
    }
}
```

在 add 方法中，参数是一个接口，需要确认具体的数据类型，通过前文可以确认这里的类型是 ConverterAdapter。第一行代码 converter.getConvertibleTypes()；是从 ConverterAdapter 中获取它的 convertibleTypes 信息，在前文提到的测试用例中这里会得到 String 和 Product。convertibleTypes 信息如图 17.3 所示。

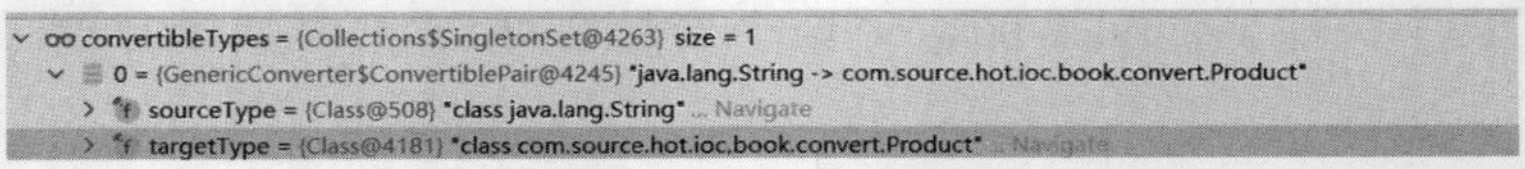

图 17.3 convertibleTypes 数据信息

此时对于获取得到的 convertibleTypes 信息会有以下两种不同操作。

（1）当 convertibleTypes 不存在的时候会将这个转换器放在全局的转换服务容器中（globalConverters）。

（2）当 convertibleTypes 存在的时候会将转换对象的信息和转换器放入 converters 容器中。

前文提供的测试用例会执行第二种操作，具体处理代码如下。

```
private ConvertersForPair getMatchableConverters(ConvertiblePair convertiblePair) {
    //缓存中获取
    ConvertersForPair convertersForPair = this.converters.get(convertiblePair);
    if (convertersForPair == null) {
        //创建一个空对象
        convertersForPair = new ConvertersForPair();
        this.converters.put(convertiblePair, convertersForPair);
    }
    return convertersForPair;
}
```

在 getMatchableConverters 中的操作就是一个数据缓存的操作，细节不多。在执行完成 getMatchableConverters 之后，Spring 会将对应关系放在 converters 容器中，但是现在 value 还没有数据。value 的设置依赖 convertersForPair.add(converter) 方法，这个方法是一个简单的 add 操作，执行后会向 value-> converters 中加入 StringToProdut 这个转换接口。添加后的数据结果如图 17.4 所示。

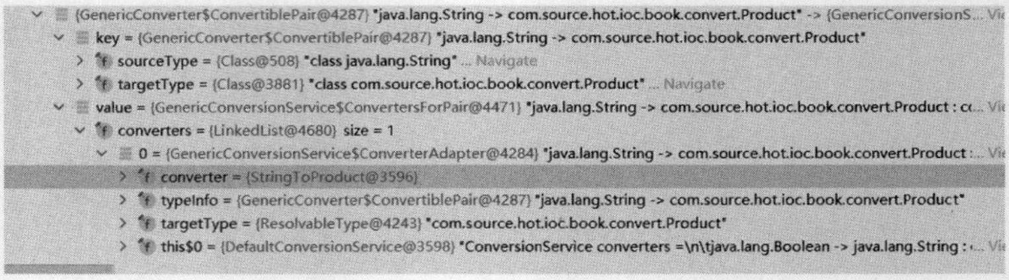

图 17.4 StringToProdut 加入容器后数据信息

从图 17.4 中可以发现，容器中已经放入 StringToProdut 对象就进入到等待执行阶段。在处理完成容器数据后还需要进行缓存处理，缓存处理代码如下。

```
private void invalidateCache() {
    this.converterCache.clear();
}
```

这里所谓缓存处理其本质就是将 map 对象清空，converterCache 的存储结构为 key 存储转换前对象的描述和转换目标的对象描述，value 存储转换服务。

4. 转换器注册方式 4——ConverterFactory 注册

下面将介绍最后一种注册方式——ConverterFactory 注册，具体处理代码如下。

```
@Override
public void addConverterFactory(ConverterFactory<?, ?> factory) {
    //获取类型信息
    ResolvableType[] typeInfo = getRequiredTypeInfo(factory.getClass(), ConverterFactory.class);
    //判断 factory 是不是 DecoratingProxy
    if (typeInfo == null && factory instanceof DecoratingProxy) {
        //其中 DecoratingProxy 可以获取 class
        typeInfo = getRequiredTypeInfo(((DecoratingProxy) factory).getDecoratedClass(), ConverterFactory.class);
    }
    if (typeInfo == null) {
        throw new IllegalArgumentException("Unable to determine source type <S> and target type <T> for your " +
            "ConverterFactory [" + factory.getClass().getName() + "]; does the class parameterize those types?");
    }
    //添加转换器
    addConverter(new ConverterFactoryAdapter(factory,
        new ConvertiblePair(typeInfo[0].toClass(), typeInfo[1].toClass())));
}
```

需要重点关注这段代码的 ConverterFactoryAdapter 对象，该对象有两个成员变量，具体信息见表 17.4。

表 17.4　ConverterFactoryAdapter 成员变量

变量名称	变量类型	变量说明
converterFactory	ConverterFactory<Object,Object>	存储转换器工厂
typeInfo	ConvertiblePair	存储源数据类型和目标类型的对象

注册的具体实现和第三种注册方式一模一样，下面将对 NumberToNumberConverterFactory 转换器的注册做一些调试来对该方法做一个简单理解。首先需要关注的变量是 ConverterFactoryAdapter，具体信息如图 17.5 所示。

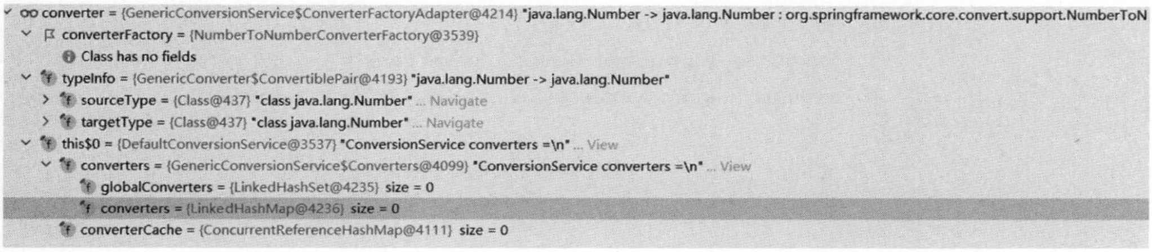

```
∨ oo converter = {GenericConversionService$ConverterFactoryAdapter@4214} "java.lang.Number -> java.lang.Number : org.springframework.core.convert.support.NumberToN
    ∨ ⌦ converterFactory = {NumberToNumberConverterFactory@3539}
        ⓘ Class has no fields
    ∨ ⌦ typeInfo = {GenericConverter$ConvertiblePair@4193} "java.lang.Number -> java.lang.Number"
        > ⌦ sourceType = {Class@437} "class java.lang.Number" ... Navigate
        > ⌦ targetType = {Class@437} "class java.lang.Number" ... Navigate
    ∨ ⌦ this$0 = {DefaultConversionService@3537} "ConversionService converters =\n" ... View
        ∨ ⌦ converters = {GenericConversionService$Converters@4099} "ConversionService converters =\n" ... View
            > ⌦ globalConverters = {LinkedHashSet@4235} size = 0
            > ⌦ converters = {LinkedHashMap@4236} size = 0
            > ⌦ converterCache = {ConcurrentReferenceHashMap@4111} size = 0
```

图 17.5　ConverterFactoryAdapter 数据信息

在得到 ConverterFactoryAdapter 数据对象后进行注册，注册到 converters 容器中，此时的数据情况如图 17.6 所示。

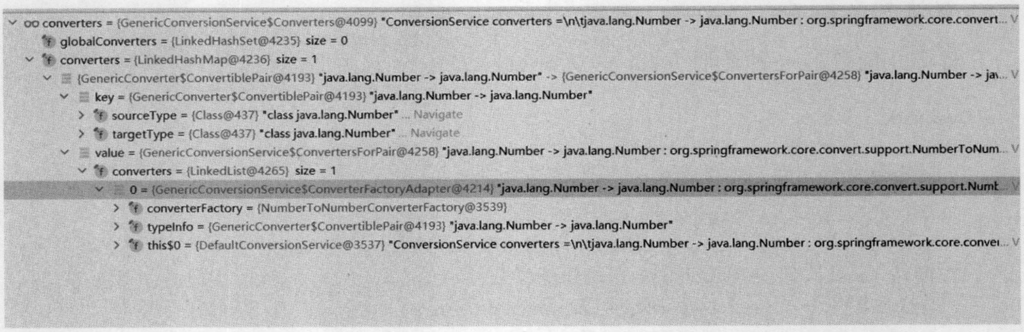

图 17.6　ConverterFactoryAdapter 添加到容器后 converters 的数据情况

在图 17.6 中可以看到已经注册成功，当各类转换器相关的内容都注册到 converters 容器后即可完成 createConversionService 方法的调用。

17.2.4　ConversionServiceFactory.registerConverters 分析

接下来回到 ConversionServiceFactoryBean 类中的 afterPropertiesSet 方法，下面将对 ConversionServiceFactory.registerConverters 方法进行分析，该方法是将非 Spring 原生提供的转换服务进行注册，例如，在测试用例中所编写的 com.source.hot.ioc.book.convert.StringToProduct。在配置文件中有下面这段配置。

```
<bean
    id="conversionService" class="org.springframework.context.support.ConversionServiceFactoryBean">
    <property name="converters">
        <list>
            <bean class="com.source.hot.ioc.book.convert.StringToProduct"/>
        </list>
    </property>
</bean>
```

在这段配置文件中所设置的 converters 属性会进行注册，有关注册的代码如下。

```
public static void registerConverters(@Nullable Set<?> converters, ConverterRegistry registry) {
    if (converters != null) {
        for (Object converter : converters) {
            if (converter instanceof GenericConverter) {
                registry.addConverter((GenericConverter) converter);
            }
            else if (converter instanceof Converter<?, ?>) {
                registry.addConverter((Converter<?, ?>) converter);
            }
            else if (converter instanceof ConverterFactory<?, ?>) {
                registry.addConverterFactory((ConverterFactory<?, ?>) converter);
```

```
            }
            else {
                throw new IllegalArgumentException ( " Each converter object must
implement one of the " + "Converter, ConverterFactory, or GenericConverter interfaces");
            }
        }
    }
}
```

在这段代码中首先需要理解两个参数,第一个参数是 converters,该参数通过外部配置进行设置,可以填写 GenericConverter、Converter 和 ConverterFactory 的实现类;第二个参数是 ConverterRegistry,该参数负责注册。在配置文件中配置的 com.source.hot.ioc.book.convert.StringToProduct 数据会被映射到第一个参数中,然后进行数据注册。有关 converters 的数据如图 17.7 所示。

图 17.7　配置文件中 converters 转换后结果

在得到图 17.7 中的数据后就会进行注册,注册的相关内容可以查看前文介绍。

17.3　转换过程分析

接下来将进入到转换过程的分析,在测试用例中转换服务使用了下面的代码。

```
ConversionService bean = context.getBean(ConversionService.class);
Product convert = bean.convert("product;10.00", Product.class);
```

在这个代码中需要关注的重点就是 ConversionService,从代码上可以看到它是转换的入口。

17.3.1　ConversionService 分析

首先需要认识 ConversionService 对象,在 SpringXML 配置阶段并没有对 ConversionService 对象进行配置,配置的内容是 ConversionServiceFactoryBean 类,进入 ConversionServiceFactoryBean 类找到 FactoryBean 相关的实现方法,具体如下。

```
@Override
@Nullable
public ConversionService getObject() {
    return this.conversionService;
}
```

```java
@Override
public Class<? extends ConversionService> getObjectType() {
    return GenericConversionService.class;
}

@Override
public boolean isSingleton() {
    return true;
}
```

从这段代码中可以看到 GenericConversionService 是作为 getObjectType 的返回值，在 Spring 中获取一个对象通过 FactoryBean 获取依靠的方法是 getObject，可以发现代码中对于 getObject 的返回值是 conversionService 对象，因此开发者可以通过 context.getBean(ConversionService.class) 操作进而得到 conversionService 对象。这个对象是通过 afterPropertiesSet 方法进行创建的。

在了解 ConversionService 对象后就可以进行最终的转换方法分析，ConversionService 对象的实际类型是 GenericConversionService，在它的方法列表中提供了转换方法，具体处理代码如下。

```java
@Override
@Nullable
public Object convert(@Nullable Object source, @Nullable TypeDescriptor sourceType, TypeDescriptor targetType) {
    Assert.notNull(targetType, "Target type to convert to cannot be null");
    if (sourceType == null) {
        Assert.isTrue(source == null, "Source must be [null] if source type == [null]");
        //处理 sourceType 为空的转换
        return handleResult(null, targetType, convertNullSource(null, targetType));
    }
    //数据验证
    if (source != null && !sourceType.getObjectType().isInstance(source)) {
        throw new IllegalArgumentException("Source to convert from must be an instance of [" +
            sourceType + "]; instead it was a [" + source.getClass().getName() + "]");
    }
    //获取转换器接口
    GenericConverter converter = getConverter(sourceType, targetType);
    if (converter != null) {
        //通过工具获得转换结果
        Object result = ConversionUtils.invokeConverter(converter, source, sourceType, targetType);
        return handleResult(sourceType, targetType, result);
    }
    //处理找不到 convert 的转换结果
    return handleConverterNotFound(source, sourceType, targetType);
}
```

在这个处理方法中涵盖下面四种处理形式。

（1）原类型描述不存在。

（2）原对象存在但是原类型和原类型的描述对象中的类型不同源。
（3）在转换器容器中存在对应的转换器。
（4）在转换器容器中不存在对应的转换器。
为了处理这四种形式 Spring 提供了一些处理方法，下面对这些方法进行分析。

17.3.2　handleResult 分析

下面对 handleResult 方法进行分析，该方法是用来处理返回结果的方法，具体代码如下。

```java
@Nullable
private Object handleResult(@Nullable
TypeDescriptor sourceType, TypeDescriptor targetType, @Nullable Object result) {
    if (result == null) {
        //判断 target type
        assertNotPrimitiveTargetType(sourceType, targetType);
    }
    return result;
}

private void assertNotPrimitiveTargetType(@Nullable
TypeDescriptor sourceType, TypeDescriptor targetType) {
    if (targetType.isPrimitive()) {
        throw new ConversionFailedException(sourceType, targetType, null,
            new IllegalArgumentException("A null value cannot be assigned to a primitive type"));
    }
}
```

注意，在 handleResult 方法处理过程中可能返回 null。

这段代码中主要逻辑是关于原始类型和目标类型的验证，验证方式是判断 targetType.isPrimitive 是否为 true，如果为 true 就会抛出异常，否则将对象原封不动返回。

17.3.3　getConverter 分析

接下来对 getConverter 方法进行分析，在这个方法中会得到一个 Converter 接口，具体处理代码如下。

```java
@Nullable
protected GenericConverter getConverter(TypeDescriptor sourceType, TypeDescriptor targetType) {
    ConverterCacheKey key = new ConverterCacheKey(sourceType, targetType);
    GenericConverter converter = this.converterCache.get(key);
    if (converter != null) {
        return (converter != NO_MATCH ? converter : null);
    }
```

```java
        //找出 converter 对象
        converter = this.converters.find(sourceType, targetType);
        if (converter == null) {
            //获取默认的 converter
            converter = getDefaultConverter(sourceType, targetType);
        }

        if (converter != null) {
            //设置缓存
            this.converterCache.put(key, converter);
            return converter;
        }
        //设置缓存
        this.converterCache.put(key, NO_MATCH);
        return null;
    }
```

从这段代码中可以看到它对使用过的内容进行了缓存,抛开缓存层面的处理更为重要的是 find 方法,在 find 方法中才是真正意义上的核心,它能够根据传入的原始类型和目标类型找到转换接口,具体处理代码如下。

```java
@Nullable
public GenericConverter find(TypeDescriptor sourceType, TypeDescriptor targetType) {
    //找到 source 类型的类关系和接口关系
    List<Class<?>> sourceCandidates = getClassHierarchy(sourceType.getType());
    //找到 target 类型的类关系和接口关系
    List<Class<?>> targetCandidates = getClassHierarchy(targetType.getType());
    //循环 source 的类列表和 target 的类列表
    for (Class<?> sourceCandidate : sourceCandidates) {
        for (Class<?> targetCandidate : targetCandidates) {
            //创建 ConvertiblePair 对象
            ConvertiblePair convertiblePair = new ConvertiblePair(sourceCandidate, targetCandidate);
            //获取 source + target 的转换接口
            GenericConverter converter = getRegisteredConverter(sourceType, targetType, convertiblePair);
            if (converter != null) {
                return converter;
            }
        }
    }
    return null;
}
```

在这个方法中首先需要理解 getClassHierarchy 方法的作用,它能够将一个类上的所有类接口都提取出来包括父类。以 String 类举例,它会找到 String、Comparable、Serializable、CharSequence 和 Object,注意 Object 是隐式的,不管是什么类都会被加入这个信息。第二个需要关注的是 getClassHierarchy 方法返回值的顺序,它的顺序是自己本身→父类→接口→Object。

在得到 source 和 target 的类关系后会做一个笛卡儿积的搜索。将 source 中的类和 target 循环一个个组装成 ConvertiblePair，然后在容器中寻找 GenericConverter 对象。如果找到了就直接返回，如果找不到就返回 null。在进行笛卡儿积结果搜索时会使用到 getRegisteredConverter 方法，确认具体的转换接口，相关代码如下。

```java
@Nullable
private GenericConverter getRegisteredConverter(TypeDescriptor sourceType,
        TypeDescriptor targetType, ConvertiblePair convertiblePair) {

    //从 map 中获取
    ConvertersForPair convertersForPair = this.converters.get(convertiblePair);
    if (convertersForPair != null) {
        //获取 GenericConverter
        GenericConverter converter =
convertersForPair.getConverter(sourceType, targetType);
        if (converter != null) {
            return converter;
        }
    }
    for (GenericConverter globalConverter : this.globalConverters) {
        if (((ConditionalConverter) globalConverter).matches(sourceType, targetType)) {
            return globalConverter;
        }
    }
    return null;
}
```

在 getRegisteredConverter 方法之中需要重点关注它的搜索顺序。首先进行 converter 的搜索，其次进行全局转换器（globalConverter）的搜索。

通过这样一个过程 find 方法有可能找到了转换接口，也有可能没有找到转换接口，如果找到，则直接返回；如果没有找到，则执行下面这段代码。

```java
if (converter == null) {
    //获取默认的 converter
    converter = getDefaultConverter(sourceType, targetType);
}
```

更深入挖掘 getDefaultConverter，详细代码如下。

```java
protected GenericConverter getDefaultConverter(TypeDescriptor
sourceType, TypeDescriptor targetType) {
    return (sourceType.isAssignableTo(targetType) ? NO_OP_CONVERTER : null);
}
```

在这个找不到的情况下会返回 null 或者没有任何处理的转换接口实现类（转换前后数据对象未做改变），有关 NoOpConverter 的代码可以查看下面的内容。

```java
private static class NoOpConverter implements GenericConverter {

    private final String name;
```

```java
public NoOpConverter(String name) {
    this.name = name;
}

@Override
public Set<ConvertiblePair> getConvertibleTypes() {
    return null;
}

@Override
@Nullable
public Object convert(@Nullable Object source,
        TypeDescriptor sourceType, TypeDescriptor targetType) {
    return source;
}

@Override
public String toString() {
    return this.name;
}
```

17.3.4 ConversionUtils.invokeConverter 分析

接下来将分析 ConversionUtils.invokeConverter 方法,该方法是用来执行转换接口得到最终的转换结果。具体处理代码如下。

```java
@Nullable
public static Object invokeConverter(GenericConverter converter, @Nullable Object source,
        TypeDescriptor sourceType, TypeDescriptor targetType) {

    try {
        //converter 方法调用
        return converter.convert(source, sourceType, targetType);
    }
    catch (ConversionFailedException ex) {
        throw ex;
    }
    catch (Throwable ex) {
        throw new ConversionFailedException(sourceType, targetType, source, ex);
    }
}
```

从这段代码中可以非常直观地看到这里只是一个转换。但是这个转换能力的提供者 GenericConverter 是一个接口,此时还需要明确这个接口在实际运算中的类型。在本章的前面提到过 ConverterAdapter 和 ConverterFactoryAdapter,在这两个类中有具体的转换处理。

1. ConverterAdapter 的转换

下面是 ConverterAdapter 中关于转换处理的代码。

```
@Override
@Nullable
public Object convert(@Nullable Object source, TypeDescriptor sourceType, TypeDescriptor targetType) {
    if (source == null) {
        //空对象转换
        return convertNullSource(sourceType, targetType);
    }
    //转换接口调用
    return this.converter.convert(source);
}
```

在这段代码中可以看到两种处理情况,第一种是需要转换的对象不存在,当需要转换的对象不存在时会做一次判断是不是 Optional 类型,如果是则返回 Optional.empty(),不是的话直接返回 null,具体实现代码如下。

```
@Nullable
protected Object convertNullSource(@Nullable TypeDescriptor sourceType, TypeDescriptor targetType) {
    if (targetType.getObjectType() == Optional.class) {
        return Optional.empty();
    }
    return null;
}
```

第二种是需要转换的对象存在,执行转换接口的转换方法,具体处理代码如下。

```
this.converter.convert(source)
```

2. ConverterFactoryAdapter 的转换

下面将介绍 ConverterFactoryAdapter 中的转换过程,具体代码如下。

```
@Override
@Nullable
public Object convert(@Nullable Object source, TypeDescriptor sourceType, TypeDescriptor targetType) {
    if (source == null) {
        return convertNullSource(sourceType, targetType);
    }
    //从工厂中获取一个 convert 接口进行转换
    return this.converterFactory.getConverter(targetType.getObjectType()).convert(source);
}
```

在这个转换过程中有以下两种处理模式。

(1) 需要转换的对象不存在,当需要转换的对象不存在时会做一次判断是不是 Optional 类型,如果是则返回 Optional.empty(),不是的话直接返回 null。

(2) 需要转换的对象存在,从 converterFactory 中找到转换接口再进行转换,假设现在正在进行 NumberToNumberConverterFactory 的转换，在这个方法中它的调用流程如下：从 NumberToNumberConverterFactory 中调用 getConverter 方法找到 Converter 接口，再进行 convert 方法的调用。

17.3.5 handleConverterNotFound 分析

下面将介绍 handleConverterNotFound 方法，该方法是负责处理 GenericConverter 对象找不到的情况，具体代码如下。

```java
private Object handleConverterNotFound(
        @Nullable Object source, @Nullable TypeDescriptor sourceType, TypeDescriptor targetType) {

    if (source == null) {
        assertNotPrimitiveTargetType(sourceType, targetType);
        return null;
    }
    if ((sourceType == null || sourceType.isAssignableTo(targetType)) &&
            targetType.getObjectType().isInstance(source)) {
        return source;
    }
    throw new ConverterNotFoundException(sourceType, targetType);
}
```

在这个方法中处理了以下三种情况。

（1）souce 对象不存在，通过 targetType.isPrimitive 验证返回 null，没有通过 targetType.isPrimitive 抛出异常 ConversionFailedException。

（2）source 对象不存在或者原类型和目标类型可以转换就返回 source 本身。

（3）抛出转换异常 ConverterNotFoundException。

17.4 脱离 Spring 实现转换服务

前文对于 Spring 中关于转换服务的内容做了比较充分的分析，接下来将要脱离 Spring 来实现转换服务。通过这个简单的实现过程来理解 Spring 中对于转换服务做的细节工作。

首先需要编写一个接口，该接口是转换的核心，接口名为 CommonConvert，具体代码如下。

```java
public interface CommonConvert<S, T> {
    /**
     * 转换
     * @param source 原始对象
     * @return 转换结果对象
     */
    T convert(S source);
}
```

在完成接口定义后需要定义一个存储容器将所有的转换接口进行统一管理，这里可以借鉴 Spring 做一个 Map 对象将所有数据收集起来，具体定义如下。

```
static Map<ConvertSourceAndTarget, CommonConvert> convertMap
    = new ConcurrentHashMap<>(256);
```

在这个容器中 key 是 ConvertSourceAndTarget，它的作用是存储 ConvertSourceAndTarget 的泛型信息，具体代码如下。

```
public class ConvertSourceAndTarget {

    /**
     * {@link CommonConvert} 中的 S
     */
    private Class<?> sourceTypeClass;

    /**
     * {@link CommonConvert} 中的 T
     */
    private Class<?> targetTypeClass;
}
```

容器准备完成，下面就需要定义注册方法，可以定义以下两种注册方法。

（1）通过实例进行注册，void register(CommonConvert<?,?> commonConvert)。

（2）通过 class 进行注册，void register(Class<? extends CommonConvert> convert) throws IllegalAccessException，InstantiationException。

这两种注册方式的具体实现代码如下。

```
@Override
public void register(Class<? extends CommonConvert> convert)
throws IllegalAccessException, InstantiationException {
    if (convert == null) {
        log.warn("当前传入的 convert 对象为空");
        return;
    }
    CommonConvert cv = convert.newInstance();

    if (cv != null) {
        handler(cv, convert);
    }

}

@Override
public void register(CommonConvert commonConvert ) {

    if (commonConvert == null) {
        log.warn("当前传入的 convert 对象为空");
```

```java
        return;
    }


    Class<? extends CommonConvert> convertClass = commonConvert.getClass();

    handler(commonConvert, convertClass);

}
```

这两种实现方式中需要使用到 handler 方法来做核心处理,它的代码如下。

```java
private void handler(CommonConvert commonConvert, Class<? extends CommonConvert> convertClass) {
    Type[] genericInterfaces = convertClass.getGenericInterfaces();

    for (Type genericInterface : genericInterfaces) {
        ParameterizedType pType = (ParameterizedType) genericInterface;
        boolean equals = pType.getRawType().equals(CommonConvert.class);
        if (equals) {
            Type[] actualTypeArguments = pType.getActualTypeArguments();

            if (actualTypeArguments.length == 2) {
                Type a1 = actualTypeArguments[0];
                Type a2 = actualTypeArguments[1];

                try {

                    Class<?> sourceClass = Class.forName(a1.getTypeName());
                    Class<?> targetClass = Class.forName(a2.getTypeName());

                    ConvertSourceAndTarget convertSourceAndTarget =
                        new ConvertSourceAndTarget(sourceClass,
                                                    targetClass);
                    //如果类型相同,覆盖
                    convertMap.put(convertSourceAndTarget, commonConvert);
                }
                catch (Exception e) {
                    log.error("a1=[{}]", a1);
                    log.error("a2=[{}]", a2);
                    log.error("从泛型中转换成 class 异常", e);
                }
            }
        }
    }
}
```

在这个方法中传递两个参数,第一个参数是 commonConvert,它表示转换器实例;第二个参数是 convertClass，它表示转换器类型。该方法的主要处理逻辑是提取转换器上的泛型将其转换成容器中的 KEY，再将转换器实例作为 VALUE 进行存储。

现在存储容器已经准备完成，注册方法也提供完成，还缺少一个关于统一转换的入口，下面定义一个统一转换入口（转换方法），具体转换方法代码如下。

```java
public static <T> T convert(Object source, Class<T> target) {
    if (log.isInfoEnabled()) {
        log.info("convert,source = {}, target = {}", source, target);
    }

    if (source == null || target == null) {
        throw new IllegalArgumentException("参数异常请重新检查");
    }


    ConvertSourceAndTarget convertSourceAndTarget =
 new ConvertSourceAndTarget(source.getClass(), target);

    CommonConvert commonConvert = DefaultConvertRegister.getConvertMap(convertSourceAndTarget);
    if (commonConvert != null) {

        return (T) commonConvert.convert(source);
    }
    return null;
}
```

在转换的时候需要从转换器容器中获取对应的转换器再进行转换。处理方式就是将需要转换的对象获取 class 和转换目标类型组装成 ConvertSourceAndTarget，拿着这个对象去转换器容器中获取转换器，再进行转换即可。

通过这些代码就完成了一个简易的转换服务，和 Spring 对比，这个转换服务提供的注册方式比该转换服务的注册模式单一。对比 Spring 中对于转换接口的泛型处理，本例所选择的方式比较简单，没有考虑其他细节内容，同时在依赖泛型的处理中没有考虑类的关系图（这个方案只适合没有继承的对象转换和没有实现类的对象转换）。这里所提出的转换方式是对 Spring 中转换方式的一个简单实现，读者还可以对其进行更多的加工处理来满足各种场景的需求。

小结

本章对 Spring 中的转换服务做了使用说明和源码分析，使用层面从一个转换服务开始到配置文件再到容器的使用给出了具有详细信息的例子，同时从这个例子出发围绕转换服务这一点进行了源码分析。在源码分析中分析了转换器注册、转换器搜索和转换器使用三个核心。在理解 Spring 的转换服务实现后，尝试实现了一个简单的转换服务，以此来加深读者对转换服务的理解。

第18章

MessageSource源码分析

本章将对 Spring IoC 辅助工具中的国际化相关内容的处理进行分析,在本章中将会涉及 MessageSource 测试环境搭建和 Spring 中 MessageSource 处理流程分析。

18.1 MessageSource 测试环境搭建

本节将搭建一个国际化相关的测试用例,编写这个测试用例分为下面三个步骤:①编写一个或多个数据源文件;②编写 SpringXML 配置文件;③使用相关内容。

首先编写两个数据源文件,在 Spring 中使用"前缀_[语言代码]_[国家/地区代码].properties"方式对数据源文件进行命名,这类文件需要放在 resources 文件夹下。创建第一个数据源文件,文件名为 messages_en_US.properties,文件内容如下。

```
home = Home
format_data = {0}.abc
```

创建第二个数据源文件,文件名为 messages_zh_CN.properties。如果需要使用中文作为对应值,需要使用中文转换 Unicode 的工具,如果直接使用中文填写在数据源文件中会出现乱码情况。

接下来编写 SpringXML 配置文件,文件名为 message-source.xml,具体内容如下。

```xml
<?xml version = "1.0" encoding = "UTF-8"?>
< beans xmlns = "http://www.springframework.org/schema/beans"
       xmlns:xsi = "http://www.w3.org/2001/XMLSchema-instance"
       xsi:schemaLocation = "http://www.springframework.org/schema/beans http://www.springframework.org/schema/beans/spring-beans.xsd">

    < bean id = "messageSource" class = "org.springframework.context.support.ResourceBundleMessageSource">
        < property name = "basename"
```

```xml
            <value>messages</value>
        </property>
    </bean>
</beans>
```

完成 SpringXML 编写后编写具体的测试用例,具体代码如下。

```java
@Test
void testXml() {
    ClassPathXmlApplicationContext context =
    new ClassPathXmlApplicationContext("META-INF/message-source.xml");

    String usHome = context.getMessage("home", null, Locale.US);
    assert usHome.equals("Home");
    String zhHome = context.getMessage("home", null, Locale.CHINESE);
    assert zhHome.equals("jia");

    String format_data =
    context.getMessage("format_data", new Object[] {"abc"}, Locale.US);
    System.out.println(format_data);

}
```

18.2　MessageSource 实例化

接下来将介绍 MessageSource 对象的实例化,该方法的作用就是用来完成 MessageSource 对象的实例化。具体处理代码如下。

```java
protected void initMessageSource() {
    //获取 BeanFactory
    ConfigurableListableBeanFactory beanFactory = getBeanFactory();
    //判断容器中是否存在 messageSource 这个 BeanName
    //存在的情况
    if (beanFactory.containsLocalBean(MESSAGE_SOURCE_BEAN_NAME)) {
        //获取 messageSource 对象
        this.messageSource = beanFactory.getBean(MESSAGE_SOURCE_BEAN_NAME, MessageSource.class);

        //设置父 MessageSource
        if (this.parent != null &&
        this.messageSource instanceof HierarchicalMessageSource) {
            HierarchicalMessageSource hms =
            (HierarchicalMessageSource) this.messageSource;
            if (hms.getParentMessageSource() == null) {
                hms.setParentMessageSource(getInternalParentMessageSource());
            }
        }
    }
    //不存在的情况
```

```
        else {
            //MessageSource 实现类
            DelegatingMessageSource dms = new DelegatingMessageSource();
            //设置父 MessageSource
            dms.setParentMessageSource(getInternalParentMessageSource());
            this.messageSource = dms;
            //注册 MessageSource
            beanFactory.registerSingleton(MESSAGE_SOURCE_BEAN_NAME, this.messageSource);
        }
    }
```

在这段代码中可以确认一个关键字 MESSAGE_SOURCE_BEAN_NAME,它的数据是 messageSource,这个数据和测试用例中在 SpringXML 中配置的 bean id 一致,通过这样的配置就会进入 if 而不是进入 else 了。关于 getBean 方法的分析在本章不再赘述,可以阅读第 9 章,当 getBean 方法调用后数据如图 18.1 所示。

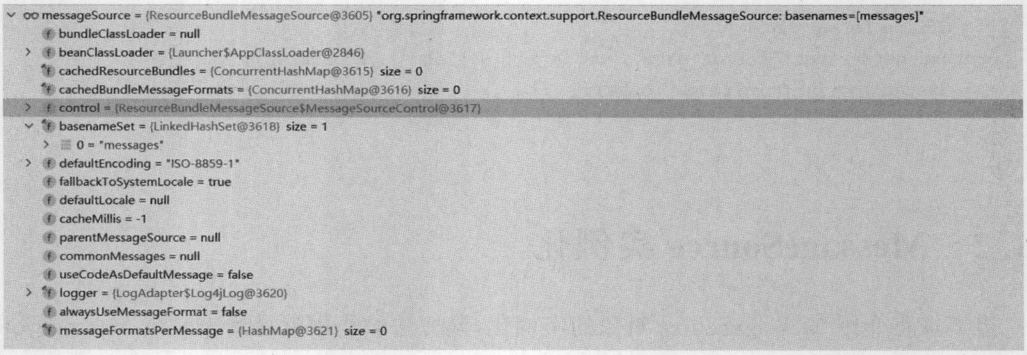

```
∨ oo messageSource = {ResourceBundleMessageSource@3605} "org.springframework.context.support.ResourceBundleMessageSource: basenames=[messages]"
    f bundleClassLoader = null
  > f beanClassLoader = {Launcher$AppClassLoader@2846}
    f cachedResourceBundles = {ConcurrentHashMap@3615} size = 0
    f cachedBundleMessageFormats = {ConcurrentHashMap@3616} size = 0
    f control = {ResourceBundleMessageSource$MessageSourceControl@3617}
  ∨ f basenameSet = {LinkedHashSet@3618} size = 1
    >   0 = "messages"
  > f defaultEncoding = "ISO-8859-1"
    f fallbackToSystemLocale = true
    f defaultLocale = null
    f cacheMillis = -1
    f parentMessageSource = null
    f commonMessages = null
    f useCodeAsDefaultMessage = false
  > f logger = {LogAdapter$Log4jLog@3620}
    f alwaysUseMessageFormat = false
    f messageFormatsPerMessage = {HashMap@3621} size = 0
```

图 18.1 messageSource 数据信息

在图 18.1 中需要关注的信息是 basenameSet,在 basenameSet 中存储的内容是在 SpringXML 中配置的 basename 属性。这个属性会为后续的读取配置文件提供帮助。

18.3 getMessage 方法分析

接下来将 getMessage 方法进行分析,该方法的方法签名是 org.springframework.context.support.AbstractMessageSource#getMessage(java.lang.String,java.lang.Object[],java.util.Locale),具体处理代码如下。

```
@Override
public final String getMessage(String code, @Nullable Object[] args, Locale locale) throws
NoSuchMessageException {
    String msg = getMessageInternal(code, args, locale);
    if (msg != null) {
        return msg;
    }
    String fallback = getDefaultMessage(code);
    if (fallback != null) {
```

```
        return fallback;
    }
    throw new NoSuchMessageException(code, locale);
}
```

在这段代码中需要关注以下三点。

（1）getMessageInternal 方法：该方法可以返回资源文件中的数据。
（2）getDefaultMessage 方法：该方法用来返回默认的消息信息。
（3）处理失败的异常信息处理。

接下来对 getMessageInternal 方法进行分析，下面是处理代码。

```
@Nullable
protected String getMessageInternal(@Nullable String code, @Nullable Object[] args,
        @Nullable Locale locale) {
    //code 不存在,返回空
    if (code == null) {
        return null;
    }
    //locale 不存在,设置默认
    if (locale == null) {
        locale = Locale.getDefault();
    }
    //需要替换的真实数据
    Object[] argsToUse = args;

    //是否需要进行消息解析,真实数据是否存在
    if (!isAlwaysUseMessageFormat() && ObjectUtils.isEmpty(args)) {

        //解析消息
        String message = resolveCodeWithoutArguments(code, locale);
        if (message != null) {
            return message;
        }
    }

    else {
        argsToUse = resolveArguments(args, locale);

        MessageFormat messageFormat = resolveCode(code, locale);
        if (messageFormat != null) {
            synchronized (messageFormat) {
                return messageFormat.format(argsToUse);
            }
        }
    }

    Properties commonMessages = getCommonMessages();
    if (commonMessages != null) {
        String commonMessage = commonMessages.getProperty(code);
        if (commonMessage != null) {
```

```
            return formatMessage(commonMessage, args, locale);
    }
}

return getMessageFromParent(code, argsToUse, locale);
}
```

在这段代码中有以下四种处理方式。

(1) code 不存在的处理。
(2) 不需要进行消息解析并且消息体为空。
(3) 需要进行消息解析并且消息体不为空。
(4) 其他情况。

在第一种情况处理中会直接返回 null，没有什么其他处理逻辑，主要关注第二种和第三种处理。

18.3.1　resolveCodeWithoutArguments 方法分析

下面介绍第二种处理方式。第二种处理方式所用到的方法是 resolveCodeWithoutArguments。方法 resolveCodeWithoutArguments 存在多个实现类，这里着重关注在 SpringXML 中配置的 ResourceBundleMessageSource 类，ResourceBundleMessageSource 类中的实现代码如下。

```
@Override
protected String resolveCodeWithoutArguments(String code, Locale locale) {
    Set<String> basenames = getBasenameSet();
    for (String basename : basenames) {
        ResourceBundle bundle = getResourceBundle(basename, locale);
        if (bundle != null) {
            String result = getStringOrNull(bundle, code);
            if (result != null) {
                return result;
            }
        }
    }
    return null;
}
```

在上述代码中有一个比较熟悉的方法 getBasenameSet，该方法将返回 SpringXML 配置中的数据，在测试用例中所配置的信息是 messages，通过 getResourceBundle(basename, locale) 方法调用可以得到 ResourceBundle 对象。下面是 getResourceBundle 方法的详细内容。

```
@Nullable
protected ResourceBundle getResourceBundle(String basename, Locale locale) {
    if (getCacheMillis() >= 0) {
        return doGetBundle(basename, locale);
    }
    else {
```

```java
        Map<Locale, ResourceBundle> localeMap = this.cachedResourceBundles.get(basename);
        if (localeMap != null) {
            ResourceBundle bundle = localeMap.get(locale);
            if (bundle != null) {
                return bundle;
            }
        }
        try {
            ResourceBundle bundle = doGetBundle(basename, locale);
            if (localeMap == null) {
                localeMap = new ConcurrentHashMap<>();
                Map<Locale, ResourceBundle> existing = this.cachedResourceBundles.putIfAbsent(basename, localeMap);
                if (existing != null) {
                    localeMap = existing;
                }
            }
            localeMap.put(locale, bundle);
            return bundle;
        }
        catch (MissingResourceException ex) {
            if (logger.isWarnEnabled()) {
                logger.warn("ResourceBundle [" + basename + "] not found for MessageSource: " + ex.getMessage());
            }
            return null;
        }
    }
}
```

在这段代码中对于 ResourceBundle 对象的获取有以下两种方式。

(1) 从缓存容器中获取,容器存储结构如下。

`Map<String, Map<Locale, ResourceBundle>> cachedResourceBundles`

(2) 通过 doGetBundle 方法进行获取。在 doGetBundle 方法中最终会调用 JDK 提供的 ResourceBundle.getBundle 方法,JDK 中的代码不做分析。抛开 JDK 的细节可以简单思考成这样一种方式:通过字符串拼接组成(组成规则:basename + lang + country + .properties),接着读取该文件的信息。

通常情况下会进入第二种。第一种情况需要有进行缓存设置才可以获取,第二种方法得到的 bundle 数据如图 18.2 所示。

图 18.2 bundle 数据信息

得到 bundle 数据对象后就可以进行映射关系转换。在测试用例中,code 传递的实际参数是 home,在资源文件中与 home 对应的数据是 Home,经过 getStringOrNull(bundle,code)方法就可以得到 Home 数据。getStringOrNull 方法的作用是通过 code 在 bundle 对象中找到对应的数据值将其返回,具体对应数据在 bundle 的 lookup 容器中。

18.3.2　resolveCode 方法分析

前文介绍了关于 code 的转换资源文件的过程，在 Spring 中还提供了另外一种方式：code＋占位符的取值方式，相关测试代码如下，后文称为用例 1。

```
String format_data = context.getMessage("format_data", new Object[] {"abc"}, Locale.US);
assert format_data.equals("abc.abc");
```

上述代码的底层核心是由 AbstractMessageSource#getMessageInternal 方法提供的，下面是有关这部分操作的代码。

```
argsToUse = resolveArguments(args, locale);

MessageFormat messageFormat = resolveCode(code, locale);
if (messageFormat != null) {
   synchronized (messageFormat) {
      return messageFormat.format(argsToUse);
   }
}
```

在这段 format 相关的代码中有以下三个处理细节。
（1）将参数 args 进行转换。
（2）通过 code 和 locala 在消息文件中将对应的内容转换成 MessageFormat。
（3）MessageFormat 和转换后的 args 解析得到最终数据。
接下来对第一个操作的处理方法 resolveArguments 进行分析，处理代码如下。

```
@Override
protected Object[] resolveArguments(@Nullable Object[] args, Locale locale) {
   if (ObjectUtils.isEmpty(args)) {
      return super.resolveArguments(args, locale);
   }
   List<Object> resolvedArgs = new ArrayList<>(args.length);
   for (Object arg : args) {
      if (arg instanceof MessageSourceResolvable) {
         resolvedArgs.add(getMessage((MessageSourceResolvable) arg, locale));
      }
      else {
         resolvedArgs.add(arg);
      }
   }
   return resolvedArgs.toArray();
}
```

如果需要进入这部分代码的调试需要编写新的测试用例，测试用例代码如下，后文称为用例 2。

```
String format_data2 =
context.getMessage("format_data", new Object[] {new MessageSourceResolvable() {
```

```
    @Override
    public String[] getCodes() {
        return new String[] {"home"};
    }
}}, Locale.US);
```

在这个测试用例中自定义了 MessageSourceResolvable 接口实现，对于 MessageSourceResolvable 的处理方式就是将每个 code 在资源文件中寻找对应值，如果不是 MessageSourceResolvable 类型则忽略处理。

用例 1 中 args 和 argsToUse 数据信息如图 18.3 所示。

用例 2 中 args 和 argsToUse 数据信息如图 18.4 所示。

图 18.3　用例 1 数据信息

图 18.4　用例 2 数据信息

第二步和第三步中的处理方法在 Spring 中做了一些接口抽象，抛开这些抽象内容可以将其理解为下面这段代码。

```
String messageFormat = MessageFormat.format("{0}.abc", "aaa");
```

这段代码使用 JDK 中 MessageFormat 类完成，具体实现方法如下。

```
public static String format(String pattern, Object ... arguments) {
    MessageFormat temp = new MessageFormat(pattern);
    return temp.format(arguments);
}
```

在 Java 中的 new MessageFormat(pattern); 对应的就是 resolveCode(code, locale)，整个处理流程如下。

（1）通过 basename + locale 找到对应的 message 文件。

（2）通过 code 在 message 文件中找到对应的字符串。

（3）将字符串创建成 MessageFormat。

（4）使用 format 方法将最终数据返回。

小结

本章对 Spring 中关于国际化的相关内容进行了分析，包含两块内容，第一块是 Spring 国际化使用，第二块是国际化相关源码分析。在 Spring 中负责国际化处理的核心类有两个，一个是 MessageSource，它是消息源接口，用来获取消息；另一个是 MessageSourceResolvable，它是消息源解析接口。此外，还对消息的两种处理模式进行了分析，第一种模式是直接通过 code 获取消息文件中对应的数据；第二种模式是通过 code 加替换参数获取消息文件中对应的数据并根据占位符进行替换。

第19章

资源解析器

本章将对 Spring IoC 辅助工具中的资源解析器进行分析,具体涉及的对象是 ResourcePatternResolver。在 SpringXML 模式下资源解析器通常用来解析 SpringXML 配置文件,在本章将会对资源解析器的处理流程进行分析。

19.1 资源解析器测试环境搭建

在 org.springframework.context.annotation.ClassPathScanningCandidateComponentProvider #scanCandidateComponents 方法中可以看到下面这段代码。

```
String packageSearchPath = ResourcePatternResolver.CLASSPATH_ALL_URL_PREFIX +
    resolveBasePackage(basePackage) + '/' + this.resourcePattern;
Resource[] resources = getResourcePatternResolver().getResources(packageSearchPath);
```

根据这段代码可以进行一些修改,从而得到下面的测试用例。

```
public class ResourcePatternResolverTest {
    @Test
    void testResourcePatternResolver() throws IOException {
        ResourcePatternResolver resourcePatternResolver =
new PathMatchingResourcePatternResolver();
        String basePackage = "com.source.hot.ioc.book.ioc";
        String packageSearchPath =
ResourcePatternResolver.CLASSPATH_ALL_URL_PREFIX +
            resolveBasePackage(basePackage) + "/" + "**/*.class";
        Resource[] resources = resourcePatternResolver.getResources(packageSearchPath);
        System.out.println();
    }
}
```

```java
@Test
void testGetResource(){
    ResourcePatternResolver resourcePatternResolver =
new PathMatchingResourcePatternResolver();
    String basePackage =
resolveBasePackage("com.source.hot.ioc.book.ioc.ResourcePatternResolverTest.class");
    String packageSearchPath =
ResourcePatternResolver.CLASSPATH_ALL_URL_PREFIX +
            basePackage;
    Resource resource = resourcePatternResolver.getResource(packageSearchPath);
    System.out.println();
}

private String resolveBasePackage(String basePackage){
    return basePackage.replace(".", "/");
}
}
```

19.2　ResourcePatternResolver 类图分析

接下来对 ResourcePatternResolver 类图进行分析，类图信息如图 19.1 所示。

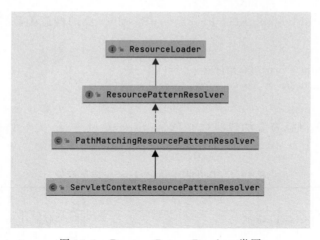

图 19.1　ResourcePatternResolver 类图

Spring 中除了 PathMatchingResourcePatternResolver 和 ServletContextResourcePatternResolver 以外，还有一些 ApplicationContext 也同时继承了 ResourcePatternResolver，这里就不列出来了。在 Spring IoC 中主要使用到的是 PathMatchingResourcePatternResolver。首先查看 ResourceLoader 代码，具体信息如下。

```java
public interface ResourceLoader {

    String CLASSPATH_URL_PREFIX = "classpath:";

    Resource getResource(String location);

    @Nullable
```

```
    ClassLoader getClassLoader();
}
```

ResourceLoader 定义了两个方法：getResource 方法通过文件地址获取对应的资源对象，getClassLoader 方法用来获取类加载器。

继续查看 ResourcePatternResolver 的代码，完整代码如下。

```
public interface ResourcePatternResolver extends ResourceLoader {

    String CLASSPATH_ALL_URL_PREFIX = "classpath*:";

    Resource[] getResources(String locationPattern) throws IOException;

}
```

ResourcePatternResolver 接口提供了可以根据路径通配符的方式获取 Resource 对象的方法。

19.3　PathMatchingResourcePatternResolver 构造器分析

接下来将对 PathMatchingResourcePatternResolver 类的构造函数进行分析，具体处理代码如下。

```
public PathMatchingResourcePatternResolver() {
    this.resourceLoader = new DefaultResourceLoader();
}
```

该构造函数创建默认资源加载器对象 DefaultResourceLoader，关于 DefaultResourceLoader 的成员变量见表 19.1。

表 19.1　DefaultResourceLoader 成员变量

变量名称	变量类型	变量说明
classLoader	ClassLoader	类加载器
protocolResolvers	Set<ProtocolResolver>	协议解析器集合，ProtocolResolver 定义了从字符串到 Resource 的解析函数，ProtocolResolver 是一个接口
resourceCaches	Map<Class<?>,Map<Resource,?>>	资源缓存

在 DefaultResourceLoader 的成员变量表中关键变量是 protocolResolvers，它提供了各类协议的解析处理规则，它生成的数据会放入到 resourceCaches 变量中。

19.4　getResource 方法分析

通过构造器将这些数据进行初始化后即可进行方法调用，接下来将对 getResource 方法进行分析，具体的处理代码如下。

```java
public Resource getResource(String location) {
    return getResourceLoader().getResource(location);
}
```

在这段代码中可以看到它使用的是 ResourceLoader 做具体的解析操作，在此之前通过构造函数了解到 ResourceLoader 具体是 DefaultResourceLoader，DefaultResourceLoader 中的 getResource 方法如下。

```java
@Override
public Resource getResource(String location) {
    Assert.notNull(location, "Location must not be null");

    //获取协议解析器列表,循环
    for (ProtocolResolver protocolResolver : getProtocolResolvers()) {
        Resource resource = protocolResolver.resolve(location, this);
        if (resource != null) {
            return resource;
        }
    }

    //路径地址是以/开头
    if (location.startsWith("/")) {
        return getResourceByPath(location);
    }
    //地址路径是以 classpath: 开头
    else if (location.startsWith(CLASSPATH_URL_PREFIX)) {
        return new ClassPathResource(location.substring(CLASSPATH_URL_PREFIX.length()), getClassLoader());
    }
    else {
        try {
            //尝试将 location 转换成 URL 进行读取
            URL url = new URL(location);
            return (ResourceUtils.isFileURL(url) ? new FileUrlResource(url) : new UrlResource(url));
        }
        catch (MalformedURLException ex) {
            return getResourceByPath(location);
        }
    }
}
```

在这个方法中有以下四种处理方式。

（1）依靠协议解析接口（ProtocolResolver）进行处理。

（2）待解析的资源路径是以"/"开头的处理，创建 ClassPathContextResource 对象返回。

（3）待解析的资源路径是以"classpath:"开头的处理，创建 ClassPathResource 对象返回。

（4）尝试创建 URL 对象。创建成功：如果 URL 对象是文件类型的就返回 FileUrlResource 对象，否则就返回 UrlResource 对象。创建失败：返回 ClassPathContextResource 对象。

带入测试用例来查看这里的处理结果，测试用例中传递的数据是"classpath*:com/source/hot/ioc/book/ioc/ResourcePatternResolverTest/class"，这个参数会直接进入第四种处理模式并且进入失败阶段，因此会创建 ClassPathContextResource 对象，创建结果如图 19.2 所示。

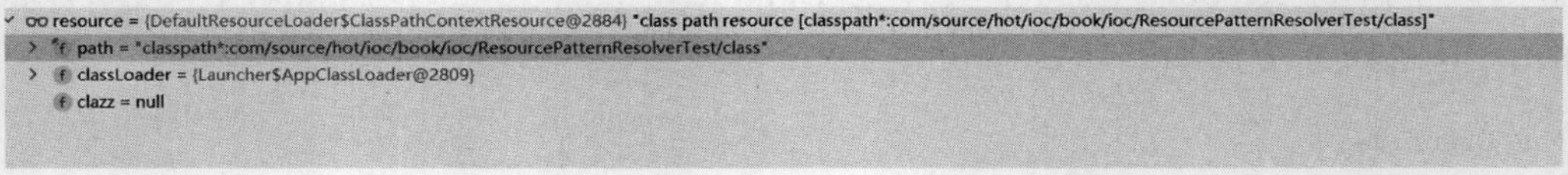

```
oo resource = {DefaultResourceLoader$ClassPathContextResource@2884} "class path resource [classpath*:com/source/hot/ioc/book/ioc/ResourcePatternResolverTest/class]"
  f path = "classpath*:com/source/hot/ioc/book/ioc/ResourcePatternResolverTest/class"
  f classLoader = {Launcher$AppClassLoader@2809}
  f clazz = null
```

图 19.2　测试用例中的 ClassPathContextResource 数据对象

19.5　getResources 方法分析

接下来将分析 getResources 方法，具体处理代码如下。

```
@Override
public Resource[] getResources(String locationPattern) throws IOException {
    Assert.notNull(locationPattern, "Location pattern must not be null");
    //是否以 classpath*: 开头
    if (locationPattern.startsWith(CLASSPATH_ALL_URL_PREFIX)) {
        //正则验证是否通过
        //判断是否包含 * 和 ?
        if (getPathMatcher().isPattern(locationPattern.substring(CLASSPATH_ALL_URL_PREFIX.length()))) {
            //解析 locationPattern 转换成资源接口
            return findPathMatchingResources(locationPattern);
        }
        else {
            //寻找所有资源路径
            return findAllClassPathResources(locationPattern.substring(CLASSPATH_ALL_URL_PREFIX.length()));
        }
    }
    else {
        //前缀寻找模式
        //查询 war*: 三种
        int prefixEnd = (locationPattern.startsWith("war:") ? locationPattern.indexOf("*/") + 1 : locationPattern.indexOf(':') + 1);
        if (getPathMatcher().isPattern(locationPattern.substring(prefixEnd))) {
            //寻找匹配的资源路径
            return findPathMatchingResources(locationPattern);
        }
        else {
            //通过 ResourceLoader 解析得到这资源列表
            return new Resource[] {getResourceLoader().getResource(locationPattern)};
        }
    }
}
```

这段代码的操作流程如图 19.3 所示。

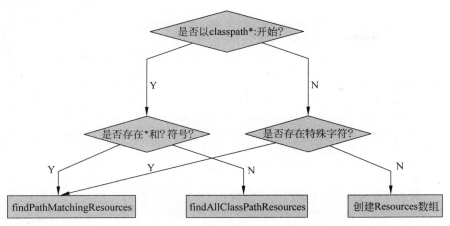

图 19.3 getResources 处理流程

图 19.3 中的特殊字符指代 war:、*/和:三个字符串。在这个处理过程中具体的实现类是 AntPathMatcher。isPattern 的核心代码如下。

```
@Override
public boolean isPattern(@Nullable String path) {
    if (path == null) {
        return false;
    }
    boolean uriVar = false;
    for (int i = 0; i < path.length(); i++) {
        char c = path.charAt(i);
        if (c == '*' || c == '?') {
            return true;
        }
        if (c == '{') {
            uriVar = true;
            continue;
        }
        if (c == '}' && uriVar) {
            return true;
        }
    }
    return false;
}
```

在这段代码中会判断是否存在 *、?、{ 和 },只要存在四个字符中的一个就会返回 true（花括号需要成对出现）。

19.5.1 findPathMatchingResources 方法分析

接下来对 findPathMatchingResources 方法进行分析,该方法的作用是将 locationPattern 转换成多个 Resource 对象,具体处理代码如下。

```java
protected Resource[] findPathMatchingResources(String locationPattern) throws IOException {
    //确定 root 路径地址
    String rootDirPath = determineRootDir(locationPattern);
    //切除 root 地址留下真实地址
    String subPattern = locationPattern.substring(rootDirPath.length());

    //第二部分
    //获取 root 下的资源
    Resource[] rootDirResources = getResources(rootDirPath);
    Set<Resource> result = new LinkedHashSet<>(16);
    //循环处理 root 下的资源对象
    for (Resource rootDirResource : rootDirResources) {
        //解析资源
        //目前而言 Spring 没有做额外处理,属于预留方法
        rootDirResource = resolveRootDirResource(rootDirResource);
        URL rootDirUrl = rootDirResource.getURL();
        //判断协议是不是 bundle
        if (equinoxResolveMethod != null && rootDirUrl.getProtocol().startsWith("bundle")) {
            //equinoxResolveMethod 方法执行
            URL resolvedUrl = (URL) ReflectionUtils.invokeMethod(equinoxResolveMethod, null, rootDirUrl);
            if (resolvedUrl != null) {
                rootDirUrl = resolvedUrl;
            }
            //解成 UrlResource 对象
            rootDirResource = new UrlResource(rootDirUrl);
        }
        //协议是否是 vfs
        if (rootDirUrl.getProtocol().startsWith(ResourceUtils.URL_PROTOCOL_VFS)) {
            //添加 vfs 相关资源
            result.addAll(VfsResourceMatchingDelegate.findMatchingResources(rootDirUrl, subPattern, getPathMatcher()));
        }
        //协议是否是 jar
        else if (ResourceUtils.isJarURL(rootDirUrl) || isJarResource(rootDirResource)) {
            //添加 jar 相关资源
            result.addAll(doFindPathMatchingJarResources(rootDirResource, rootDirUrl, subPattern));
        }
        else {
            //文件协议处理
            result.addAll(doFindPathMatchingFileResources(rootDirResource, subPattern));
        }
    }
    if (logger.isTraceEnabled()) {
        logger.trace("Resolved location pattern [" + locationPattern + "] to resources " + result);
    }
    return result.toArray(new Resource[0]);
}
```

在这段代码中首先需要关注 determineRootDir 方法,该方法会将地址正确解析,解析过程:切除 root 地址留下真实地址。真实地址可以是一个需要进行匹配的字符串,也可以是一个全路径地址,具体处理代码如下。

```java
protected String determineRootDir(String location) {
    int prefixEnd = location.indexOf(':') + 1;
    int rootDirEnd = location.length();
    while (rootDirEnd > prefixEnd
            && getPathMatcher().isPattern(location.substring(prefixEnd, rootDirEnd))) {
        rootDirEnd = location.lastIndexOf('/', rootDirEnd - 2) + 1;
    }
    if (rootDirEnd == 0) {
        rootDirEnd = prefixEnd;
    }
    return location.substring(0, rootDirEnd);
}
```

该方法的主要操作就是进行字符串切割,将/符号作为切割符进行切割,直到下一个符号是"*"或者"?"就结束切割并将其返回。以测试用例为例,locationPattern 数据是"classpath*:com/source/hot/ioc/book/ioc/**/*.class",rootDirPath 数据是"classpath*:com/source/hot/ioc/book/ioc/",subPattern 数据是"**/*.class"。当得到 rootDirResources 对象后会经过 getResources 方法进行处理,处理结果是多种 Resource 的集合。在这个集合基础上还需要进一步进行处理,单个 Resource 的处理代码如下。

```java
{
    //解析资源
    //目前而言 Spring 没有做额外处理,属于预留方法
    rootDirResource = resolveRootDirResource(rootDirResource);
    URL rootDirUrl = rootDirResource.getURL();
    ///判断协议是不是 bundle
    if (equinoxResolveMethod != null && rootDirUrl.getProtocol().startsWith("bundle")) {
        //equinoxResolveMethod 方法执行
        URL resolvedUrl = (URL) ReflectionUtils.invokeMethod(equinoxResolveMethod, null, rootDirUrl);
        if (resolvedUrl != null) {
            rootDirUrl = resolvedUrl;
        }
        //解成 UrlResource 对象
        rootDirResource = new UrlResource(rootDirUrl);
    }
    //协议是否是 vfs
    if (rootDirUrl.getProtocol().startsWith(ResourceUtils.URL_PROTOCOL_VFS)) {
        //添加 vfs 相关资源
        result.addAll(VfsResourceMatchingDelegate.findMatchingResources(rootDirUrl, subPattern, getPathMatcher()));
    }
    //协议是否是 jar
    else if (ResourceUtils.isJarURL(rootDirUrl) || isJarResource(rootDirResource)) {
        //添加 jar 相关资源
```

```
                result.addAll(doFindPathMatchingJarResources(rootDirResource,
rootDirUrl, subPattern));
            }
            else {
                //文件协议处理
                result.addAll(doFindPathMatchingFileResources(rootDirResource, subPattern));
            }
        }
```

关于单个 Resource 的处理有两个操作，第一个操作是数据对象的前置准备，这部分处理代码如下。

```
rootDirResource = resolveRootDirResource(rootDirResource);
URL rootDirUrl = rootDirResource.getURL();
```

在这段代码中 resolveRootDirResource 方法目前是直接将参数返回，通过 rootDirResource 对象得到 URL 对象，得到该对象后会进行第二个操作，按照各类协议进行处理，处理的协议有下面四种。

（1）bundle 协议。
（2）vfs 协议。
（3）jar 相关协议。
（4）其他协议。

1. bundle 协议处理

接下来对 bundle 协议处理进行分析，相关处理代码如下。

```
if (equinoxResolveMethod != null && rootDirUrl.getProtocol().startsWith("bundle")) {
    //equinoxResolveMethod 方法执行
    URL resolvedUrl = (URL) ReflectionUtils.invokeMethod(equinoxResolveMethod, null,
rootDirUrl);
    if (resolvedUrl != null) {
        rootDirUrl = resolvedUrl;
    }
    //解成 UrlResource 对象
    rootDirResource = new UrlResource(rootDirUrl);
}
```

在处理 bundle 协议时使用了 equinoxResolveMethod 变量，该变量的实际类型是 PathMatchingResourcePatternResolver，对于这个类型需要重点关注它的静态代码块，具体代码如下。

```
@Nullable
private static Method equinoxResolveMethod;

static {
    try {
        Class<?> fileLocatorClass =
ClassUtils.forName("org.eclipse.core.runtime.FileLocator",
            PathMatchingResourcePatternResolver.class.getClassLoader());
```

```
        equinoxResolveMethod = fileLocatorClass.getMethod("resolve", URL.class);
        logger.trace("Found Equinox FileLocator for OSGi bundle URL resolution");
    }
    catch (Throwable ex) {
        equinoxResolveMethod = null;
    }
}
```

在这个静态代码块中回去寻找具体的 Method 赋值给 equinoxResolveMethod,这里不一定每次都会寻找到。假设现在有 equinoxResolveMethod 对象,那么在处理 bundle 协议的时候会将这个方法反射调用得到 URL 对象并赋值给 rootDirUrl,同时创建新的 rootDirResource 对象。

2. vfs 协议处理

接下来对 vfs 协议的处理进行分析,处理代码如下。

```
if (rootDirUrl.getProtocol().startsWith(ResourceUtils.URL_PROTOCOL_VFS)) {
    //添加 vfs 相关资源
    result.addAll(VfsResourceMatchingDelegate.findMatchingResources(rootDirUrl, subPattern,
getPathMatcher()));
}
```

上述代码为主要入口,具体核心处理在下面的代码中。

```
private static class VfsResourceMatchingDelegate {

    public static Set<Resource> findMatchingResources(
            URL rootDirURL, String locationPattern, PathMatcher pathMatcher) throws IOException {

        Object root = VfsPatternUtils.findRoot(rootDirURL);
        PatternVirtualFileVisitor visitor =
                new PatternVirtualFileVisitor(VfsPatternUtils.getPath(root), locationPattern,
pathMatcher);
        VfsPatternUtils.visit(root, visitor);
        return visitor.getResources();
    }
}
```

在处理 vfs 协议中重点对象是 PatternVirtualFileVisitor 和 VfsPatternUtils,在这个处理过程中可以简单理解在 PatternVirtualFileVisitor 中有需要的资源对象列表。具体的获取方式是通过反射获取,方法细节如下。

```
static void visit(Object resource, InvocationHandler visitor) throws IOException {
    Object visitorProxy = Proxy.newProxyInstance(
            VIRTUAL_FILE_VISITOR_INTERFACE.getClassLoader(),
            new Class<?>[] {VIRTUAL_FILE_VISITOR_INTERFACE}, visitor);
    invokeVfsMethod(VIRTUAL_FILE_METHOD_VISIT, resource, visitorProxy);
}
```

3. jar 相关协议处理

接下来将对 jar 相关协议的处理进行分析,具体处理代码如下。

```
//协议是否是 jar
else if (ResourceUtils.isJarURL(rootDirUrl) || isJarResource(rootDirResource)) {
    //添加 jar 相关资源
    result.addAll(doFindPathMatchingJarResources(rootDirResource,
rootDirUrl, subPattern));
}
```

上述代码中需要了解 jar 相关协议的具体内容，Spring 中通过 isJarURL 方法来判断是否属于 jar 相关协议，具体处理代码如下。

```
public static boolean isJarURL(URL url) {
    String protocol = url.getProtocol();
    return
(URL_PROTOCOL_JAR.equals(protocol) || URL_PROTOCOL_WAR.equals(protocol) ||
URL_PROTOCOL_ZIP.equals(protocol) || URL_PROTOCOL_VFSZIP.equals(protocol) ||
URL_PROTOCOL_WSJAR.equals(protocol));
}
```

在 isJarURL 方法中可以确认 jar 相关协议有 jar、war、zip、vfszip 和 wsjar。判断完成是否属于 jar 协议后会进行处理，处理方法是 doFindPathMatchingJarResources，具体代码如下。

```
protected Set<Resource> doFindPathMatchingJarResources(Resource rootDirResource, URL rootDirURL, String subPattern)
        throws IOException {

    URLConnection con = rootDirURL.openConnection();
    JarFile jarFile;
    String jarFileUrl;
    String rootEntryPath;
    boolean closeJarFile;

    //是不是 jar 连接
    if (con instanceof JarURLConnection) {
        JarURLConnection jarCon = (JarURLConnection) con;
        ResourceUtils.useCachesIfNecessary(jarCon);
        jarFile = jarCon.getJarFile();
        jarFileUrl = jarCon.getJarFileURL().toExternalForm();
        JarEntry jarEntry = jarCon.getJarEntry();
        rootEntryPath = (jarEntry != null ? jarEntry.getName() : "");
        closeJarFile = !jarCon.getUseCaches();
    }
    else {
        String urlFile = rootDirURL.getFile();
        try {
            //确定是否是 war
            int separatorIndex = urlFile.indexOf(ResourceUtils.WAR_URL_SEPARATOR);
            if (separatorIndex == -1) {
                //确定是不是 jar
                separatorIndex = urlFile.indexOf(ResourceUtils.JAR_URL_SEPARATOR);
            }
```

```java
            //jar、war 都不是的处理
            //将 url 直接转换成 jarFile 对象
            if (separatorIndex != -1) {
                jarFileUrl = urlFile.substring(0, separatorIndex);
                rootEntryPath = urlFile.substring(separatorIndex + 2);
                jarFile = getJarFile(jarFileUrl);
            }
            else {
                jarFile = new JarFile(urlFile);
                jarFileUrl = urlFile;
                rootEntryPath = "";
            }
            closeJarFile = true;
        }
        catch (ZipException ex) {
            if (logger.isDebugEnabled()) {
                logger.debug("Skipping invalid jar classpath entry [" + urlFile + "]");
            }
            return Collections.emptySet();
        }
    }

    try {
        if (logger.isTraceEnabled()) {
            logger.trace("Looking for matching resources in jar file [" + jarFileUrl + "]");
        }
        if (!"".equals(rootEntryPath) && !rootEntryPath.endsWith("/")) {
            rootEntryPath = rootEntryPath + "/";
        }
        Set<Resource> result = new LinkedHashSet<>(8);
        //循环 jar file 得到资源对象
        for (Enumeration<JarEntry> entries = jarFile.entries(); entries.hasMoreElements();) {
            JarEntry entry = entries.nextElement();
            String entryPath = entry.getName();
            if (entryPath.startsWith(rootEntryPath)) {
                String relativePath = entryPath.substring(rootEntryPath.length());
                if (getPathMatcher().match(subPattern, relativePath)) {
                    result.add(rootDirResource.createRelative(relativePath));
                }
            }
        }
        return result;
    }
    finally {
        if (closeJarFile) {
            jarFile.close();
        }
    }
}
```

在这个处理方法中有三个操作对其进行处理，具体操作如下。

(1) 将 URL 对象转换成 JarFile 对象。

(2) 将 JarFile 下的所有资源提取出来(提取出来的对象是 JarEntry),对每个 JarEntry 进行判断,判断是否和切分后的真实路径相匹配。

(3) 将操作二中匹配通过的创建出资源对象 Resource。

有关 jar 协议的处理可以将其进行模拟,制作一个简单版本的 JarFile 资源搜索的方法,具体代码如下。

```
@Test
void testJarFile() throws IOException {
    File f =
new File("D:\jar_repo\org\springframework\spring-core\5.2.3.RELEASE\spring-core-5.2.3.
RELEASE.jar");

    JarFile jarFile = new JarFile(f);

    for (Enumeration<JarEntry> entries = jarFile.entries(); entries.hasMoreElements(); ) {
        JarEntry entry = entries.nextElement();
        String entryPath = entry.getName();
        if (match(entryPath)) {
            System.out.println(entryPath);
        }
    }
}

private boolean match(String entryPath) {
    return entryPath.endsWith(".class");
}
```

在这个例子中采用 File 来创建 JarFile,在 Spring 源码中使用的是 URL,通过 match 方法来判断是否符合资源匹配要求。例子中只要是以 .class 结尾就是符合的。在 Spring 中对于 match 存在两层判断:第一层,前置路径符合;第二层,后置路径符合。前置路径和后置路径分别对应 findPathMatchingResources 处理过程中最开始的两个变量 rootDirPath 和 subPattern,通过这两层判断就会被加入资源集合中。

4. 其他协议处理

接下来将对其他协议进行分析,主要处理代码如下。

```
protected Set<Resource> doFindPathMatchingFileResources(Resource rootDirResource, String
subPattern)
        throws IOException {

    File rootDir;
    try {
        //绝对路径文件
        rootDir = rootDirResource.getFile().getAbsoluteFile();
    }
    catch (FileNotFoundException ex) {
        if (logger.isDebugEnabled()) {
```

```
            logger.debug("Cannot search for matching files underneath " + rootDirResource +
                    " in the file system: " + ex.getMessage());
        }
        return Collections.emptySet();
    }
    catch (Exception ex) {
        if (logger.isInfoEnabled()) {
            logger.info("Failed to resolve " + rootDirResource + " in the file system: " + ex);
        }
        return Collections.emptySet();
    }
    //查询匹配的文件系统资源
    return doFindMatchingFileSystemResources(rootDir, subPattern);
}
```

在这段代码中会根据 rootDir 对象进行资源搜索，具体的处理方法是 doFindMatching-FileSystemResources，相关代码如下。

```
protected Set < Resource > doFindMatchingFileSystemResources(File rootDir, String subPattern)
        throws IOException {
    if (logger.isTraceEnabled()) {
        logger.trace("Looking for matching resources in directory tree [" + rootDir.getPath()
                + "]");
    }
    //搜索匹配文件
    Set < File > matchingFiles = retrieveMatchingFiles(rootDir, subPattern);
    Set < Resource > result = new LinkedHashSet<>(matchingFiles.size());
    for (File file : matchingFiles) {
        //系统文件资源添加到结果容器
        result.add(new FileSystemResource(file));
    }
    return result;
}
```

在 doFindMatchingFileSystemResources 方法中处理事项有以下两个。

（1）将 rootDir 和 subPattern 进行匹配得到匹配的文件列表。

（2）将匹配的文件列表转换成 Resource，具体是 FileSystemResource。

首先对匹配方法 retrieveMatchingFiles 进行分析，在这个方法中会有以下四种匹配情况。

（1）文件不存在返回空集合。

（2）文件不是目录返回空集合。

（3）文件不可读返回空集合。

（4）解析全路径后进行匹配。

在上述四个情况中重点关注第四个，在第四个情况中有 fullPattern 方法，该方法对数据进行了转换，转换步骤有以下四步。

（1）提取当前 rootDir 的绝对地址。

（2）将绝对地址中的文件分隔符替换成"/"。

(3) 将 pattern(实际上就是 subParttern)中的文件分隔符替换成"/"。

(4) 将第(2)步和第(4)步的结果相加得到完整的匹配符号。

将测试用例带入到这个转换过程中,第(1)步和第(2)步的处理结果是 D:/desktop/git_repo/spring-ebk/spring-framework-read/spring-source-hot-ioc-book/build/classes/java/test/com/source/hot/ioc/book/ioc,第(3)步和第(4)步的处理结果是 D:/desktop/git_repo/spring-ebk/spring-framework-read/spring-source-hot-ioc-book/build/classes/java/test/com/source/hot/ioc/book/ioc/ ** / * . class。有关匹配过程的代码如下。

```java
protected Set<File> retrieveMatchingFiles(File rootDir, String pattern) throws IOException {
    //目录不存在
    if (!rootDir.exists()) {
        if (logger.isDebugEnabled()) {
            logger.debug("Skipping [" + rootDir.getAbsolutePath() + "] because it does not exist");
        }
        return Collections.emptySet();
    }
    //不是目录
    if (!rootDir.isDirectory()) {
        if (logger.isInfoEnabled()) {
            logger.info("Skipping [" + rootDir.getAbsolutePath() + "] because it does not denote a directory");
        }
        return Collections.emptySet();
    }
    //不可读
    if (!rootDir.canRead()) {
        if (logger.isInfoEnabled()) {
            logger.info("Skipping search for matching files underneath directory [" + rootDir.getAbsolutePath() +
                    "] because the application is not allowed to read the directory");
        }
        return Collections.emptySet();
    }
    //转换成全路径
    String fullPattern = StringUtils.replace(rootDir.getAbsolutePath(), File.separator, "/");
    if (!pattern.startsWith("/")) {
        fullPattern += "/";
    }
    //转换成全路径
    fullPattern = fullPattern + StringUtils.replace(pattern, File.separator, "/");

    //最终结果容器
    Set<File> result = new LinkedHashSet<>(8);
    //搜索匹配的文件
    doRetrieveMatchingFiles(fullPattern, rootDir, result);
    return result;
}
```

通过前面的操作得到了一个需要进行匹配的文件路径，最后就需要进行搜索，具体的方法是 doRetrieveMatchingFiles，详细代码如下。

```
protected void doRetrieveMatchingFiles(String fullPattern, File dir, Set < File > result)
throws IOException {
    if (logger.isTraceEnabled()) {
        logger.trace("Searching directory [" + dir.getAbsolutePath() +
            "] for files matching pattern [" + fullPattern + "]");
    }
    //循环文件夹下的所有文件
    for (File content : listDirectory(dir)) {
        //当前需要处理的文件路径
        String currPath = StringUtils.replace(content.getAbsolutePath(), File.separator, "/");
        //当前文件是不是目录
        //当前文件是否是以根路径开头
        if (content.isDirectory() && getPathMatcher().matchStart(fullPattern, currPath +
"/")) {
            if (!content.canRead()) {
                if (logger.isDebugEnabled()) {
                    logger.debug("Skipping subdirectory [" + dir.getAbsolutePath() +
                        "] because the application is not allowed to read the directory");
                }
            }
            else {
                //递归查询文件夹下的文件
                doRetrieveMatchingFiles(fullPattern, content, result);
            }
        }
        //匹配放入数据
        if (getPathMatcher().match(fullPattern, currPath)) {
            result.add(content);
        }
    }
}
```

在这里需要着重关注的是 match 这个方法，该方法是判断当前文件路径是否符合匹配路径的重点。真正的处理方法是 org.springframework.util.AntPathMatcher#doMatch，在这个方法中的处理逻辑如下（和 AntPathMatcher 略有差异，主体含义相同）。

首先进入方法时拥有下面两个变量：①fullPattern，数据信息为"D:/desktop/git_repo/spring-ebk/spring-framework-read/spring-source-hot-ioc-book/build/classes/java/test/com/source/hot/ioc/book/ioc/**/*.class"；②currPath，数据信息为"D:/desktop/git_repo/spring-ebk/spring-framework-read/spring-source-hot-ioc-book/build/classes/java/test/com/source/hot/ioc/book/ioc/A.class"。

在这两个变量基础上有以下两个操作。

(1) 将 fullPattern 中存在通配符的切割出来得到"D:/desktop/git_repo/spring-ebk/spring-framework-read/spring-source-hot-ioc-book/build/classes/java/test/com/source/hot/ioc/book/ioc/"和"**/*.class"。

(2) 将 currPath 切分成两组，切分依据是第一步得到的结果"D:/desktop/git_repo/spring-ebk/spring-framework-read/spring-source-hot-ioc-book/build/classes/java/test/com/source/hot/ioc/book/ioc/"，切分成功后会得到"D:/desktop/git_repo/spring-ebk/spring-framework-read/spring-source-hot-ioc-book/build/classes/java/test/com/source/hot/ioc/book/ioc/" 和 "A.class"。在这个切分结果中主要比较的是第二部分，在这里"*"会忽略比较直到出现非"*"字符，在这个例子中字符是"."匹配到"A.class"中的"."，再继续往下比较 class，此时通过比较就是符合匹配条件的数据；反之不是符合条件的数据。

19.5.2　findAllClassPathResources 方法分析

接下来对另一种获取 Resource 的方式进行分析，主要处理代码如下。

```java
protected Resource[] findAllClassPathResources(String location) throws IOException {
    String path = location;
    //是不是以/开头.如果是去掉第一个/
    if (path.startsWith("/")) {
        path = path.substring(1);
    }
    //查询方法
    Set<Resource> result = doFindAllClassPathResources(path);
    if (logger.isTraceEnabled()) {
        logger.trace("Resolved classpath location [" + location + "] to resources " + result);
    }
    return result.toArray(new Resource[0]);
}
```

在 findAllClassPathResources 方法中的重点是 doFindAllClassPathResources，在做该数据处理前会做一个以"/"为首的切割，即去掉第一个"/"符号。真正处理的方法是 doFindAllClassPathResources，具体代码如下。

```java
protected Set<Resource> doFindAllClassPathResources(String path) throws IOException {
    //结果对象
    Set<Resource> result = new LinkedHashSet<>(16);
    //获取类加载器
    ClassLoader cl = getClassLoader();
    //将 path 解析成 URL
    Enumeration<URL> resourceUrls = (cl != null ?
        cl.getResources(path) : ClassLoader.getSystemResources(path));
    while (resourceUrls.hasMoreElements()) {
        URL url = resourceUrls.nextElement();
        //url 对象转换成 UrlResource 放入结果集合
        result.add(convertClassLoaderURL(url));
    }
    if ("".equals(path)) {
        //jar 相关资源处理
        addAllClassLoaderJarRoots(cl, result);
    }
```

```
        return result;
}
```

在 doFindAllClassPathResources 方法中存在两种 Resource 的处理方式。

（1）将资源地址中的数据解析成为 UrlResource。

（2）资源地址为空字符串的情况下处理成为 UrlResource。

处理方式一中的核心方法是 convertClassLoaderURL，该方法的本质是 new UrlResource(url)。处理方式二的本质也是创建 UrlResource 对象，但是和处理方式一的差异是参数。下面是第二种处理方式的相关代码。

```
UrlResource jarResource
    = (ResourceUtils.URL_PROTOCOL_JAR.equals(url.getProtocol()) ?
        new UrlResource(url) : new
UrlResource(ResourceUtils.JAR_URL_PREFIX + url
    + ResourceUtils.JAR_URL_SEPARATOR));
```

处理方式二中还有另外一种 UrlResource 的创建，进行这种模式创建需要满足下面的条件：当前 classLoader 和 Spring 中提供的系统类加载器相同。有关 UrlResource 的创建代码如下。

```
UrlResource jarResource = new UrlResource(ResourceUtils.JAR_URL_PREFIX +
        ResourceUtils.FILE_URL_PREFIX +
filePath + ResourceUtils.JAR_URL_SEPARATOR);
```

小结

本章主要分析 PathMatchingResourcePatternResolver 中的实现，在对它的分析中对一些协议和对应协议的处理方式进行分析，以及间接了解了 Resource 的一些实现类。

第20章

BeanName生成策略

本章将对 Spring IoC 辅助工具中的 BeanName 生成策略进行分析，主要讨论的是 Spring 中的 BeanName 属性。

20.1 AnnotationBeanNameGenerator 分析

在开始分析之前需要先找到负责 BeanName 生成的接口，在 Spring 中负责这项工作的接口是 BeanNameGenerator，有关它的类图如图 20.1 所示。

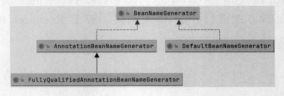

图 20.1　BeanNameGenerator 类图

在 Spring 中对 BeanNameGenerator 接口提供了三个相关实现，首先对 AnnotationBeanNameGenerator 实现进行分析，具体处理代码如下。

```
@Override
public String generateBeanName(BeanDefinition definition, BeanDefinitionRegistry registry) {
    if (definition instanceof AnnotatedBeanDefinition) {
        //从注解中获取 BeanName
        //获取注解的 value 属性值
        String beanName = determineBeanNameFromAnnotation((AnnotatedBeanDefinition) definition);
        if (StringUtils.hasText(beanName)) {
            //如果存在直接返回
            return beanName;
        }
}
```

```
        }
        //默认 BeanName
        //类名,首字母小写
        return buildDefaultBeanName(definition, registry);
    }
```

在这段代码中有以下两种处理。

(1) BeanDefinition 是 AnnotatedBeanDefinition 类型。

(2) BeanDefinition 不是 AnnotatedBeanDefinition 类型,或者第一种处理情况得不到 BeanName。

20.1.1 AnnotatedBeanDefinition 类型的 BeanName 生成策略

接下来将对 AnnotatedBeanDefinition 类型的 BeanName 生成策略进行分析,具体的处理代码如下。

```
@Nullable
protected String
determineBeanNameFromAnnotation(AnnotatedBeanDefinition annotatedDef) {
    //获取注解信息
    AnnotationMetadata amd = annotatedDef.getMetadata();
    //所有注解
    Set<String> types = amd.getAnnotationTypes();
    String beanName = null;
    for (String type : types) {
        //注解属性 map 对象
        AnnotationAttributes attributes = AnnotationConfigUtils.attributesFor(amd, type);
        if (attributes != null
&& isStereotypeWithNameValue(type, amd.getMetaAnnotationTypes(type), attributes)) {
            Object value = attributes.get("value");
            if (value instanceof String) {
                String strVal = (String) value;
                if (StringUtils.hasLength(strVal)) {
                    if (beanName != null && !strVal.equals(beanName)) {
                        throw new IllegalStateException("Stereotype annotations suggest inconsistent " + "component names: '" + beanName + "' versus '" + strVal + "'");
                    }
                    beanName = strVal;
                }
            }
        }
    }
    return beanName;
}
```

在 determineBeanNameFromAnnotation 方法中有以下五项处理。

(1) 提取 BeanDefinition 中的注解元数据。

(2) 从注解元数据中提取所有的注解类名。

(3) 循环注解类名,从注解元数据中提取对应的注解属性。
(4) 从注解属性中尝试获取 value 对应的值。
(5) 将 value 对应值判断是否是 String 类型,如果是那么就会被作为 BeanName。

在代码中第一步是获取 AnnotationMetadata 对象,在得到该对象后需要进行数据提取,根据 AnnotationConfigUtils.attributesFor(amd,type)可以依次获取在类上(方法上)标记的注解数据,AnnotationAttributes 类定义如下。

```java
public class AnnotationAttributes extends LinkedHashMap<String, Object> {

    private static final String UNKNOWN = "unknown";

    @Nullable
    private final Class<? extends Annotation> annotationType;

    final String displayName;

    boolean validated = false;
}
```

通过上面这段代码可以知道,AnnotationAttributes 是一个 Map 结构的对象,并在其中存有一个注解类信息字段 annotationType,这个字段数据不会显式存在,是允许为空的。以@Component 注解举例,在单纯使用的情况下(不添加属性值)会得到如图 20.2 所示的结果。

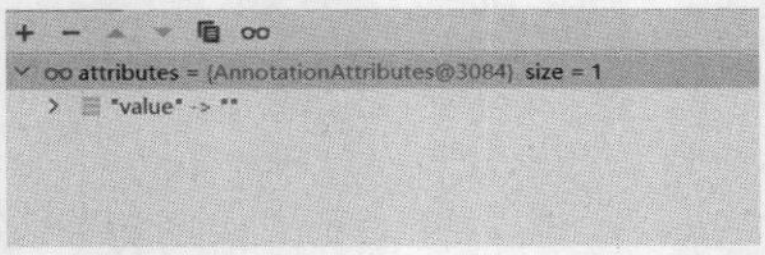

简而言之,AnnotationAttributes 存储了单个注解的所有属性,当拥有了 AnnotationAttributes 数据对象后,就需要从属性表中提取 value 的数据,如果 value 不为空同时不是空字符串的情况下,会将 value 的数据作为 BeanName。

图 20.2 Component 注解中属性对应的数据

20.1.2 非 AnnotatedBeanDefinition 类型的 BeanName 生成策略

接下来将对非 AnnotatedBeanDefinition 类型的 BeanName 生成策略进行分析,注意如果在 value 提取数据为空或者空字符串的情况下也会进行该操作,具体处理代码如下。

```java
protected String buildDefaultBeanName(BeanDefinition definition) {
    //获取 class name
    String beanClassName = definition.getBeanClassName();
    Assert.state(beanClassName != null, "No bean class name set");
    //获取短类名
    String shortClassName = ClassUtils.getShortName(beanClassName);
    //首字母小写
    return Introspector.decapitalize(shortClassName);
}
```

在这段代码中将 BeanName 的处理进行了以下三步操作。
(1) 提取全类名。

(2) 全类名中提取类名称(短名)。

(3) 短类名首字母小写处理。

20.2 FullyQualifiedAnnotationBeanNameGenerator 分析

接下来将对 FullyQualifiedAnnotationBeanNameGenerator 类进行分析，这个类也是进行 BeanName 生成的类，它是 AnnotationBeanNameGenerator 的子类，它重写了 buildDefaultBeanName 方法，具体方法如下。

```java
@Override
protected String buildDefaultBeanName(BeanDefinition definition) {
    String beanClassName = definition.getBeanClassName();
    Assert.state(beanClassName != null, "No bean class name set");
    return beanClassName;
}
```

在这个重写方法中将 BeanClass 直接返回，从而来改变处理结果，从原来的短类名变成了全类名。

20.3 DefaultBeanNameGenerator 分析

接下来将对 DefaultBeanNameGenerator 类进行分析，这个类是 Spring 中 BeanName 的默认生成方法，具体处理代码如下。

```java
public static String generateBeanName(
        BeanDefinition definition, BeanDefinitionRegistry registry, boolean isInnerBean)
        throws BeanDefinitionStoreException {

    //获取 bean class 的名称
    //Class.getName()
    String generatedBeanName = definition.getBeanClassName();
    if (generatedBeanName == null) {
        //父类名称是否存在
        if (definition.getParentName() != null) {
            generatedBeanName = definition.getParentName() + "$child";
        }
        //工厂 BeanName 是否为空
        else if (definition.getFactoryBeanName() != null) {
            generatedBeanName = definition.getFactoryBeanName() + "$created";
        }
    }
    if (!StringUtils.hasText(generatedBeanName)) {
        throw new BeanDefinitionStoreException("Unnamed bean definition specifies neither " +
                "'class' nor 'parent' nor 'factory-bean' - can't generate bean name");
    }

    String id = generatedBeanName;
```

```
        if (isInnerBean) {
            //组装名称
            //生成名称 + ＃ +十六进制的一个字符串
            id = generatedBeanName + GENERATED_BEAN_NAME_SEPARATOR + ObjectUtils.getIdentityHexString
(definition);
        }
        else {
            //唯一 BeanName 设置
            //BeanName + ＃ + 序号
            return uniqueBeanName(generatedBeanName, registry);
        }
        return id;
    }
```

这段代码的处理逻辑如图 20.3 所示。

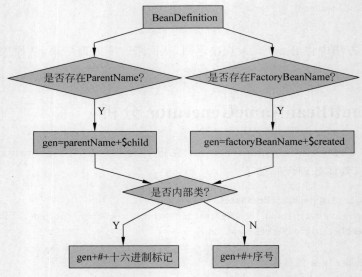

图 20.3　DefaultBeanNameGenerator 处理逻辑

总结图 20.3 的生成策略有下面四种。

（1）ClassName ＋"＃"＋ 十六进制字符。

（2）parentName ＋"＄child"+"＃"＋ 十六进制字符。

（3）factoryBeanName ＋"＄created"+"＃"＋ 十六进制字符。

（4）beanName ＋"＃"＋ 序号。

小结

本章主要对 BeanName 的生成接口 BeanNameGenerator 进行分析，对 BeanNameGenerator 接口的三个实现类进行源码分析，对比了部分细节差异，同时对 BeanName 的生成策略有了明确的认知。

第21章

条 件 注 解

本章将对 Spring IoC 辅助工具中的条件注解进行分析,包含条件注解测试环境搭建和条件注解处理流程分析。

21.1 条件注解测试环境搭建

为了进行条件注解的源码分析,首先需要搭建一个测试环境,第一步编写一个接口,接口名为 InterFunc,具体代码如下。

```
public interface InterFunc {
    String data();
}
```

第二步编写 Condition 接口的实现类,总共有两个实现,一个是 LinuxCondition,具体代码如下。

```
public class LinuxCondition implements Condition {

    @Override
    public boolean matches(ConditionContext context, AnnotatedTypeMetadata metadata) {
        return context.getEnvironment().getProperty("os.name").contains("Linux");
    }

}
```

另一个是 WindowsCondition,具体代码如下。

```
public class WindowsCondition implements Condition {

    @Override
    public boolean matches(ConditionContext context, AnnotatedTypeMetadata metadata) {
```

```
        return context.getEnvironment().getProperty("os.name").contains("Windows");
    }
}
```

接下来编写 Spring 配置类,配置类类名为 ConditionBeans,具体代码如下。

```
@Configuration
public class ConditionBeans {

    @Bean
    @Conditional(LinuxCondition.class)
    public InterFunc linux() {
        return new InterFunc() {
            @Override
            public String data() {
                return "linux";
            }
        };
    }

    @Bean
    @Conditional(WindowsCondition.class)
    public InterFunc windows() {
        return new InterFunc() {
            @Override
            public String data() {
                return "windows";
            }
        };
    }

}
```

最后编写测试用例,具体代码如下。

```
public class ConditionTest {
    @Test
    void testCondition() {
        AnnotationConfigApplicationContext context =
 new AnnotationConfigApplicationContext(ConditionBeans.class);
        InterFunc bean = context.getBean(InterFunc.class);
        assert bean.data().equals("windows");
    }
}
```

21.2 条件注解分析

在 Spring 注解模式启动过程中 doRegisterBean 方法有一段关于条件注解处理的代码,具体代码如下。

```
//和条件注解相关的函数
if (this.conditionEvaluator.shouldSkip(abd.getMetadata())) {
   return;
}
```

doRegisterBean 操作是针对 Spring Configuration Bean 进行的，对于普遍的 Bean Spring 处理操作可以在 loadBeanDefinitionsForBeanMethod 方法中看到处理逻辑，具体代码如下。

```
private void loadBeanDefinitionsForBeanMethod(BeanMethod beanMethod) {
    ConfigurationClass configClass = beanMethod.getConfigurationClass();
    MethodMetadata metadata = beanMethod.getMetadata();
    String methodName = metadata.getMethodName();

    if (this.conditionEvaluator.shouldSkip(metadata, ConfigurationPhase.REGISTER_BEAN)) {
        configClass.skippedBeanMethods.add(methodName);
        return;
    }
    if (configClass.skippedBeanMethods.contains(methodName)) {
        return;
    }
    //省略后续代码
}
```

loadBeanDefinitionsForBeanMethod 方法的调用层级如下。

（1）org.springframework.context.annotation.ConfigurationClassBeanDefinitionReader#loadBeanDefinitions。

（2）org.springframework.context.annotation.ConfigurationClassBeanDefinitionReader#loadBeanDefinitionsForConfigurationClass。

（3）org.springframework.context.annotation.ConfigurationClassBeanDefinitionReader#loadBeanDefinitionsForBeanMethod。

在 loadBeanDefinitionsForBeanMethod 方法中需要进行分析的方法是 conditionEvaluator.shouldSkip，在 doRegisterBean 中也可以找到类似的处理代码 conditionEvaluator.shouldSkip。在这两个处理过程中都使用 conditionEvaluator，实际类型是 org.springframework.context.annotation.ConditionEvaluator。继续深入源码找到 shouldSkip 的核心处理方法。

```
public boolean shouldSkip(@Nullable AnnotatedTypeMetadata metadata, @Nullable ConfigurationPhase phase) {
    //注解元数据不存在或者注解元数据中不包含 Conditional 注解
    if (metadata == null || !metadata.isAnnotated(Conditional.class.getName())) {
        return false;
    }

    //配置解析阶段处理
    if (phase == null) {
        if (metadata instanceof AnnotationMetadata &&
```

```java
                    ConfigurationClassUtils.isConfigurationCandidate((AnnotationMetadata)
    metadata)) {
        return shouldSkip(metadata, ConfigurationPhase.PARSE_CONFIGURATION);
    }
    return shouldSkip(metadata, ConfigurationPhase.REGISTER_BEAN);
}

//需要处理的 Condition, 数据从注解 Conditional 中来
List<Condition> conditions = new ArrayList<>();
//获取注解 Conditional 的属性值
for (String[] conditionClasses : getConditionClasses(metadata)) {
    for (String conditionClass : conditionClasses) {
        //将注解中的数据转换成 Condition 接口
        //从 class 转换成实例
        Condition condition =
            getCondition(conditionClass, this.context.getClassLoader());
        //插入注解列表
        conditions.add(condition);
    }
}

//对 Condition 进行排序
AnnotationAwareOrderComparator.sort(conditions);

//执行 Condition 得到验证结果
for (Condition condition : conditions) {
    ConfigurationPhase requiredPhase = null;
    //如果类型是 ConfigurationCondition
    if (condition instanceof ConfigurationCondition) {
        requiredPhase = ((ConfigurationCondition) condition).getConfigurationPhase();
    }

    //matches 进行验证
    if ((requiredPhase == null || requiredPhase
        == phase) && !condition.matches(this.context, metadata)) {
        return true;
    }
}

return false;
}
```

这个方法是整个条件注解处理的核心。首先理解这个方法的两个参数,第一个参数是 metadata,类型是 AnnotatedTypeMetadata,表示注解元数据,存储了注解的数据信息;第二个参数是 phase,类型是 ConfigurationPhase,表示配置解析阶段枚举存在两种状态,第一种状态是 PARSE_CONFIGURATION 配置解析阶段,第二种状态是 REGISTER_BEAN Bean 注册阶段。

方法 shouldSkip 的具体处理流程如下。

(1) 注解元数据不存在或者注解元数据中不存在 Conditional 注解,返回 false。

（2）阶段信息的补充。如果阶段信息为空，需要做如下操作，如果注解元数据的类型是 AnnotationMetadata 并且注解元数据中存在 Bean、Component、ComponentScan、Import 和 ImportResource 注解，将设置阶段信息为 PARSE_CONFIGURATION，表示配置解析阶段；反之则将阶段信息设置为 REGISTER_BEAN，表示 Bean 注册阶段。以测试用例为例，doRegister 方法中传递的是 null，但是注解元数据解析符合条件会被设置为 PARSE_CONFIGURATION，当进入 loadBeanDefinitionsForBeanMethod 方法后对单个 Bean 的注册，传递的是一个明确的变量 ConfigurationPhase.REGISTER_BEAN。

（3）提取注解 Conditional 中 value 的属性并将其转换成实例对象，得到当前 Conditional 中所有的 Condition 后进行排序，排序与 Ordered 有关。

（4）执行排序后的 Condition，如果满足下面的条件就会返回 true。

① requiredPhase 不存在或者 requiredPhase 的数据等于参数 phase。

② 方法!condition.matches(this.context,metadata) 执行结果为 true。

在条件一中 requiredPhase 数据的获取是通过 ConfigurationCondition 接口获取的，该接口也是 Condition 的子接口，条件二所执行的 matches 方法是通过 Condition 接口提供的，也是开发者自定义的验证方法。

小结

本章对条件注解的使用及原理进行分析，主要围绕条件解析接口 ConditionEvaluator 进行源码分析。条件注解的本质是执行条件注解中对应的条件接口，判断是否返回 true，如果返回 true 就进行 Bean 初始化；反之则不进行 Bean 初始化。

第22章

Spring 排序注解

本章将对 Spring IoC 辅助工具中的排序注解 Order 展开,对其进行分析,本章包含排序注解测试环境搭建和排序注解的处理流程分析。

22.1 排序注解测试环境搭建

在开始分析源码之前,首先需要搭建一个可以用来进行测试的环境。首先编写一个 Spring 配置类,在配置类中将对一些注解进行使用,用到的注解有 Order 和 Component 等,配置类类名为 OrderedBeans,具体代码如下。

```
@Configuration
@ComponentScan(basePackageClasses =
{OrderedBeans.OrderBeanOne.class, OrderedBeans.OrderBeanTwo.class})
public class OrderedBeans {
    @Order(2)
    @Component
    public class OrderBeanTwo {
    }

    @Order(1)
    @Component
    public class OrderBeanOne {
    }

}
```

完成 Spring 配置类编写后继续编写对应的单元测试,测试类类名为 OrderedTest,具体测试方法如下。

```
public class OrderedTest {
```

```
@Test
void orderedTest() {
    AnnotationConfigApplicationContext context =
new AnnotationConfigApplicationContext(OrderedBeans.class);
    context.close();
}
```

测试用例编写完成，接下来需要确认分析目标，在之前的内容中有提到在 org.springframework.context.annotation.ConfigurationClassParser#processMemberClasses 方法中有一段排序操作 OrderComparator.sort(candidates)，这段代码就是分析目标。

22.2　OrderComparator.sort 方法分析

前文确定了分析目标，下面是需要分析的完整代码。

```
private void processMemberClasses(ConfigurationClass configClass, SourceClass sourceClass)
		throws IOException {
    //找到当前配置类中存在的成员类
    Collection<SourceClass> memberClasses = sourceClass.getMemberClasses();
    //成员类列表不为空
    if (!memberClasses.isEmpty()) {
        List<SourceClass> candidates = new ArrayList<>(memberClasses.size());
        for (SourceClass memberClass : memberClasses) {
            //成员类是否符合配置类候选标准，Component、ComponentScan、Import、ImportResource 注
            //解是否存在
            //成员类是否和配置类同名
            if (ConfigurationClassUtils.isConfigurationCandidate(memberClass.getMetadata()) &&
                    !memberClass.getMetadata().getClassName().equals(configClass.getMetadata().getClassName())) {

                candidates.add(memberClass);
            }
        }
        //排序候选类
        OrderComparator.sort(candidates);

        //省略其他
    }
```

在这段代码中需要关注参数 candidates，可以看到在这段方法中分为两个步骤：第一个步骤是将需要进行排序的对象从配置类中提取出来，第二个步骤是将提取出来的对象进行排序，这两步需要操作的对象都是 candidates。下面进入调试阶段观察 candidates 对象的数据信息，如图 22.1 所示。

在图 22.1 中可以发现这两个元素是在配置类中标记了 Order 注解的对象，此时它们通过了筛选被选出等待后续的操作，这也是第一步得到的数据结果。在这个数据结果中排序关键的信息可以在 metadata 对象中找到，具体信息如图 22.2 所示。

图 22.1　candidates 数据信息

图 22.2　metadata 中关于 Order 注解的信息

继续向下深入分析找到最重要的代码 OrderComparator.sort，该方法的代码如下。

```
public static void sort(List<?> list) {
    if (list.size() > 1) {
        list.sort(INSTANCE);
    }
}
```

在这个方法中需要关注 INSTANCE 对象，这个对象是一个比较器对象（java.util.Comparator），也就是说，这里对于对象 list 的排序依靠 INSTANCE 进行，在代码中可以找到 INSTANCE 的定义信息如下。

```
public static final OrderComparator INSTANCE = new OrderComparator();
```

此时可以明确 INSTANCE 的类型是 OrderComparator。OrderComparator 对象必然实现类 Comparator 接口，需要重点分析的方法就是 Comparator 接口的实现方法 compare，具体代码如下。

```
@Override
public int compare(@Nullable Object o1, @Nullable Object o2) {
    return doCompare(o1, o2, null);
}
```

在 compare 方法中还有一层调用指向了 doCompare 方法，具体代码如下。

```
private int doCompare(@Nullable Object o1, @Nullable Object o2, @Nullable OrderSourceProvider sourceProvider) {
    boolean p1 = (o1 instanceof PriorityOrdered);
    boolean p2 = (o2 instanceof PriorityOrdered);
```

```
        if (p1 && !p2) {
            return -1;
        }
        else if (p2 && !p1) {
            return 1;
        }

        int i1 = getOrder(o1, sourceProvider);
        int i2 = getOrder(o2, sourceProvider);
        return Integer.compare(i1, i2);
}
```

在这段代码中可以看到两种比较模式,第一种是关于是否实现了 PriorityOrdered 接口进行比较,第二种是只根据 Ordered 接口进行比较。第一种比较模式就是代码中的 if 和 else if 部分代码,需要重点关注的是下面的 getOrder 方法。在 getOrder 方法中会得到一个数据,这个数据就是排序号。关于排序号的获取有下面三种方式,第一种是从接口 Ordered 中获取,第二种是从 PriorityOrdered 中获取(默认最高),第三种是从注解 Order 中获取。由于第二种会被单独讨论,通常情况下不会去进行获取操作。对于这个方法的分析就转到对 getOrder 方法的分析,下面是具体处理代码。

```
private int getOrder(@Nullable Object obj, @Nullable OrderSourceProvider sourceProvider) {
    Integer order = null;
    if (obj != null && sourceProvider != null) {
        Object orderSource = sourceProvider.getOrderSource(obj);
        if (orderSource != null) {
            if (orderSource.getClass().isArray()) {
                Object[] sources = ObjectUtils.toObjectArray(orderSource);
                for (Object source : sources) {
                    order = findOrder(source);
                    if (order != null) {
                        break;
                    }
                }
            }
            else {
                order = findOrder(orderSource);
            }
        }
    }
    return (order != null ? order : getOrder(obj));
}
```

在这段代码中核心方法是 findOrder,在这里需要寻找排序号所有的方法,最终都会指向到 findOrder 方法中,下面是 findOrder 的处理代码。

```
@Nullable
protected Integer findOrder(Object obj) {
    return (obj instanceof Ordered ? ((Ordered) obj).getOrder() : null);
}
```

可以看到在这个处理代码中它会判断是否实现了 Ordered 接口，如果实现了，那么直接从接口获取，否则就会返回 null。跳出这个方法回到方法入口，在调用方法时传递参数 candidates，这个对象的数据类型是 SourceClass。findOrder 方法的参数和 SourceClass 之间存在一种关系，这个关系是通过 Ordered 接口建立的。这一点信息可以通过 SourceClass 类图得知，如图 22.3 所示。

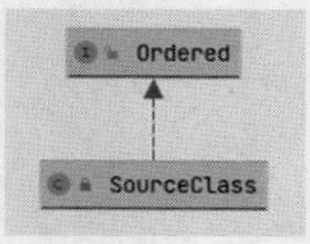

图 22.3　SourceClass 类图

在 SourceClass 中关于 Ordered 接口的实现代码如下。

```
@Override
public int getOrder() {
    Integer order = ConfigurationClassUtils.getOrder(this.metadata);
    return (order != null ? order : Ordered.LOWEST_PRECEDENCE);
}
```

在这段代码中可以看到关于数据提取依赖 ConfigurationClassUtils.getOrder 方法，具体的提取方式如下。

```
@Nullable
public static Integer getOrder(AnnotationMetadata metadata) {
    Map<String, Object> orderAttributes =
        metadata.getAnnotationAttributes(Order.class.getName());
    return (orderAttributes != null ? ((Integer) orderAttributes.get(AnnotationUtils.VALUE)) :
        null);
}
```

在这段代码中可以理解为直接从 metadata 中根据注解名称获取对应的数据结果，将数据结果作为返回对象即可，具体的数据信息可以查看图 22.2。至此，得到 Order 数据信息即可完成所有的数据操作。

小结

Spring 排序相关内容主要有两个，第一个是排序号的获取方式，第二个是 Ordered 接口和 PriorityOrdered 的优先级。排序号可以通过注解获取，也可以通过 Ordered 接口获取。PriorityOrdered 的优先级是最高的。